F. A. Shutov

Integral / Structural Polymer Foams

Integral/Structural Polymer Foams

Technology, Properties and Applications

by

Prof. Dr. Fyodor Anatoljevitsch Shutov
Moscow

(English by F. A. Shutov)

With 127 Figures and 95 Tables

Akademie-Verlag Berlin
1986

Die Originalausgabe erscheint im Springer-Verlag
Berlin · Heidelberg · New York · Tokyo

Vertrieb ausschließlich für die DDR und die sozialistischen Länder

ISBN 3-540-15038-2 Springer-Verlag Berlin · Heidelberg · New York · Tokyo
ISBN 0-387-15038-2 Springer-Verlag New York · Heidelberg · Berlin · Tokyo

ISBN 3-05-500138-9

Erschienen im Akademie-Verlag Berlin, DDR-1086 Berlin, Leipziger Straße 3—4
Lizenznummer: 202 · 100/497/86
Printed in the German Democratic Republic
Gesamtherstellung: VEB Druckerei „Thomas Müntzer“, 5820 Bad Langensalza
Umschlaggestaltung: Karl Salzbrunn
LSV 1275
Bestellnummer: 763 565 2 (6946)
19800

To my beloved parents:
Natali and Anatol

Preface

Integral, or structural, foams are one of the most remarkable materials that have been developed over the last fifteen years.

As with all rapidly growing fields, the terminology seems to have grown even faster. Thus there are two names for the material structure itself. In the United States and in Japan the term for these plastics is *Structural Foams*, whereas in Europe and the USSR the term used is usually *Integral Foams*. We have adhered to the European term in the text and hope our colleagues will bear with us.

Integral foams have a specific structure: a cellular core that gradually turns into a solid skin. The skin gives the part its form and stiffness, while the cellular core contributes to the very high strength-to-weight values of the material. These are higher than those of some unfoamed plastics and metals.

The sandwich-like structure with its unique mechanical properties was prompted by nature. Wood and bone are strong and light-weight natural materials having a cellular structure.

Since the sandwich-like structure of the integral foams resembles that of natural wood, the foams are often referred to as *artifical wood* or *plastic wood*, thereby emphasizing not only the formal structural similarity of these materials, but also one of the main functional applications of integral foams — replacement of wooden articles in various fields of engineering and construction.

There is another very important aspect to the use of foamed plastics. These materials can help resolve two current global crises, i.e. the depletion of non-renewable organic raw materials (oil and gas) and the energy crisis. Foamed plastics use less polymer (than unfoamed plastics) because a considerable part of the volume is plastics taken up by a gas (air). As to the energy crisis, the cellular structure produces a very low thermal conductivity making foamed excellent thermal insulators.

It is the physics of manufacturing these materials that is unique. Indeed, the integral structure is created by the interplay of the main physical parameters — temperature and pressure — on the foaming polymer melts or solutions. This is why I have devoted a considerable portion of the book to a discussion of the technology and equipment of integral foams and to an analysis of the optimum processing parameters.

This book covers in detail the fundamental correlations between the morphology and properties of integral foams and the formulation, equipment, and processing parameters.

Various applications of the products are discussed together with a review of the design and marketing problems. In addition, I have included an economic analysis of the commercial processes, for most of the materials.

I have tried to write a book that will be useful for:

- chemists and physicists, who create the integral foams,
- technologists and engineers, who manufacture them,
- consumers and managers, who apply these plastics, and
- students and post-graduates, who study the science, technology and applications of polymer materials.

I tried to write not just a scientific monograph, but a manual and handbook as well.

The references are surveyed trough Summer 1985.

I should be extremely grateful for any comments or criticism you may care to send me.

Moscow, July 1985 Fyodor A. Shutov

Acknowledgements

Scientists from many countries have been "invisible" coauthors to this book, for it is their efforts that have resulted in the remarkable scientific and practical developments in the field of integral foamed plastics. I would like to express my deep gratitude to my colleagues in different countries who kindly took part in the discussions of this manuscript, or sent me copies of their papers and reports.

I have the pleasure to acknowledge the contributions of the following workers in the field:

Argentina: R. J. J. Williams;
Austria: L. Golser, H. Hubeny, P. J. Schmidt;
Belgium: J. L. Lambert, P. Stachel, R. H. Young;
Bulgaria: N. Popov, S. Semerdjiev;
Canada: C. C. Elliott;
England: G. E. Anderton, R. H. Burton, K. T. Collington, P. R. Hornsly, D. R. Moore, G. Woods;
France: Ch. Bonfillon, P. Jentet;
Federal Republic of Germany: H.-J. Barth, A. Bauer, W. Becker, G. Blay, F. L. Boschke, H. Börger, H. Eckardt, V. Faroga, H. Hauf, J. Härting, W. Kleber, A. Malburg, G. Menges, H. Müller, U. Osinski, K. Pontius, H. Reichstein, H. Sadlon, R. Schaffrath, E. Schirlbauer, K. Schlüter, E. Zahn, H. Zehender;
Holland: G. Hesse, P. Salwiczek;
Italy: C. Fiorentini, L. Nicolais, O. Orlandi, R. Pernice;
Japan: K. Ashida, H. Okamoto, Y. Oyanagi;
Mexico: A. Rios Sanchez;
Switzerland: B. Baumberger, G. H. Hull, J. P. Sormani, A. Sternfield, W. K. Veith;
USA: J. Ahnemiller, M. Amon, R. G. Angell, E. W. Archer, W. E. Becker, D. L. Bernard, W. R. Burk, N. Chessin, M. Cook, R. Drake, J. A. Gribens, C. D. Han, G. T. Harrick, R. L. Heck, V. Herbert, S. Hettinga, S. Hobbs, E. Hunerberg, J. Klemm, D. L. Leis, L. T. Luft, R. J. Manno, G. L. Nelson, F. R. Nissel, R. C. Progelhof, D. Reithoffer, C. Rosis, M. Rubenstein, D. L. Smith, C. D. Storms, P. M. Thomas, J. L. Throne, H. P. Torner, C. S. Wang, B. C. Wendle, A. C. Werner, J. E. Widdel, and M. G. Wilson.

I owe thanks to my colleagues from the USSR who kindly agreed to look through various chapters of this manuscript and who came up with many useful remarks and comments; they are: I. Chaikin, Yu. Esipov, G. Kusmin, Yu. Lozhechko, and A. Zukerman.

Nevertheless, my main coauthor has been my loving, devoted and understanding wife Helena, whose faith in my work has been constant.

Table of Contents

Section B: Integral Polymer Foam Technology

Abbreviations

ABS	Acrylonitrile-butadiene-styrene copolymer
ACA	Azodicarbonamide
BA	Blowing agent
BMC	Bulk molding compound
CBA	Chemical blowing agent
CCIM	Counter-current impingement mixing
CFE	Controlled foam extrusion
CIM	Conventional injection molding
CTE	Coefficient of thermal expansion
DMCHA	Dimethylcyclohexylamine
DMEA	Dimethylethanolamine
DNPMTA	Dinitrosopentamethylenetetramine
DUDEG	Diurethane-diethylene glycol
DWV	Drain, waste and vent pipe
EMI	Electromagnetic interference plastic
EMT	Electromagnetic transparent plastic
ESD	Electrostatic discharge (dissipative) effect
FFE	Free foaming extrusion
FR	Fire-resistant material
FRP	Fiber reinforced plastic
GCP	Gas counter pressure process
GSE	Gas-structural element
HDPE	High-density polyethylene
HF	High-frequency heating element
HIPS	High-impact polystyrene
HM	High modulus material
HP	High pressure process
HPE	High performance elastomeric material
HPIM	High pressure impingement mixing
HV	High velocity process
IF	Integral foam
IM	Injection molding
IM-HP	High pressure injection molding
IM-LP	Low pressure injection molding
IPN	Interpenetrating polymer network
LIM	Liquid injection molding
LRIM	Liquid reaction injection molding

LP	Low pressure process
LPF	Low pressure foaming
LRM	Liquid reaction molding
MDI	Methylene-diisocyanate
MMD	Molecular mass distribution
PA	Polyamide
PAPI	Polymethylenc-polyphenylisocyanate
PBA	Physical blowing agent
PBE	Private branch exchange
PBT	Poly-(butylene terephthalate)
PC	Polycarbonate
PE	Polyethylene
PF	Phenol-formaldehyde resin
PIC	Polyisocyanate
PMMA	Poly(methyl methacrylate)
PO	Polyolefin
PP	Polypropylene
PPMS	Poly-para-methylstyrene
PPO	Poly(phenylene oxide)
PS	Polystyrene
PUR	Polyurethane
PVA	Poly(vinyl alcohol)
PVC	Poly(vinyl chloride)
RFI	Radio frequency interference plastic
RIM	Reaction injection molding
	Reinforced reaction injection molding
RSG	Reaction "Schaum Guss"
SAN	Styrene-acrylonitrile copolymer
SF	Structural foam
SIN	Simultaneous interpenetrating network
SMC	Sheet molding compound
SW	Structural web
TCF	Two component foam
TCM	Thermoplastic cellular molding process
TDI	Tolylene diisocyanate
TEA	Triethylamine
TEC	Thermal expansion coefficient
TSG	Thermoplast "Schaum Guss"
UV	Ultraviolet

Chemical and Physical Symbols

A	coefficient; width; work
B	coefficient; thickness
C	concentration; shape factor
d, D	diameter
D	diffusion coefficient
D_ϱ	density gradient

E	elasticity modulus
E_0	modulus of unfoamed polymer
E_c	compression modulus; core modulus
E_{ds}	dielectric strength
E_f	flexural modulus
E_s	skin modulus
E_t	tensile modulus
G	shear modulus
H	hardness; roughness; thickness
I	moment of inertia
I_i	isocyanate index
k	Boltzmann constant; solubility coefficient
K	curvature; foaming coefficient
K_F	Fikentscher constant
L	length; coefficient
L/D	length-to-diameter ratio
n, N	coefficients
n_{OH}	hydroxyl group concentration
N	Avogadro number
P	pressure; weight
Q	shear force
r	radius; δ_c/δ_a
R	gas constant
s_y	static moment
S	surface area
$\operatorname{tg} \alpha$	dielectric losses (dissipation factor)
T	reduced moment of inertia; temperature; tight
T_{dec}	decomposition temperature
T_m	maximum temperature; molding temperature
T_{mold}	molding temperature
W	moment of resistance; width
x	distance
y	deflection; distance; height
Z	distance
α	angle; coefficient of thermal expansion
δ	thickness
δ_a	article thickness
δ_c	core thickness
δ_{iz}	intermediate zone thickness
δ_s	skin thickness
ε	dielectric constant; deformation
ϑ_p	volume fraction of polymer phase
$\varkappa$	shrinkage
$\varkappa_s$	surface resistance
$\varkappa_v$	volume resistance
λ	thermal conductivity; wave length
λ_0	thermal conductivity of air

λ_g	thermal conductivity of gas
λ_p	thermal conductivity of unfoamed polymer
ν	Poisson number
ϱ	density; apparent density; specific density; local density; volume weight
ϱ_0	density of umfoamed polymer
ϱ_a	article density
ϱ_c	core density
ϱ_m	overall density
ϱ_p	density of unfoamed polymer
ϱ_r	relative density
ϱ_s	skin density
ϱ_{st}	density of structure
$\Delta\varrho$	differential density
σ	strength; surface tension
σ_c	compressive strength; core strength
σ_f	flexural strength
σ_p	strength of unfoamed polymer
σ_t	tensile strength
σ_{tr}	tearing strength
τ	time; cycle
τ_c	cooling time
τ_d	dwelling (molding) time
τ_φ	conversion period
φ	degree of conversion
φ_ϱ	integral density distribution

Units Conversion Chart

Length	1 mm	= 0.0394 in
	1 m	= 3.2808 ft
Area	1 mm^2	= 0.0016 sq in
	1 m^2	= 10.764 sq ft
Volume	1 m^3	= 35.3144 ft^3
	1 l	= 0.26417 gal
Density	1 kg/m^3	= 0.0624 pcf
Mass	1 kg	= 2.2046 lb
Force	1 kg	= 9.80665 N $\approx$ 10 N
Strength/Modulus	1 kg/cm^2	= 0.0981 N/mm^2 $\approx$ 0.1 N/mm
	1 MPa	= 1 N/mm^2
	1 MPa	= 145 psi
Pressure	1 at	= 1 kg/cm^2 = 0.9806 bar $\approx$ 1 bar
	1 Pa	= 10^{-5} bar
Thermal Conductivity	1 W/Km	= 0.86 kcal/m h °C
	1 W/Km	= 0.579 BTU/ft h °F
	1 W/Km	= 6.95 BTU in/ft^2 h
Heat Quantity	1 kcal	= 4.187 kJ
Dynamic Viscosity	1 cP	= 1 mPa · s
Temperature	°F	= °C · 1.8 + 32

Section A:
General Information

1 General Description of Integral (Structural) Foams

1.1 Definition of Integral Foam

An important achievement of plastics technology is the development of *integral, or structural foams*. These foams possess a peculiar structure which is distinguished by a skin, whose density equals or is close to that of the unfoamed plastic, and a porous core (Figs. 1.1 and 1.2).

Integral foams (IF) should not be confused with laminated foams (laminates) which have a similar structure, or with sandwich-panels. Laminates are produced by joining (gluing, melting, welding, etc.) together pre-fabricated "parts", that is the external coverings are joined to the cores; and sandwich-panels are made by foaming different compositions between pre-molded coverings or sheets (Fig. 1.3).

The distinguishing technological feature of integral foams is that they are fabricated in a *single manufacturing cycle*, while their distinguishing "morphological" feature is their unified structure which *integrates* two types of plastics (unfoamed and foamed) and an "intermediate density" zone (intermediate zone). In this zone the density gradually increases from that of the core to that of the high-density unfoamed skin (Figure 1.3d and e). The skin and intermediate zones are therefore both *integral parts* of any IF article (Fig. 1.1, 1.2).

Since this structure resembles that of natural wood, IF's are often referred to as *artifical wood*, thereby not only emphasizing the formal structural similarity of the

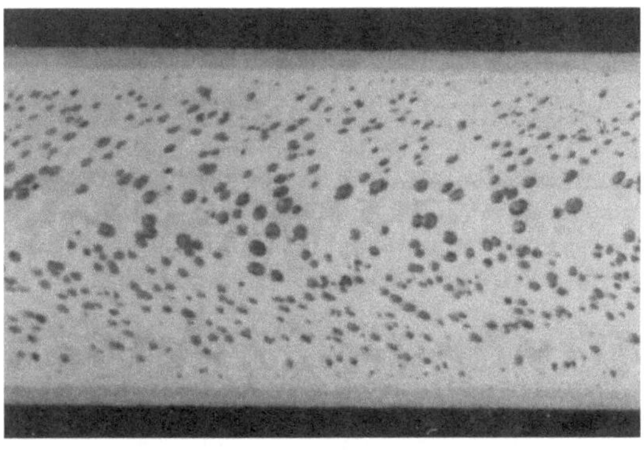

Fig. 1.1. Cross Section of rectangular integral foam molding fabricated from Lexan ® polycarbonate resin, showing real skin-core morphology (photo submitted by S. Y. Hobbs, General Electric Company, Schenectady, USA)

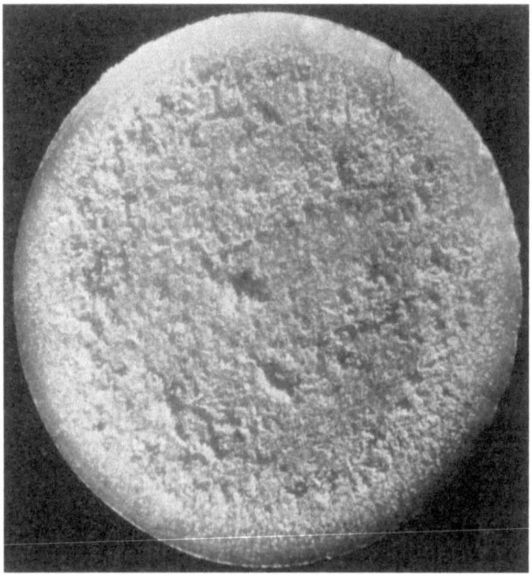

Fig. 1.2. Cross Section of round integral foam extruded article fabricated from poly(vinyl chloride) resin, showing real skin-core morphology (photo submitted by P. Jentet, Ugine Kuhlmann, Paris, France)

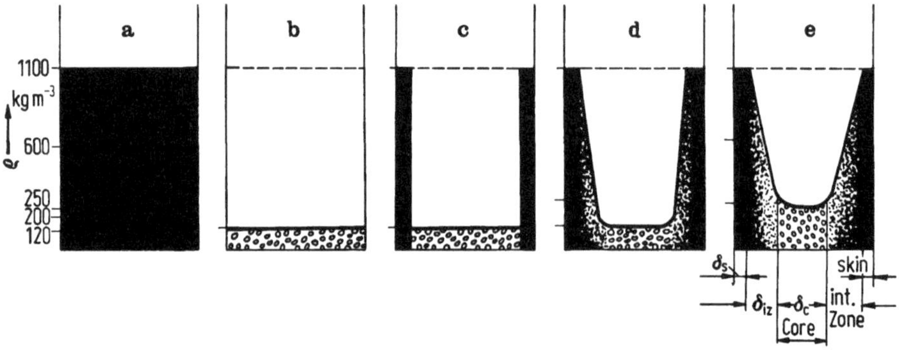

Fig. 1.3a–e. The density distribution over a cross-section for different plastic materials having the same thickness; **a**: unfoamed (compact) plastic, density $\varrho = 1{,}100$ kg/m^3; **b**: non-integral (isotropic) foamed plastic, $\varrho = 120$ kg/m^3; **c**: glued sandwich of unfoamed plastic sheets, $\varrho = 1{,}100$ kg/m^3, and a non-integral foamed plastic sheet, $\varrho = 120$ kg/m^3; **d**: integral plastic foam with an average density $\varrho = 200$ kg/m^3 (core density $\varrho = 120$ kg/m^3); **e**: integral plastic foam with an average density $\varrho = 600$ kg/m^3 (core density $\varrho = 250$ kg/m^3); diagrams d and e show the three zones of integral structures

materials but also indicating the main functional application of integral foams which is to replace wooden products and structures in various areas of engineering and construction.

A typical cross-section of an IF part is approximately 6.4 mm thick with a cellular core and an outer skin that is from 0.25 to 1 mm thick. Recently, it has been demon-

Table 1.1. Comparison of integral and non-integral (conventional) polyurethane foams [1]

Properties	Integral foam	Non-integral foam
Density, kg/m^3	432	416
Flexural strength, MPa	15.05	10.0
Flexural modulus, MPa	441	311
Tensile strength, MPa	8.1	6.4

strated that IF parts can be molded as thin as 3.2 mm (so called thin-wall parts, see Chapt. 22).

The main difference between integral foams and conventional (non-integral) foamed plastics resides in their different ratios of skin thickness to overall part thickness. This ratio is the key to explaining the better structural properties of IF's compared to conventional foams, even though their skin densities and overall densities are about the same (see Chapt. 21). For example, an IF article based on polyurethane has mechanical properties that are about 50% higher than those of an article made from a conventional non-integral polyurethane foam with the same density and formulation (Table 1.1) [1].

1.2 Terminology

Currently, there is no common terminology for these materials in the literature. We use the term "integral foams", not only because it enjoys a wider usage, but also because it is a fairly accurate definition of the physical structure and technology of these materials. On the other hand, "structural" foams, which are sometimes also referred to as "construction" foams, are believed to include almost all foam-based materials having construction applications, i.e. filled and reinforced foams, foam laminates and sandwich panels, extruded profiles, partially foamed (pre-foamed) plastics, and even conventional (non-integral) foams with the very thin compact ("technological") skin that exists at the external surface of any foamed article (before cutting or secondary treatment) [2].

According to Throne [3], "structural foam" is a misleading term implying that the foamed plastic has, somehow, better physical properties (strength, for example) than unfoamed plastics.

The problem of terminology for these plastics is still under discussion, and some years ago the debate became quite heated. Wendle remembers [4] that "back in 1969 at a Society of Plastics Industry, Cellular Plastics Division meeting some 30 men from various segments of the plastics industry had a knockdown battle over the definition of 'structural foam'. Before the meeting was over, half of the group got up and in anger left the meeting because no one could really define this new plastic form".

The term "structural foam" is now used primarily for describing materials based on thermoplastic polymers, and is used to emphasize that they are stronger than conventional foamed plastics. Another term is "*self-skinning foam*", which is generally used in English-speaking countries for describing integral foams based on polyurethane.

Today the FRG and the USA have accepted standards for IF's. For example, the new version of the DIN-7726 standard "Foams, terms and classification", which defines these materials, is used in West Germany. The definition given (DIN Draft 53432) is: "Integral skin foam (also known as structural foam) is a foam in accordance with DIN-7726, having a skin zone which, as a result of the molding processes, has a higher density (almost the density of the polymer skeleton substance), than the core" [5].

In the USA the Structural Foam Division of the Society of Plastics Industry defined this material as a plastic product having integral skins, a cellular core and a high enough strength-to-weight ratio to be classed as structural. This definition leaves much to be desired because many foamed plastics meet these criteria [4a].

1.3 Distinctive Properties

The integral structure of the materials concerned gives them their extremely high rigidity and bending/shear strength. Their specific rigidities and specific strengths (per unit mass) not only exceed those of unfoamed plastics but also exceed the values typical of some metals and wood as well (Tables 1.2 and 1.3).

Table 1.2. Comparison of different integral foams and metals [6]

Material	Relative stiffness (A)	Relative weight (B)	A/B	A/B to article price ratio
Acetal copolymer[a], unfoamed	1.0	1.0	1.0	1.0
Nylon 6,6 (+33% glass fiber)	13.5	1.4	9.6	9.5
Polycarbonate (+5% glass fiber)	7.1	1.2	5.9	5.9
Poly(phenylene oxide)	5.2	1.1	4.7	5.9
Polyethylene	3.1	1.1	2.8	9.3
Polypropylene	2.8	0.9	3.1	12.9
Polystyrene, high impact	4.7	1.1	4.3	14.3
ABS copolymer	4.3	0.7	6.1	15.6
Steel sheet[b]	8.9	2.7	3.3	24.4
Aluminum sheet[c]	5.0	1.1	4.5	9.0
Aluminum, die cast[a]	27.5	1.9	14.5	29.0

Note: all materials are 6.3 mm thick, except [a] (3.1 mm), [b] (1.45 mm), and [c] (1.8 mm)

Table 1.3. Comparison of weight-stiffness ratios for different woods, plastics and metals [7]

Material	Flexural modulus of elasticity, MPa	Specific gravity, kg/m^3	Weight-equal stiffness ratio
White Pine	7,910	370	0.307
Elm	9,310	550	0.400
Red Oak	12,460	660	0.480
Plywood (exterior)	6,300	560	0.515
Integral ABS copolymer	630	520	1.000
Aluminum	70,000	2,800	1.120
Filled plastic	5,110	1,310	1.206
Steel	203,000	7,830	2.270

These properties, in turn, mean that IF's can be used as *light construction materials* capable of withstanding considerable mechanical stresses, and as substitutes for wood and metal.

However, in many applications stiffness or rigidity is not the only structural requirement and depending upon the type of stress induced on the part, a balance of mechanical properties may be required. For example, the automobile industry must consider the flexural modulus, tensile and elongation properties as well as the heat distortion temperature, impact strength and product design life when choosing a material. If an integral foam is to be considered as a replacement for metal, in applications requiring strength and good impact properties, then a high Young's modulus coupled with a high elongation-before-break (high work to rupture) will be required [8].

Integral foam articles have the following advantages:

the weight is less than that of compact (unfoamed) articles of the same volume and configuration;

they are very rigid due to the fact that very thick walls can be produced;

finished IF items contain only small frozen-in stresses and hence they have little or no distortion;

very thick walls and very abrupt changes in the wall thickness can be achieved;

little or no provision for ribs is necessary and hence smooth-surfaced designs can be employed;

due to their cellular structure the finished articles can be machined well;

surface finishing presents no problems although this expense must be weighed against required product quality.

There are two limitations to the IF process: the surface appearance of the moldings and the increased cooling time. Integral foams characteristically have a "silvered" swirled surface finish. For the majority of technical applications this is not a major problem, but where appearance is important some degree of finishing is normally required. Various techniques have been developed to yield moldings with improved surface finishes and these are discussed later. Because a wall section of at least 6 mm is normally required to produce IF parts with a worthwhile density reduction, cooling times are often longer than with some other injection moldings [9].

1.4 Comparison with Other Materials

An interesting comparison between different material and process alternatives was made by Hazneci [10] (Table 1.4). As to sheet metals, even though these materials have been used extensively in the past, raw material and labor costs have recently increased, thus making this choice economically prohibitive. Metals can also be dented or damaged, resulting in the complete loss of the part; in addition fully integrated parts cannot be made with a metal fabrication process. The *sheet molding compound* (SMC) approach is often practical, but is labor intensive for large parts with complex configurations. The *reaction injection molding* (RIM) process utilizing polyurethane is used more often and its appeal is increasing (see Chapt. 8), although economically it sometimes does not yet have a clear benefit.

Table 1.4. Comparison of integral foams and various other methods and materials[10]

IF's vs Sheet Metals	IF's vs Die Casting	IF's vs Sheet Molding Compounds	IF's vs unfoamed RIM
ADVANTAGES			
1. Fabrication economy: less assembly time; less final product inspection time 2. Fewer parts required for assembly 3. Dent resistance 4. Greater design freedom 5. Better sound damping 6. Reduced tooling costs for complex configuration	1. Much lower tooling costs 2. Longer tooling life 3. No trim dies required 4. Lighter weight 5. Better impact resistance 6. Better sound damping	1. Uniform physical properties 2. Warping and sink marks reduced or eliminated 3. No resin-rich areas to cause configuration problems 4. Greater design freedom 5. Large parts more economical 6. Better sound damping 7. Lower tooling costs	1. Lower tooling costs 2. Better large part capability 3. Better sound damping 4. Lower internal stresses 5. Sink marks reduced or eliminated
DISADVANTAGES			
1. Smaller variety of finishes available, such as chrome or baked enamel 2. Creep 3. Low impact resistance at −5 °C 4. No grounding capabilities 5. Thicker wall	1. No heat sink capabilities 2. No grounding capabilities 3. Higher finishing costs 4. Thicker walls 5. Possible internal voids	1. Increased finishing costs (surface swirl) 2. Heat distortion 3. Creep 4. Thicker wall 5. Lower physical properties 6. Possible internal voids	1. Poorer surface finish 2. Longer cycle time 3. Thicker walls 4. Poorer high-volume economics 5. Less equipment available for various shot sizes

1.4.1 Integral Foam *versus* Wood

The first major success of IF came as a wood substitute, mainly for furniture, and this application is still its major use. The continuing high prices and uncertain supply of hard woods (such as mahogany, maple and pecan) ensures that IF's will go on replacing wood in many applications, such as structural members, carving detail, and large sections for modular cabinet units (Table 1.3).

Wood is 5 to 10 times stiffer than IF, but proper design can often solve this disparity. In all other respects, IF equals or exceeds wood's performance. It does not rot or absorb water nor is it a source of nourishment for insects. Large surface areas will not warp, split or splinter. IF's have the appearance, feel, weight and sound of wood.

IF parts can be fastened using the same methods as with wood. Screws, staples and nails have approximately the same retention properties as in wood. Joining elements such as tongues, grooves, dovetails, as well as mortice and tenon can be molded into the mating parts, saving assembly operations and providing excellent joint strength.

IF can be made to look very much like wood with color and graining molded in, and the surface can be finished with stains, waxes, varnish, etc. Unlike wood, an IF does not possess a porous surface and so it resists discoloration [11].

1.4.2 Integral Foam *versus* Unfoamed Plastics

The design of injection molded solid plastic parts is limited by the process. These limitations affect part size and wall thickness, and require design adjustments to preclude internal stresses and sink marks. Integral foams relax the dimensional limitations considerably, and the low material density reduces the chance of warpage, shrinkage and sink marking during cooling, even for thick sections. The absence of internal stresses removes sites of possible subsequent failure due to cracking and crazing.

IF obviously outperforms unfoamed plastic for large or thick-section parts. Increasing the thickness may raise the rigidity of the part three- or four-fold for the same weight of material.

Injection molding of solid plastics also allows functional items like bosses to be included, but these must be carefully designed to avoid sink marks. Using IF, large bosses can be designed, thus reducing the number of fastening points. Large unfoamed plastics often require long and heavy ribbing to achieve the desired rigidity, and this increases tooling costs; IF often has the necessary rigidity without any ribbing [11].

Integral foams clearly have substantial advantages over unfoamed plastic materials, and these advantages can be achieved with articles having wall thickness of 4 to 5 mm. Although technically there are no upper limits to the wall thickness of an IF item, in most cases economic reasons dictate a wall thickness of 5 to 10 mm.

1.4.3 Integral Foam *versus* Metals

IF has a higher stiffness-to-weight ratio than steel and aluminum (Table 1.2), possessing also good impact resistance and dimensional stability. It is therefore strong enough to challenge sheet metal and die cast construction. The first application in which IF's are challenging the more traditional materials is for the housing and components of business machines and computers, where considerable savings in fabrication and assembly costs can be achieved. The advantage is that several parts can be combined into one large molding, including all necessary hardware mountings. Bosses, supports, mounting pins, hinges, guides, handles and other items can all be designed into the molding, and these details will be faithfully reproduced by a low pressure process. Tolerances up to 0.08 mm over a span of 300 mm can be achieved.

The surface appearance of these metal-replacement housings and panels is excellent with pebbling and other textural effects being possible. The surface readily accepts painting, and several thermoplastics can be electroplated. The components are dent, crack, and rust resistant whilst the cellular-core structure gives the added advantages of sound dampening and thermal insulation for business appliances [11].

1.4.4 Integral Foam *versus* Concrete

The transportation and handling costs for concrete components are very high because of the weight involved; foamed parts offer an economic alternative. IF's are as strong,

Fig. 1.4. This six-compartment utility cable ducting weights 8.6 kg per 92 cm section (in concrete the unit weighs 182 kg); the exterior ribs provide stability during back-filling and increasing stiffness (Molder: Phone-Ducs, Inc.; plastic: HDPE. Courtesy of Union Carbide Corporation.)

inert, impact resistant and durable as required — over a wide temperature range — and so they can replace concrete in rugged outdoor uses (Fig. 1.4).

IF is being used to replace concrete for underground utilities, ducting and housing, and appears to be suitable for such massive parts as casket vaults. It also competes with decorative concrete parts like birdbaths, planters, and statuary, and IF items are very weatherable and will neither rot nor mildew. Their ultraviolet resistance depends upon the plastic used, and can be improved with additives or surface coatings [11].

1.4.5 General Comparison

Among the many reasons underlying the extremely rapid development of the integral foam industry (these reasons will be discussed in more detail in Chapter 23) the following are particularly important: IF's bridge the gap between articles produced by molding solid low-performance polymers and those produced from sophisticated engineering polymers. Moreover, in many areas IF has overcome the problems that previously had been associated with changing over from conventional materials such as wood, metals and concrete to plastics, and hence new markets have been opened up to the foam molder. At the same time, one of the pioneers of IF-technology, J. Hendry [12], pointed out that integral foams "will not replace injection molding, nor will they compete against reaction injection molding, sheet molding, or vacuum forming. If the geometry is flat, dull and uninteresting such as flat panels, it could go sheet molding or vacuum forming, or if soft pliability is required, reaction injection molding could be used. But structural foam will compete where multiple assemblies or sub-assemblies can be married into one part" (Fig. 1.5). We shall analyse the design of these assemblies in Chapter 22.

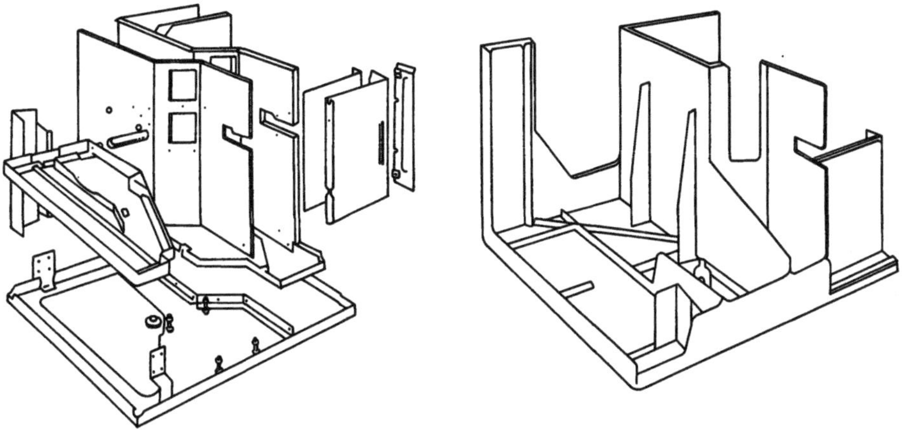

Fig. 1.5. Comparison between a conventional metal assembly and an IF molding assembly. The advantages of the latter are more economic manufacture and substantial weight savings (Courtesy of General Electric Plastics B.V., Bergen op Zoom, Holland)

Generally speaking, even in countries where wood and metal are cheap and where energy is not yet a critical problem, producers and consumers are very seriously considering solutions which use the most economic materials; moreover they are trying to save as much plastic material and labor as possible for assembling, finishing and sometimes maintaining the products. Integral foamed plastics are already serving these purposes in many applications.[17,18]

1.5 Classification

Integral foams can be divided into two types depending upon the quality of the finished article's surface: the first type includes articles with a *smooth constant-thickness skin*, and the second type includes articles with an *embossing variable-thickness skin* that resembles the bark of a tree. In other words, integral foam items can be divided into articles with *smooth-mirror surfaces* and those with *rough, dull surfaces*[13,14]. The advent of methods for producing integral foams from different polymers by molding together two unlike polymers (one for the unfoamed skin and the other for the foamed core), has made it necessary to distinguish between *one-component* and *two-component* IF. Finally, there are numerous classifications that depend on the type of equipment used to produce the integral foam, or on the conditions under which it is made: standard or special-purpose machines; low or high-pressure methods; slow or fast mixing; constantly or cyclically heated molds; with or without back pressure, etc. [15].

We believe that the technology underlying the production of an integral foam should be basic to its classification. All the existing manufacturing processes used for IF's can be reduced to: *injection molding*, *extrusion*, and *rotational molding* [16]

1.6 References

1. Horvath, M.: J. Cell Plast. *12*, 289 (1976)
2. N.N.: Plast. Eng. *37*, No. 10, 16 (1981)

3. Throne, J. L.: Principles of Thermoplastic Structural Foam Moulding, In: Proc. Intern. Conf. Polymer Process, 1977, MIT Press, 1979, pp. 77–131
4. Wendle, B. C.: Future of Structural Foam. In: Engineering Guide to Structural Foams. Wendle, B. C. (ed.) Westport: Technomic, 1976, pp. 224–228
4a. Wendle, B. C.: Intern. Conf. Foamed Plastics, Düsseldorf FRG, 1983
5. Botsch, H.: Europ. J. Cell. Plast. *3*, 115 (1980)
6. Wendle, B. C.: Materials. In: Engineering Guide to Structural Foams. Wendle, B. C. (ed.). Westport: Technomic, 1976, pp. 76–89
7. Rider, L. B.: Plast. Eng. *31*, No 11, 19 (1975)
8. Domenick, G.: J. Elast. Plast. *11*, No 2, 133 (1979)
9. Norgan, M. R.: Cell. Polymers, *1*, No 2, 161 (1982)
10. Hazneci, N.: 6th Conf. Structural Foam, Bal Harbour USA, 1978
11. Structural Foam: Union Carbide Corporation, New York USA, F-44669B, 11/74-10M
12. Hendry, J.: 6th Conf. Structural Foam, Bal Harbour USA, 1978
13. Berlin, A. A., Shutov, F. A.: Strengthened Gas-Filled Polymers. Moscow: Khimia, 1980 (in Russian)
14. Shutov, F. A.: USSR Conf. Gas-Filled Plastics, Suzdal, USSR, 1982
15. Colangelo, M.: Plast. Technology *29*, No 6, 41 (1983)
16. Shutov, F. A.: Intern. Symposium on Plastics in Building, Liège Belgium, 1984
17. Shutov, F. A.: 4th Conf. Mechanics and Technology of Composites, Varna Bulgaria, 1985
18. Shutov, F. A.: 4th Conf. Polymer Science and Technology, Mar-del-Plata Argentina, 1985

2 Starting Materials

The first patents in IF technology were granted between 1962 and 1964 to its two inventors, R. Angell (USA) and T. P. Engel (France). These patents pertained to integral polystyrene foam [1], and were later used as a basis in Europe and the USA for the production of the first commercial grades. Integral polystyrene foam was followed by integral polyolefin and poly(vinyl chloride) foams.

2.1 High Molecular Weight Polymers and Reactive Oligomers

Commercial IF's are mostly made up from high molecular weight polymers, namely *polystyrene* (PS), the *polyolefins* (PO), i.e., *polyethylene* (PE) and *polypropylene* (PP), *poly(vinyl chloride)* (PVC), *polycarbonate* (PC), *poly(phenylene oxide)* (PPO), *ABS-copolymers* (ABS), *polyamides* (PA), thermoplastic polyesters, etc.

Until recently, *polyurethanes* (PUR) and urethane compositions modified by other polymers were the predominant thermosetting resins used for integral foams. Later, however, commercial IF's based on phenol-formaldehyde resins (PF), and epoxy and polyisocyanurate resins have appeared.

Taking into consideration the favorable opportunities provided by the "oligomer" route, we should expect IF production based on polymerizable oligomers to expand rapidly in the near future, especially in the *Reaction Injection Molding* (RIM) of polyurethanes (see Chapter 8) [2].

An extremely interesting and promising area of polymer science surrounds two new closely related classes of polymer systems, viz., simultaneous interpenetrating networks (SIN) and interpenetrating polymer networks (IPN). These systems have recently been applied to produce integral foam components by RIM technology.

The manufacturing technology, properties, applications, as well as economic and marketing positions will be considered in separate chapters.

In this chapter we shall discuss problems pertinent to any IF, namely the problems of blowing agents and additives, because without these substances it is impossible to produce a foamed polymer.

2.2 Blowing Agents

The properties and production economics of an IF can vary significantly, depending on the *blowing agent* (BA). The "ideal" blowing agent is one that is compatible with the base resin, produces a void-free compact outer skin, ensures a uniform transition to the foam core, and gives the best blowing efficiency for the minimum concentration and price.

Both *physical blowing agents* (PBA) and *chemical blowing agents* (CBA), as well as mixtures of them are used to produce integral foams.

The composition used to produce an IF must take into account the compatibility of the BA with the starting polymer or oligomer, as well as the effect of the composition components upon the thermal decomposition (the effect of variations in the composition on temperature and gas numbers) of the chemical BA involved. Thus, a small water content would not only affect (reduce) the activity of either a PBA or a CBA, but it would also render a CBA and the polymer incompatible. At the same time, many flame-extinguishing admixtures, fillers and nucleating agents (e.g., titanium oxide and talc) tend to affect the foaming activity of a CBA adversely.

2.3 Physical Blowing Agents

The most frequently used PBA's are the freons, nitrogen, hydrogen and compressed air [3]. Despite the high price of *freons*, their application is justified by the 10–30% lower densities of the final IF parts compared with the same compositions foamed by nitrogen [4]. Hence the unit costs are the same due to feedstock savings and reduced production period. Recently General Electric Plastics reported[5] some interesting results by substituting Freon-22 for nitrogen to make PPO integral foams. According to the report Freon reduced swirl, upgraded physical properties, cut cycle time, and saved money (see Chapter 18).

Freons and foaming systems based on *isocyanate water mixtures* are largely used to produce integral PUR, whereas the other PBA's are useful for foaming thermoplastics.

It should be noted that in the USA *nitrogen* is by far the most widely used PBA for manufacturing IF's, but CBA's predominate in Western Europe, the main reason for this difference being the type of equipment available. Union Carbide Corporation is leading in the promotion of the foam process which uses N_2 as the foaming agent. Meanwhile some equipment has been introduced in USA which uses only CBA's and some machines allow the option of either N_2 or CBA, or both.

There is a widespread opinion that the cost of using N_2 is less than that of using CBA's, but the point is arguable since there has been no direct comparison between the two for identical machines, tools and conditions over an extended run. Of course, the comparative cost of blowing agents is not going to be the decisive factor as to which process or machinery to purchase. The finishing cost and reject rates will most likely depend on the surface finish and the uniformity of cell structure, and it is known that CBA's provide a better cell structure and superior surface than those provided by nitrogen.

PBA's can be introduced into a foaming system either as *volatile liquids* or as *compressed gases*. Liquid agents such as the fluorocarbons are either introduced directly into the plastic melt in a barrel, or added beforehand to the plastic granules. Gaseous blowing agents (N_2, CO_2, etc.) are introduced into the melt in the barrel at a pressure higher than that applied to the melt.

2.4 Chemical Blowing Agents

2.4.1 Types of Agent

As for the CBA's, the most commonly used are *azodicarbonamide* (ACA) and the bicarbonates of the alkali metals; 5-phenyltetrazole and sulfonylhydrazides are also used. When thermoplastic IF's are being produced, the CBA should be introduced

Table 2.1. Commercial organic CBA's for integral foams [6)]

Chemical	Decomposition Temperature, °C	Temperature Range used for IF's, °C[a]	Typical Usage Level for IF, %[b]	Trademark*	Manufacturer[c]
Azodicarbonamide	204	177–232	0.5	Azocel Celogen AZ Ficel AC Kempore Porofor ADC Vinyfor AC	F UC FIC NPD B E
Modified Azodicarbonamide	204	177–232	0.5	Celogen AZNP Ficel EP Kempore MC	UC FIC NPD
p, p'-oxy-bis (benzene sulfonyl hydrazide)	157	149–177	0.75	Celogen OT Ficel OB Neocellborn Nitropore OBSH	UC FIC E NPD
Proprietary Hydrazide	199	177–232	0.75	Celogen CB	UC
P-Toluenesulfonylsemicarbazide	227	204–260	0.75	Celogen RA	UC
5-Phenyltetrazole	249	249–316	0.3	Expandex 5-RT	NPO
Trihydrazine Triazine Type	265	260–316	0.4	Ficel THT	FIC
Hydrazine Derivative	288	260–343	0.3	Celogen HT-550	UC
Aromatic 300		154–232	0.6	Kemtec 300	S
Aromatic 500		271–315	0.8	Kemtec 500	S
Aromatic 500 HP		271–315	0.8	Kemtec 500 HP	S
Aromatic 550		287–330	0.8	Kemtec 550	S

* Registered U.S. Patent Office.

[a] At the lower end of range activation may be required;

[b] Actual use level will depend on part thickness, part design, desired density, molding conditions and other factors;

[c] Code: B — Bayer, FRG; E — Ewia Chemical Ind., Japan through A & S Corp., USA; F — Fairmont Chemical Co., USA; FIC — Fisons Ind. Chemical, England through Sobin Chemical Co., USA; NPD — National Polychemical Div., Stephan Chemical Co., USA; S — Sherwin Williams, USA; UC — Uniroyal Chemical, Div. of Uniroyal Inc., USA

Table 2.2. Celogen® CBA selection guide for different IF [6)]

Type	Features	Process		Tooling		Polymers
		Low Press.	High Press.	All Except Be-Cu	Be-Cu	
Celogen OT	Low temperature, no ammonia	Yes	Yes	Yes	Yes[a]	LDPE*, EVA
Celogen AZ						
Grade 130	Low cost, most efficient	Yes	No	Yes	No	Acetal, HIPS, ABS, HDPE, PP, Thermoplastic Rubber, (LDPE, EVA*, Vinyl* with activation), Noryl
Grade 190	Slower decomposition rate, low cost, most efficient	Yes	Yes	Yes	No	Same as above
Celogen AZHP						
Grade 130	Non-plate out type	Yes[a]	No	Yes	No	Acetal*, HIPS*, ABS*, HDPE, PP*, Thermoplastic Rub.*, Noryl (LDPE, EVA, Vinyl with activation)
Grade 199	Non-plate out type, slower decomposition rate	Yes	Yes*	Yes	No	Same as above
Celogen CB	No ammonia	Yes[a]	—	Yes	Yes[a]	Acetal, HIPS, ABS, HDPE, PP, Vinyl
Celogen RA	Intermediate to high temperature	Yes[a]	Yes	Yes	—	Acetal, ABS, HIPS, HDPE, PP, Thermoplastic Rub., Noryl*, Nylon*

[a] The most common recommendation with respect to polymer, tooling, and process

just prior to composition plastication, thereby preventing any possible BA losses and providing more accurate component ratios (Tables 2.1 and 2.2).

The decomposition temperatures of many CBA's can be lowered by the addition of special additives, called activators or "kickers", such as zinc stearate.

A serious problem is *polymer-CBA incompatibility*. Although there does not seem to be an equivalent polymer-PBA incompatibility, the question of N_2 embrittlement at low foaming levels has not been fully answered [7)]. One of the main reasons for the polymer-CBA incompatibility apparently stems from chemical reactions between CBA byprodukts and the polymer. A typical example of this is the polycarbonate azodicarbonamide system. Here, NH_3 (the byproduct of the CBA) cleaves the PC backbone over long residence times leaving the PC chalk-like and friable [7)]. It has been shown that CBA's which generate H_2O as a byproduct are unsuitable for many condensation polymers such as Nylon-66 and poly(butylene terephthalate) [8], and that small amounts of CO will reduce the strength of polyoxymethylene-based resins.

The *amount of CBA* required for the foaming process depends upon the physical nature of the polymer to be foamed, namely, upon its crystallinity. Thus, amorphous polymers require far less CBA to be foamed than crystalline polymers do. For instance, in order to produce integral polystyrene using ACA, a concentration as low as 0.2% mass ACA is sufficient (30%-foaming) [9]. Under actual process conditions, the BA concentrations generally prove to be somewhat higher since possible gas leakage during the injection and mold opening, and via ventilation vents must be considered.

2.4.2 Forms of Agents

Commercial CBA's are supplied as free-flowing powders, pelletized resin/CBA concentrates, or liquid dispersions (pastes) [10-12].

CBA's in the form of powder or granular resin concentrates are added to the resin pellets, either before plastication and outside of the machine in a slow running mixer (dry tumbler) or during plastication by pouring it directly into the resin hopper, using an attached metering device (Fig. 2.1) [10].

CBA in form of pastes (usually at concentrations of 40–50%) are introduced just

Fig. 2.1. Mixing and metering equipment for introducing CBA powders or concentrates; different metering devices are used for powder and for concentrate [10] (Courtesy of Battenfeld Maschinenfabriken GmbH, Meinerzhagen, FRG)

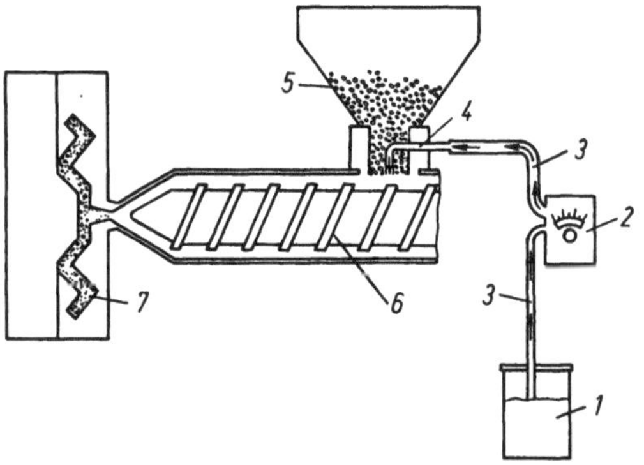

Fig. 2.2. Feeding in a CBA in the form of a liquid dispersion (paste) using a hose pump; **1**: CBA paste, **2**: metering pump, **3**: flexible tubing, **4**: metal tubing, **5**: polymer, **6**: injection machine, **7**: IF part

before and above the screw by pumping them in. The rate of pump feed is determined by the rotation velocity of the screw (Fig. 2.2).

A recent development comprises the use of a composite liquid that carries a standard CBA and is metered via a liquid pump. This system promotes faster cycles, better surface finishes and tighter short control. A liquid blowing agent can be pumped at temperatures from −20 °C to +50 °C and is available in 1 to 50-gal. drums. It is expected that it will add 2 c/lb to the price of an IF compound, although it does eliminate the need for wetting agents and a mixing step [10].

The concentrates contain the CBA incorporated in a compatible thermoplastic carrier. By using small quantities of these concentrates, the custom molder can foam-mold a variety of resins (with or without fillers, reinforcements, or other additives) using his existing equipment and handling procedures. At a ratio of 1: 200 these concentrates have proven effective in controlling the foaming coefficient and appearance of sink marks in a wide variety of resins and composites [13].

It should be noted that blowing agent concentrates can be used in any of the low pressure systems now used (see below). They are mixed with the resin before it is placed in the hopper, the applied ratio depending on the particular resin system and desired apparent density of the IF article. Theberge and Cloud [13] give the following as a guide to determine a starting point for the letdown ratio of concentrates:

	Letdown Ratio
Injection Molding	
minor sink control	200/1
major sink control, light foaming	50/1
major foaming	50/1
Extrusion	
major foaming	100/1

For example, LNP Corp., USA, manufactures some concentrates for IF's: Grade Kon-22 for PS, PPO and thermoplastic polyester (PBT), and Grade Kon-23 for PP, PE, PA, PC, polysulfones and poly(phenylene sulfides). When the CBA's are applied as liquids or granulated concentrates, decomposition activators, plasticizers, pigments, etc. can also be incorporated. Such concentrates help avoid inevitable losses that arise using powdered CBA's when preparing a composition, and considerably improve the efficiency of the equipment due to the high-speed pneumatic and vacuum transfer of the concentrates into the working zone of the processing machines. Concentrates containing up to 15–20% CBA are generally prepared in extruders and mixed with the carrier polymer in ratios between 1:10 and 1:20. The novel concentrates intended for integral PS's and ABS-plastics contain PBA-CBA mixtures. The PBA serves, as a rule, to provide the foam while the CBA is acting as a nucleating agent coming into action first (in the plasticizer) and, hence, reducing the size of the cells.

The *commercial granulated CBA concentrates* that are produced in the United States have a wide decomposition temperature range, i.e. 150° to 290 °C [12]. These concentrates are based on polypropylene and poly(butylene terephthalate). The former are intended for PO production and the CBA content needed for this application is between 5 and 50 per cent; the latter are designed for polymers which are processed at higher temperatures and contain 5 to 10 per cent CBA. These concentrates help reduce IF density by 40 per cent (for PP) and also promote shorter molding cycles, e.g. 22 sec instead of 60 sec in the case of PS; 90 sec instead of 240 sec in the case of PP, and 35 sec instead of 90 sec in the case of ABS IF articles.

2.5 Effect of Blowing Agents

Regardless of the process technology, the blowing agent can significantly affect the physical properties of an IF article. This is especially true for CBA's in low pressure processes because of the agent's additional effect on the surface quality of the molding. Eckardt [10, 11] studied the effects of using a CBA in various forms and concentrations on product physical properties and found that the effects included a wide range of changes in density, foam structure, shrinkage quality, and cooling time. Since this study is very important for any IF technology, we shall discuss the data in more detail. In the study the concentration and form of CBA (azodicarbonamide) were varied (powder, paste, concentrate), and $8.9 \times 40 \times 640$ mm plaques were molded from *high impact polystyrene* (HIPS) resin at 200 °C and 250 °C (Fig. 2.3). The concentration of CBA referred to henceforth reflects the actual concentration of pure blowing agent.

2.5.1 Density

Increasing the concentration of CBA generally reduces the density. However, there are limits beyond which there is no practical further reduction in density. This effect is shown in Fig. 2.4 for three forms of CBA and two stock temperatures. At 200 °C, lower densities (as low as 0.65 g/cm^3) can be achieved with the concentrate or paste blowing agent as compared to the powder. This means that for the same final density less CBA is required if a concentrate or paste is used. The relative advantage of larger density reductions may be due to lower losses of foaming gas during melting. At 250 °C, the differences in performance among the three forms are substantially less, though the trend remains.

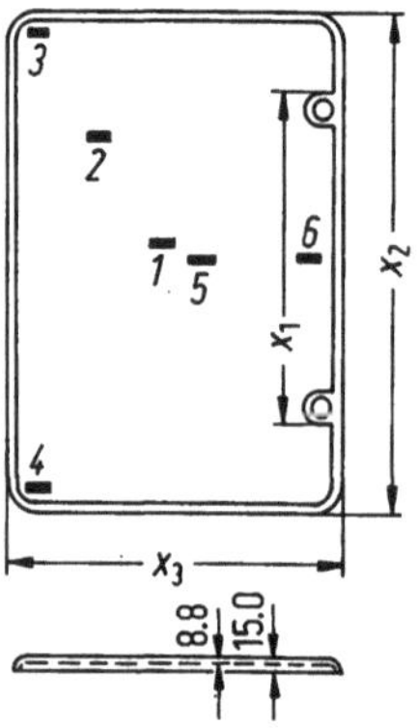

Fig. 2.3. Experimental IF plaque made from HIPS of size 640 ×40 × 8.8 mm; 1 to 6: samples taken to determine the density (Fig. 2.4) and roughness (Table 2.3); X_1 to X_3 are the distance measures to determine the shrinkage (Fig. 2.6) [10]

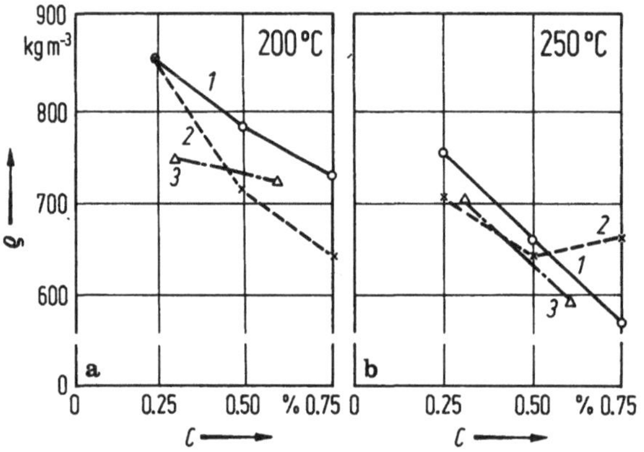

Fig. 2.4. Effect of the type (1, 2, 3) and concentration (C) of the CBA on the minimum achievable density ϱ of the IF part (Fig. 2.3) for two stock temperatures, (a) 200° and (b) 250 °C; 1: powder, 2: concentrate, and 3: paste [10]

2.5.2 Cellular Structure

The most uniform cells were obtained when the CBA was uniformly dispersed within the melt prior to injection into the mold. Owing to the various forms in which they are available (powder, pellets, liquids) and the particular way they act, CBA's are particularly well-suited to achieving uniform dispersions. In real processes, the concentration of the CBA can be adjusted to obtain the best and most uniform cellular structure. This effect is shown in Fig. 2.5 in which the concentration was varied to achieve the lowest possible density consistent with uniform cellular structure. It can be seen that although heavy loadings of CBA (1% by weight) produced a low density (250 kg/m³) and good quality surfaces, excessive gas pressure generated by too much CBA formed large bubbles and voids in the interior of the molding (Fig. 2.5, right). By cutting CBA concentration in half, the large voids were eliminated and a uniform cell structure was achieved with only a minor reduction in surface quality (Fig. 2.5, left).

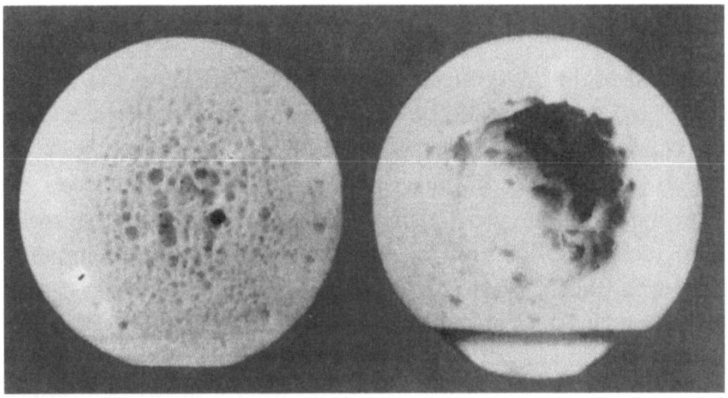

Fig. 2.5. Cross Section through an IF ball made of PE, diameter 125 mm, average density 250 kg/m³. Produced using 1% CBA (right) and 0.5% CBA (left) [10] (Courtesy of Battenfeld Maschinenfabriken GmbH, Meinerzhagen, FRG)

2.5.3 Shrinkage

According to Eckardt [10, 11] CBA's can be used in injection foam molding to reduce the in-mold shrinkage of the IF articles as they cool. The gas generated by a CBA approximates the effect of the holding pressure in conventional non-foam injection molding. The effect of increasing the amount of blowing agent (powder) to reduce even further the shrinkage at low pressure is shown in Fig. 2.6 for molded plaques based on HIPS integral foam (Fig. 2.3). At 200 °C and 0.25% CBA loading there are pronounced differences among the three sections of the IF plaque. The two sections adjacent to the core pins shrank about 0.6% whilst the transverse section shrank 0.7%. If the amount of CBA was increased to 0.5%, or the temperature was increased to 250 °C, the differences became very slight. Higher CBA loadings designed to fill the mold cavity did not reduce the shrinkage.

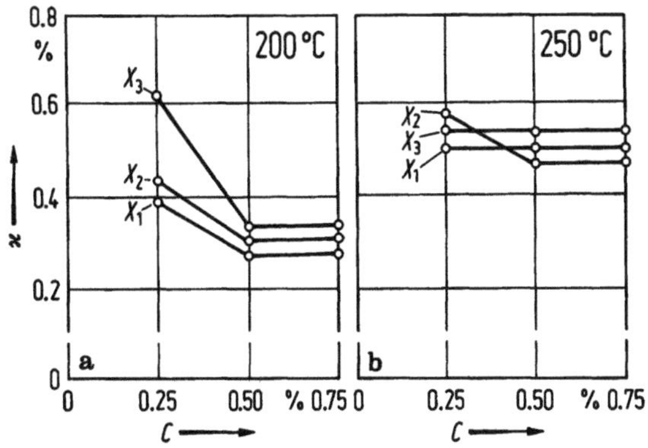

Fig. 2.6. Effect of the concentration (C) of CBA (powder) on the shrinkage ϰ of an IF part (Figure 2.3) for two stock temperatures (a) 200° and (b) 250 °C, and an injection time of 0.55 sec, average density 850 kg/m³ [10]

2.5.4 Surface Quality

The surface quality of integral foam components is highly dependent on the nature and amount of foaming gas, mold geometry, wall thickness, base resin, and various other processing parameters. For example, one of the several methods of improving of surface quality (see Chapter 12) is simply to increase the CBA concentration, or to use one that has a higher yield of foaming gas. Venting of the mold cavity must be matched to the particular blowing agent used or the surface quality will suffer. Because of the rapid injection rate used in foam injection molding, cavity air must be permitted to escape rapidly from the mold cavity. At the same time, the amount of foaming gas that is allowed to escape must be minimized. The surface of an IF article is required to be as flaw-free as possible to simplify subsequent after-treatment. The surface roughness (peak-to-valley height) of IF components is dependent on the distance from the sprue, and the roughness is less at the upper temperature. Table 2.3 shows also that no effect of the form of the CBA on roughness is detectable.

Table 2.3. Effect of form of CBA on surface roughness (μm) for a molded plaque based on polystyrene IF (see Fig. 2.3) at different stock temperatures

T, °C	Point on the plaque	Concentration CBA, % by weight								
		Powder			Concentrate			Paste		
		0.25	0.5	0.75	0.25	0.5	0.75	0.31	0.62	0.94
200	1	4	3	8	3	4	6	4	4	3
	2	4	15	15	7	9	13	9	4	24
	3	44	48	41	23	28	25	36	35	17
	4	45	>50	43	40	46	52	46	36	22
	5	3	3	10	4	14	15	4	3	3
	6	6	12	20	5	24	30	24	21	28
250	1	2	3	6	3	5	5	2	4	3
	2	4	7	20	11	15	13	14	24	28
	3	32	26	20	23	25	23	36	27	27
	4	29	23	22	20	35	22	33	31	34
	5	2	3	4	2	4	5	3	4	3
	6	15	17	20	16	17	22	22	22	21

2.5.5 Cooling Time

Increasing the concentration of CBA without also increasing the cooling time will result in the post-expansion of the molding. If a component is released from its mold too early, heat is still being dissipated from it (from the foam core to the solid outer walls). This results in higher outer-skin temperatures and can cause the outer surfaces of the molding to distort under the residual internal pressure of the foaming gas. Excessive CBA not only increases the cost of the molding materials but also increases the cooling time and thus the machine costs for the job. Eckardt [10] calculated that increasing the concentration of the CBA from 0.25% to 0.75% in a HIPS IF component molded at 200 °C, increased the cooling time required from 2 min to 2.5 min. A similar concentration increase at 250 °C raised the cooling time from 2.5 to 3 min for the same molding. Similar molding runs have shown that the efficiency of blowing agent concentrates and pastes is substantially better than that of CBA in powder form. This applies, above all, to lower processing temperatures (see Chapters 3 and 13).

Similar changes in the physical properties of IF's result when using physical blowing agents such as freon or nitrogen gas. An important factor is the homogeneous incorporation of these gases or liquids into the melt.

Eckardt has drawn the following conclusions from his work [10, 11]:

CBA's in concentrate and paste form are more efficient for reducing the density and maintaining uniform foam structures than are dry powder agents.

Increasing the concentration of the CBA can be used to reduce directional shrinkage.

The surface roughness of a molded foam item is more a function of its shape than of the physical form of the CBA.

Excessive CBA loading will unduly increase the cooling time.

At equal densities, CBA's and PBA's generally produce foam with comparable cell uniformity.

2.6 Nucleating Agents

The following substances can be used as nucleating agents: powder-like metal oxides, citric acid (for sodium bicarbonate), inorganic powders with a large specific surface (10 to 800 m^2/g) based on *silica* and crystalline asbestos, talc, etc. The required amount of nucleating agent is determined by the polymer melt viscosity and the foaming method, and varies between 0.05 and 1% (mass). The function of a nucleating agent can also be performed by powder-like pigments, fillers, and other admixtures [14, 15, 20].

2.7 Other Additives

2.7.1 Types of Additives

Because IF articles tend to lose their advantageous physical/mechanical properties as the density is decreased, reinforcing agents are sometimes used to bring back the desired strength without affecting the foam's characteristics. In fact, any sort of particles added to a foaming system tend to act as nucleating agents forming a new cellular structure in an IF product.

As in the case of blowing agents, there are basically three methods of introducing additives into a foaming system: dry powder, concentrate, and directly into the hopper as if it were a regrind. Some of the new blending units mounted directly on the machine hopper allow each of the three types to be mixed and dispersed within the base resin [15].

The following is (an incomplete) list of additives and fillers (except blowing agents):

Reinforcing agents: glass and carbon fibers, compact and hollow glas microspheres.

Fillers and extenders: wood flour, peanut hull flour, calcium carbonate.
Release additives: silicone fluids, zinc stearate.
Flame retardants: halogens, antimony trioxide, aluminum hydroxide (usually silane coated for bonding), chlorinated waxes.
Smoke suppressants: inorganic metal complexes.
Heat stabilizers: barium-cadmium or cadmium-zinc mercaptides.
UV resistant additives.
Pigments and Colorants.
Impact improving agents.
The additive used in any particular case may depend on the region or country of manufacture, due to cost considerations.

2.7.2 Fillers and Reinforcing Agents

In order to improve the strength of an IF (its modulus of elasticity and rigidity) and to reduce the IF's price, various fillers are added to the starting compositions. *Glass fibers* of various lengths are the most often used, and an article may contain 20–40% of them. The fibers are sometimes, but not always, pre-dressed to improve their adhesion to the polymeric matrix.

Later on it will be shown that the fibers introduced into an IF orient themselves and prove to be nonuniformly distributed over the article's cross section (there are more fibers in the skin than in the core). At this point, however, we would like to mention a method for additionally orienting the fibers. This is done by applying an electromagnetic field to the mold where the foaming takes place. *Metal particles* are introduced to the polyurethane composition together with *carbon fibers* which then are oriented along the field lines. This results in highly ordered fibers in the finished article.

Other fillers used are talc, metal oxides and salts, asbestos, ceramic powders with a particle size of the order of 10 μm, saw-dust, etc. The quantity of filler varies widely – from 0.001% to 60%, and depends upon the polymer type, processing method, and final article.[19]

Fiber reinforcement can also be used to reduce molding shrinkage, to improve the creep properties and to shorten the molding cycle by enabling the article to be ejected earlier from the mold, due to its improved thermal and mechanical properties [15–18].

A new and very promising application of foamed plastics is their use as *electromagnetic interference* (EMI) and *conductive materials* [16]. It has been stated [17] that 25–30% (by weight) conductive filler is needed in an integral foam to ensure shielding and that carbon fibers in IF's can provide 20 dB to 40 dB of shielding at 10 and 1000 MHz, respectively (See also Chapter 11).

2.7.3 Pigments

In order to dye an IF article directly during its fabrication, pigments have to be added to the starting composition. It is noteworthy, however, that this problem is still not fully solved because articles dyed in this way very soon lose their fresh color and fade. Hence, it is a more frequent practice to dye the finished products (see Chapter 11).

Nevertheless, some of the pigments added directly to starting compositions are lampblack and pigments based on zinc, cadmium, lead, and barium [14]. The choice of which pigment to use requires consideration of its compatibility with the CBA used. Thus, barium pigments raise the decomposition temperature of azo compounds and, at the same time, adsorb the nitrogen liberated from the BA. Pigments based on zinc, cadmium and lead, by contrast, activate the thermal decomposition of azo and nitro-compounds, but they lose their color as they interact with the decomposition products of the CBA's.
Some fillers (talc, titanium oxide) and flame-extinguishing admixtures impair the saturation of the dyestuff in an article. The thermal decomposition products of many CBA (except ACA) tend to dye integral materials in an uncontrollable way.

Note, however, that adding too much filler or pigment would complicate and lengthen all operations involved in producing IF's as these substances contribute to the composition viscosity. Sometimes compositions with fillers can be processed provided the compounding and IF production conditions are changed by introducing plasticizers or solvents, by raising the melt and mold temperatures, by increasing the BA concentration, or by slowing down injection rates [15,182].

2.8 References

1. Ryder, L. B.: Plast. Eng. *31*, No 11, 19 (1975)
2. Shutov, F. A.: Adv. Polym. Sci. *39*, 3 (1981)
3. Litman, A., et al.: AIChE Ser. *73*, No 170, 163 (1977)
4. N.N.: Mod. Plast. Int. *10*, No 9, 49 (1980); Plast. World 39, No 9, 64 (1981)
5. N.N.: Plast. World *39*, No 8, 64 (1981)
6. LaClair, R. C.: 6th Structural Foam Conf. Bal Harbour USA 1978
7. Throne, J. L.: Principles of Thermoplastic Structural Foam Moulding. In: Proc. Intern. Conf. Polymer Process, 1977. MIT Press, 1979, pp. 77–131
8. Avery, J. A., Kromer, M.: SPE ANTEC Tech. Pap. *20*, 356 (1976)
9. Mahn, H.: Plastverarbeiter *26*, No 1, 8 (1975); Kunststoffe, *23*, No 3, 24 (1976)
10. Eckardt, H.: Possible ways of influencing product quality by means of different blowing agents. Meinerzhagen: Battenfeld Maschinenfabriken GmbH, 1979
11. Eckardt, H.: Plast. World *38*, No 4, 66 (1980)
12. N.N.: Mod. Plast. Int. *12*, No 6, 53 (1982)
13. Theberge, J. E., Cloud, P.: Structural RP Foam Molding with Foam Concentrate. In: Proc. 32nd Ann. Tech. Conf. 1977, Reinforced Plastics/Composites Institute, New York: SPI, 1977, pp. 14B/1–14B/4
14. Berlin, A. A. Shutov, F. A.: Strengthened Gas-Filled Polymers. Moscow: Khimia, 1980 (in Russian)
15. Wendle, B. C.: Additives and Fillers for Structural Foam. In: Engeneering Guide to Structural Foam. Wendle, B. C. (ed.), Westport: Technomic, 1976, pp. 115–117
16. Bodnar, D. G.: 7th Structural Foam Conf., Norfolk/USA, 1979
17. Wehrenberg, R. H.: Mater. Eng. *96*, No 3, 37 (1982)
18. Shutov, F. A.: USSR Conf. Gas-Filled Plastics, Suzdal, USSR, 1982
19. Shutov, F. A.: Adv. Polym. Sci. *73/74* (1985)
20. Shutov, F. A.: 17th Europhysics Conf. Morphology of Polymers, Prague Czechoslovakia, 1985
21. Shutov, F. A.: 5th Intern. Conf. Surface and Colloid Science, Potsdam USA, 1985

Section B:
Integral Polymer Foam Technology

3 Injection Molding: General

3.1 Classification

The most common way of producing integral foam parts is *injection molding* (IM).

Several classification systems exist, but we shall consider a classification into four groups:

(1) *Low pressure (LP) processes*, up to 20 MPa,
(2) *High pressure (HP) processes*, up to 100 MPa or more,
(3) *Gas counter pressure processes*, and
(4) *Two-(or more)-component processes.*

The last two are special processes for producing IF parts having smooth external surfaces. Apart from these four groups, reaction injection molding represents an isolated process.

Before getting down to a detailed description of the main IF processes (see Chapters 4–8) we first consider the general method of producing integral foams using injection molding and the main features of the required equipment.

3.2 Basic Principle

IF molding is similar to conventional injection molding (CIM). Plastic material in the form of pellets, powders, or beads, must be melted by a conventional screw extruder and the melt stored or accumulated in a temporary reservoir during plastication. The melt is forced into the mold cavity. Enough plastic material must be injected into the mold cavity to fill it and to provide for shrinkage of the plastic as it cools.

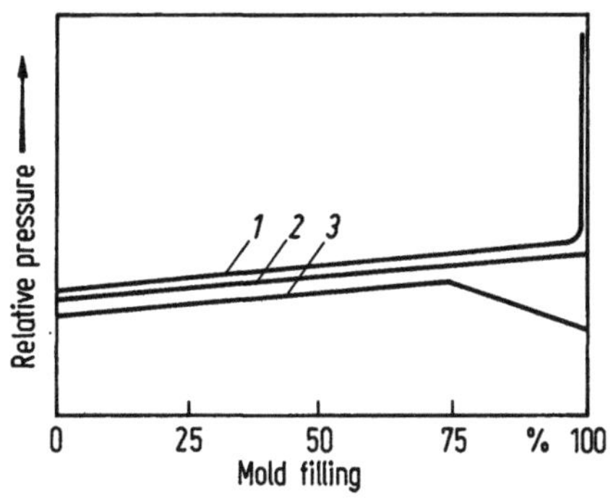

Fig. 3.1. Effect of mold filling on mold pressure during injection molding process: **1**: solid (unfoamed) article, **2**: high pressure IF article, **3**: low pressure IF article [2)]

Therefore the plastic melt must be maintained under relatively high pressure until the composition in the mold is sufficiently cool to prevent backflow. It should be emphasized that if the mold cavity is not completely packed, the article will have a poor surface quality, sink marks, voids in the thick sections, and differential distortion [1].

If the mold is not filled completely, this is called a *short shot, or low-pressure IM process* (Fig. 3.1). If the polymer melt fills the mold cavity completely, this is called a *full-shot, or high-pressure IM process*. If the polymer is mixed with a volatile substance and forced into a mold that has the same volume as it would have in a CIM process, the article appears to have been injection molded. If the mold cavity is deliberately "short shot", the volatile material forms bubbles in the melt and expands the melt until it fills the cavity. The extent of the foaming has been found to be a function of the amount and dispersion of the volatile substance added to the melt, the temperature of the melt and mold cavity, the speed of injection, and the pressure applied to the melt as it is transferred from the accumulator to the mold.

The transfer of plastic melt is common to both CIM and IF molding. During CIM it is the intention for the transfer to take place with as little flow instability or turbulence as possible, and the gates and runners are designed to minimize this sort of flow. On the other hand, the gate must perform as a check valve to prevent back-flow of the packed material from the cavity during cooling. As a result, the gate must be small and the injection speed kept low. In IF molding, however, an entirely new approach is adopted; the reason for this will be obvious once the mechanism of foaming is understood.

Later, we shall use such terms as *"low" and "high" pressures and processes* for IF technology. It should be mentioned here that this terminology is logical to those familiar with IF components. According to Burk[2], the term "high pressure" is valid when dealing with a short-shot IF process (Fig. 3.1). The pressure ratings are relative and all IF processes are low pressure processes when compared to solid (unfoamed) injection molding. Clamp pressure requirements are summarized as follows:
solid injection molding: pressure 28–56 MPa;
high pressure integral foam: pressure 7–14 MPa;
low pressure integral foam: pressure 3.5–7 MPa.

The pressure generated by these processes is a direct function of the degree of mold filling achieved. For solid injection molding the mold must be filled and packed under high pressure in order to compensate for the shrinkage which results when the part cools in the mold. The high pressure packing allows a part to be made with accurate dimensions and free of sink marks and warpage. In a high pressure IF process, the mold is just filled, but is not packed. Therefore, a high mold pressure like that used for solid injection molding is not required. The formation of the foam takes place following injection, either as a result of a mold expansion or due to egression of plastic material from the mold. In a low pressure IF process, the mold is not filled by the injection pressure, instead it is filled by the expanding (foaming) gas and so the mold pressure is very low, essentially equalling the pressure developed by the blowing agent.

3.3 Basic Commercial Processes

We try here to give a short comparative review of the main industrial processes and follow Throne's excellent survey [3].

The first process was developed by Dow Chemical Corporation in 1962, and now is available in a modified form as the Dow-TAF process (Table 3.1). Many of the

Table 3.1. Comparison of industrial IF processes according to Throne [3)]

Process name	Number inject. stations	Type of blowing agent (gas)	Mold pressure	Surface quality	Mold action	Machine action	Machine type	License cost, $	Throughput royalty
Dow	1	PBA (CH_3Cl)	Low	Poor	None	None	Mod. CIM	5,000/year	
UCC	1	N_2	Low	Fair	None	None	Special	25,000	$ 0.046/kg
UCC-Mod.	1	CBA	Low	Fair	None	None	Special	25,000	$ 0.046/kg
Variotherm	1	N_2, CBA	Low	Good	Heat-Cool	None	Special	None in Europe	
Dow-TAF	1	CBA, PBA (Freon)	Medium	Good	Motion req.-Pressurized	None	Special	5,000/year	
Allied	1	CBA	Medium	Good	None-Pressurized	Egression	Special	Not Available	
Bulgarian	1	CBA	Medium	Good	None-Pressurized	Reservoir	Special	Unknown	
USM	1	CBA	High	Good	Motion req.	None	Mod. CIM	5,000	$ 0.066/kg
ICI	2	CBA	High	Excellent	Motion req.	Sequential injection	Special	50,000	2.5% of finished price
Hanning	2	CBA	High	Excellent	Motion req.	Simultaneous injection	Special	Unknown	

processes that appeared later are modifications and some involve merely changes in the mold design. In the Dow process a polymer melt containing a dissolved gas is injected into a mold cavity. Union Carbide Corporation's process involves melting a mixture of a gas or CBA and a foamable plastic by extruding the mixture into an expanding accumulation region under positive back pressure, and then forcing this mixture from the accumulator into the mold cavity where lowered pressure permits the plastic to expand and fill the mold. The modified UCC process (UCC-Mod) is a low-pressure process, which degasses the melt as it passes from the accumulator to the mold, and adds a plastic or ceramic layer to the mold surface in order to minimize "freeze-off" and improve surface finish. Another modification of UCC's process is the "*Variotherm*". This process involves cyclic heating and cooling of the mold surface.

Researchers in Bulgaria (Institute of Metal Technology) and Allied Chemical Corporation have independently developed a gas counter-pressure, or medium pressure, process that involves pressurizing the mold cavity with inert gas and allowing the plastic to egress from the mold into a holding reservoir during a certain period of time after injection has been completed. This egression coupled with pressure control in the accumulation region offers an alternative to the Dow-TAF moving mold process.

USM Corporation, USA, has developed several mold designs that enable CIM equipment to be used to produce IF articles with good quality surfaces. The USM process consists of a normal injection molding procedure followed by moving certain mold sections outwards to allow foaming to occur in certain sections of the mold.

The best surface is achieved using a surface layer of unfoamed material that is backed with a layer of foamed material. This principle is called the *two-component foam (TCF) process*. Imperial Chemical Industries have developed a special process with a sequential injection system instead of trying to improve homogeneous skin formation. The outer unfoamed layer of the plastic is injected first and then the inner foamable plastic follows. In the ICI process it is necessary to move the mold sections in areas where low density foam is required.

The *low pressure process* was modified by Hanning GmbH in order to allow for simultaneous injection of two plastics. This process was designed to take advantage of the lower overall unit costs obtained by injecting a cheap foamable plastic (e.g. PS) into the center of more expensive plastic (e.g. PPO). It was found, however, that the simultaneous injection of unfoamed and foamed plastics yielded IF articles with nearly the same surface quality as those from the ICI process. There are some variations of this process which require two more injection stations but do not require moving molds. For example, Schloemann-Siemag designed the first Hanning machine, and Battenfeld has recently developed some equipment using the motion of a high-speed transfer valve operating between the compact and foamable accumulators.

3.4 Machines

3.4.1 General Considerations

In the early stages, integral foams were produced in standard molding machines designed for unfoamed plastics and conventional (non-integral) foams. IF articles fabricated in this equipment had high densities, uncontrolled skin thicknesses, and low-quality surfaces. Commercial experience has revealed that in order to remove these disadvantages either special-purpose machines have to be developed for integral foams or the existing standard molding machines must be modified to meet the following requirements: fast and controlled injection speed; controlled quantities of injected material; controlled injection pressure; provision of a valve to prevent the composition from foaming in the plastication cylinder and to prevent gas leakage; a small force for mold locking; and rapid but controlled cooling of the mold.

An IM machine consists basically of a clamp system, which allows the mold to open and close, and a reciprocation screw injection unit which plasticizes the polymer in a heated barrel and then injects it under pressure into the closed mold. Most of the machine's functions, except barrel heating, are normally carried out and controlled hydraulically and so a machine is fitted with electric motors driving the hydraulic pumps that provide the required hydraulic oil flows for the three main functions (clamp system, injection system, and screw drive system). The barrel heating system consists of an electric resistance heater fitted to the barrel, which ensures that the frictional heat generated by the screw yields the required melt quality. For a typical large IM machine the total installed power would be 200 kW, made up of 150 kW for the electric motors and 50 kW for the barrel heating system. The injection and screw plasticizing functions of an IM machine cycle use up the majority of the cycle's energy consumption, respectively 25.5% for injection and 58.3% for the screw processes [4]. Because of this, these two functions have been prime targets for redesign, to secure significant reductions in the total energy consumption.

It should be noted that conventional machines can be used with some modifications to produce small parts from integral foams. However, the average internal pressures needed for low pressure IF processes are rarely more than 10% of those required for CIM. Therefore, the clamp of a CIM machine is normally oversized and its platen area undersized for a given shot weight of an IF part.

Clamp Tonnage Ranges of IM Machines[5]

Low pressure machines:	
Battenfeld Corp. of America	165–2,025 tons
Desma Industrial Machines, Inc.	44– 165 tons
Ex-Cell-O Corp. Plastics Components Div.	220–1,000 tons
Hoover Universal, Inc., Structural Foam Machinery Div.	150– 750 tons
Wilmington Plastics Machinery	15– 500 tons
High pressure machines:	
Buhler-Miag Inc.	138– 880 tons
Farrel Rochester Div., USM Corp.	450–2,500 tons
Multi-component machines:	
Battenfeld Corp. of America	77–1,690 tons

Thus, even though a special purpose IF machines may be more expensive than a CIM equipment of comparable tonnage, the bigger platen gives the machine greater versatility and flexibility.

Note finally that most IF processes are only available as part of a license and royalty package, and that the most popular processes have been developed and promoted by resin suppliers (see Table 3.1) [6].

3.4.2 Types of Machines

Hazneci reviewed the main IF molding processes [7] and grouped the main machines available on the market into three categories (see Table 3.2), i.e., *standard machines; customized, or specialty single-nozzle machines; and specialty multi-nozzle machines.*

Table 3.2. Comparison of injection molding machines for producing integral foams

Parameters	Standard machine	Specialty single-nozzle machine	Specialty multi-nozzle low pressure machine	
			high speed	normal speed
PROCESS FEATURE				
Having nozzle shut-off value	No	Yes	Yes	Yes
High speed injection	No	Yes	Yes	No
Large platen area	No	Yes	Yes	Yes
Large part capacity	Limited to oz. rating	Yes	Yes	Yes
Generous platen travel (daylight opening)	No	Yes	Yes	Yes
Long flow capability	Limited	Improved	Multi-nozzle eliminates need	
Potential density reduction	Fair	Good	Good	Good
Long cycle time	Limited	Yes	Yes	Yes
Chemical blowing agents	Yes	Yes	No	No
Physical blowing agents	No	No	Yes	Yes
Typical maximum shot weight, kg	4.5	11.3	22.7 or more	
END USE REQUIREMENTS				
Small part up to 1.5 kg, appearance unessential	Good	Good	Machine oversized	
Medium part, up to 20 kg, good appearance	Poor	Best	Good	Good
Large part, over 20 kg	NA[a]	NA[a]	Good	Good
Set up with multiple molds	NA[a]	NA[a]	Good	Good
Fast set up	Good	Good	Poor	Poor
Suitable for wide range of materials	Good	Good	Poor	Poor

[a] NA — not attainable

A small part (0.5–1.5 kg), as for instance a keyboard, can be produced by any of the three process types. Most multiple nozzle machines are not suitable for small sized objects, unless they are family molded; single-nozzle machines produce the best appearance, cell structure, and physical properties, whilst the standard machines are adequate, Medium sized parts (up to 20 kg) are probably best produced in large standard, or better yet in single-nozzle, machines, which give the best cell structures and properties. A wider range of materials can be molded, with fast changeover from one material to another, or from one mold to another. Very large parts can only be molded on multi-nozzle machines and the latest generation of these machines has been greatly improved with respect to injection speed, streamlined design, material suitability, etc.

A very large monolithic structure puts many more restrictions on design, tooling, try-out, and molding than a series of modules. The main restrictions, according to Hazneci [7], are:

1) The IF equipment notionally able to handle a very large part is limited.

2) Thicker wall sections will be required for load bearing surfaces making the part heavier; thicker sections add considerably to the molding cycle time. Both these factors affect the cost of the part, especially when high cost resins and large presses are being used.

3) Engineering changes require considerably higher costs, even for minor changes such as revising core pin sizes or adding or removing bosses. Major changes may not only be expensive, but they may not be feasible due to the part's configuration or the mold design.
4) There will be some areas where long plastic flows are needed within the mold and the long path could cause poor surface qualities because the material solidifies early and the mold may not be completely filled; moreover the gas may accumulate during long journies in the mold, possibly causing voids which would weaken the area where they occur.

There are now two types of machines on the market for producing IF moldings: *one-component IM machines*, both high and low-pressure varieties, and *multi-component IM machines.*

Whilst the LP machines were developed especially for producing foamed thermoplastics, the HP machines can make compact (unfoamed) molding as well (the foamed thermoplastics, however, requiring an extra module). But HP machines are more expensive (see Chapter 12).

Multi-component IM machines are designed as HP devices but can use one of the following four process types [7]:

1) A two-component process to produce articles
(a) with a compact skin and foamed core (i.e., an IF);
(b) with a compact skin and compact core;
(c) with a foamed skin and foamed core (called laminated foamed articles); and

Table 3.3. Comparison of machines and molds required for the various processes, and the basic end properties of integral foams[7]

Process	Machine requirements	Mold requirements	Surface finish	Density
Integral foam molding	high or low pressure injection molding machines	simplified injection molds; no high quality surfaces required	wood-like finish; roughness peaks of 20–50 mm or more at very low densities	lowest density of all 3 processes
Gas-counter pressure	high or low pressure injection molding machines, but higher cavity pressure than for low pressure process; extra equipment needed for gas counter-pressure	cavity must be gas-tight and have smooth finish, as for conventional injection molding	very smooth and slightly matt surface, but not free from weld lines behind apertures	high density, as mold is filled 100%
Two-component	machine with 2 plasticizing and injection units and a specially designed nozzle with a common feed channel into the mold	as for integral foams and conventional molds; surface finish as with conventional molds	best finish of all 3 processes, high gloss	density depending on processing parameters between low pressure and gas counter-pressure

[c] Size 5,760 cm^2 (900 square inches)
[d] Size 10,240 cm^2 (1,600 square inches)

(d) with a foamed skin and compact core (called reverse integral foams).
2) Low-pressure processes.
3) Gas counter pressure processes (an extra module needed).
4) Conventional injection molding.

Thus multi-component IM machines are becoming increasingly popular because of their versatility.

Machines with only one injection unit are adequate for one-component processes, but a multi-component process necessitates a machine with two injection units. When gas counter pressure is to be used the machine must have a special unit (using electric or hydraulic control), while the LP and two-component processes use molds resembling those used in conventional molding. On the other hand, a gas counter pressure process requires the molds to be gas tight. These processes are compared in Table 3.3.

Many companies in Western Europe, the United States, and Japan are producing special-purpose machines for IF's. These include Admiral Equipment, BASF, Battenfeld, Beloit, Borg-Worner, Cannon-Afros, Cincinnati Milacron, Desma-Werke, EMCO Machines, Epco/Negri, Ex-Cell-O, Hoover Universal, Hunning, Kawagushi, Mold-Trim, Peerles Foam, Siemag, Sund-Borg, Stephan-Chemical, Sterling Extruder, and Wilmington [9,10].

An economic and technological evaluation of the various IM machines and processes will be presented in Chapter 12.

3.5 Molds

3.5.1 Mold Materials

The major requirements for the mold itself are: rigidity, corrosion resistance, repeated usability, low mass, high thermal conductivity, low heat capacity, and long-term thermostability.

The wide pressure range used to make IF's by IM (from 2 to 100 MPa) accounts for the variety of designs and materials used for the molds. Light molds can be used for the LP processes. They have thin walls of rolled aluminum sheet, zinc, steel, epoxy-resin based materials with a sublayer of quartz sand or aluminum powder, or of silicone polymers. Elastic molds with inner surfaces containing high relief can be made from vulcanized polyurethane elastomers. When moderate or high pressure are used during the molding, cast molds with thick walls of aluminum, steel, brass, or bronze are needed, the Be–Cu alloy being used most frequently. Kirksite (a high thermal conductivity alloy of Al and Zn) is another castable material that can be used for IF molds. Kirksite has a lower melting range than aluminum. Larger molds, however, tend to be heavy because of the high specific weight of kirksite, and repairs on them are difficult.

The reproducibility and quality of an IF article's skin is achieved by controlling exactly both the raw material and mold temperatures. Metal molds give better skins because of their good thermal conductivity, whereas plastic materials such as epoxy resins are poor heat conductors and consequently not as desirable when the skin quality of an IF article should be good.

A mold with high thermal conductivity and low heat capacity facilitates the rapid heating of a composition and the cooling of the finished articles. In this respect, aluminum and Be-Cu alloys are better than steel. For example, the cooling period τ_c of an integral PS foam ($\varrho = 800$ kg/m^3) in a mold of epoxy resin with a 40% aluminum powder content is 0.5 min, in a steel mold τ_c is 0.7 min., in an aluminum mold τ_c is 1 min., while in a kipsig[1] mold τ_c is 3.2 min.

Mold construction materials can affect the processing if they have poor heat transfer properties. This is particularly true for epoxy molds, even if the epoxy is aluminium filled. Epoxy tooling sometimes yields parts with "sink marks". These sink marks are not of the same type as those which occur in other IF injection moldings. While there is shrinkage during cooling of the urethane, these "sinks", which are clearly evident, appear on demolding. It is felt that the sink marks are areas which have erratic cell structures or result from gas traps that have collapsed. The sink marks can be eliminated by opening up the skin of the article; however this results in a porous article that is difficult to paint properly [12].

As will be shown later on, the mold temperature is one of the primary factors determining the structure and properties of an IF. Hence, all the heating and cooling blocks of molding machines are automatically controlled to maintain the desired temperature within 0.5 to 1 °C. Designs have been developed that include multi-section elements for heating the compositions and molds using electric furnaces, water, or steam, either in series or parallel, or both. Recent publications report machines that have infrared and high-frequency heating elements [11].

Although the temperature of the mold walls does not exceed 200 °C when producing IF's, the "lifetime" of molds is quite limited, even in the case of metal molds. This is attributed to the fact that the temperature front arising when the composition is foaming is far from uniform and the mold walls are subject to large temperature gradients giving rise to irregular local expansion and compression strains on the walls. Considering the rather complicated configurations of most IF articles and, hence, complicated mold configurations and the extra mechanical stresses that arise within the molds, it is evident that the problem of creating re-usable molds is urgent.

Aluminum molds can be re-used very often — up to 50,000 times, whereas epoxy resin molds cannot be used more than 500 times because of their rapid thermal destruction.

Today the most economic molds (for the LP process) account for about 20% of the total equipment cost. A detailed mold-cost analysis is given in Chapter 12.

3.5.2 Mold Design

During any IF molding process, the mold must be properly mounted on the platen. Reinforcing the mold with steel also lengthens mold life, this being particularly true for aluminum molds (Fig. 3.2).

Mold design should provide for sprue channels that are as short as possible, have large diameters and, desirably, round cross-sections. In some cases, in order to fabricate articles with surfaces imitating wood, point and flat-slit inlets should be provided. For reasons we shall discuss later, the minimum thickness of an IF part is of the order of 4–6 mm. [13,13a]. It is noteworthy that these values correspond to minimum spacings between the mold walls. The positions of the holes and joints as well as the mold inclination should be chosen such as to ensure uniform filling of the mold cavity and to avoid depressons in the molded article. The gap between the half-molds must not exceed 0.1 mm, otherwise the material may flow out, even at low molding pressures (0.2 to 1.0 MPa) [11]. In order to make the walls of the mold more rugged they are made thicker or external stiffening ribs and bracings are added. Aluminum molds are more resistant to corrosion than those of steel, therefore steel molds should be chrome- or nickel-plated. Be-Cu based alloys are corroded by ammonia and the decomposition residues of some CBA's, and therefore molds made from these alloys have their inner surfaces plated with nickel.

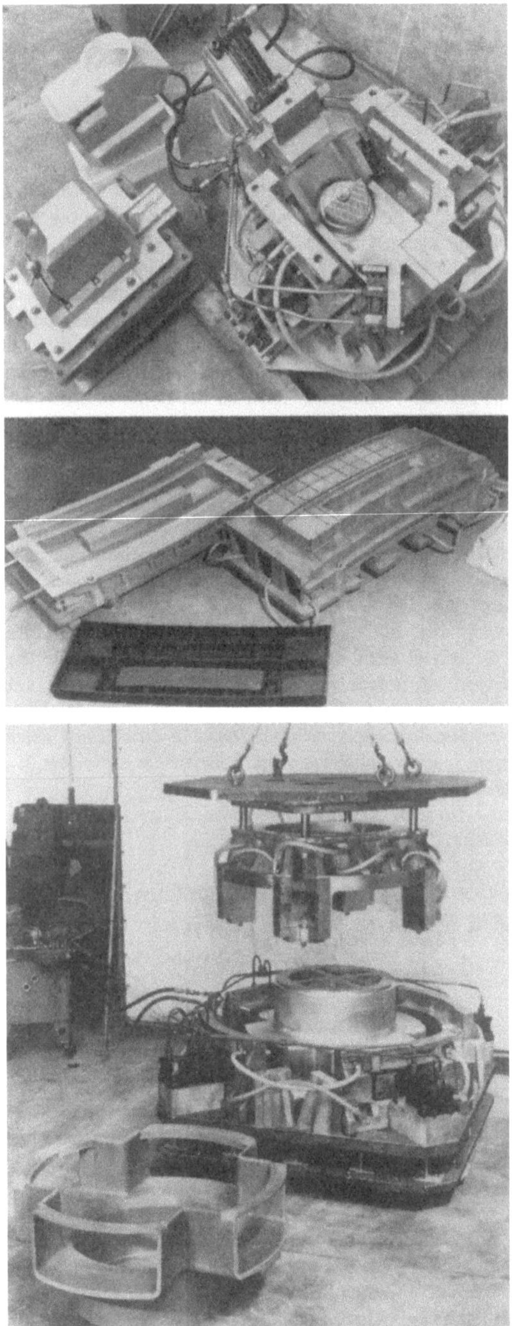

a

b

c

Fig. 3.2a–c. Various types of IF molds; **a**: 4-parted cast aluminum mold for household electric appliance casing, with foamed-in-injection-molded insert; **b**: 2-parted cast aluminum mold for bus bender centre piece of PUR elastomer; **c**: 7-parted PUR mold of "Zamak" for a drawing-room table (Courtesy of Battenfeld Maschinenfabrik GmbH, Meinerzhagen, FRG) [22]

At some point in the design procedure a prototype mold should be considered, and this is particularly true for complex articles [12a]. According to Misitano [13], 6.4 mm thick walls are the best for prototypes from wood, plastic sheet, or epoxy castings. More durable prototypes can be made from sand and plaster-cast aluminum.

An IF mold's cavity surface and the surface of the articles are closely interrelated. Nevertheless the importance of the quality of the mold surface should not be overestimated. Even mirror-finish cavities cannot produce articles having the same smoothness. For example, the surface of articles produced by the LP process is inherently somewhat uneven and not entirely smooth, and the mold's cavity surface will have little effect on the surface of an IF articles made by this process [11]. On the other hand articles produced by the Bulgarian gas counter-pressure process are even smoother than their mold's cavity surfaces (see Chapter 6).

Generally, the best reproduction of mold cavity surface is achieved using a multi-component process. However, the reproduction is still not as good as the non-foam IM process since the pressure exerted on the melt against the mold is much smaller. This explains why substantial improvements in surface reproduction are achievable by increasing the mold temperature, particularly in a two-component process. According to Eckardt [8], a mirror-finish mold heated to 60–70 °C can deliver very glossy ABS articles which often may be used without further finishing.

Gating is particularly important when processing foamed plastics. Though fundamentally the same kinds of sprues and gates can be used for foamed molding and unfoamed injection molding, there are nevertheless substantial differences regarding the importance of individual gate types, their size and configuration. The following types are commonly applied in IF technology: sprue gates, pin-point gates, fan gates, umbrella gates, and tunnel gates. Eckardt [8] and Fillman [14] have discussed the various gates very thoroughly.

RIM technology gating is quite specific and will be discussed later in Chapter 8.

3.5.3 Mold Filling

During mold filling, the flow of the plastic material should be laminar. When using closed molds, the pressure of the flow is gradually reduced between the mixing chamber and the mold cavity because the channel cross-section is increased. At the inlet of the mold cavity the sprue channel should narrow sharply for the melt to get to the mold wall in the form of a thin film. Optimum filling of the mold has been shown to depend upon the type and concentration of the BA, the temperature, the injection speed, and the geometry of the gate and the mold cavity. The thinner and less dense the article, the more difficult it is to attain an optimum filling.

For an IM process to be successful it is important that the polymer travels as short a distance and as smoothly as possible for two reasons [15]. Firstly, IF's are often based on engineering resins, or on resins containing flame retardants, reinforcement fillers, pigments and other additives. These polymers can be very sensitive (unlike general purpose polystyrene, for example) to the thermal history and the flow distance that the resin has to travel. Secondly, the BA should be under pressure, and hence in solution in the melt just before the melt is shot into an IF mold. If the pressure is prematurely reduced, the BA-gas will be allowed to expand and to escape into voids before the melt is shot into the mold. The shorter the distance the polymer has to flow, the less unwanted heat it will pick up, and the more gas stays in solution in the melt.

Wherever a molded article comes into contact with the mold's surface, it obtains a good skin and surface. This is particularly advantageous in complicated designs which normally require systems of high chemical reactivity. Polyurethane systems, such as "Baydur" for producing rigid integral foams, or "Bayflex" for flexible foams, are good examples, and are now used worldwide (Chapter 13).

A well designed mold is a major step toward the production of a commercially viable IF article. Numerous aspects must be borne in mind when designing a mold. However, there are two general mold design criteria that should apply to every IF mold, namely flow length and part thickness (Krumm and Sauers [16]).

The combination of temperature, pressure and injection speed (filling speed) required to establish the best flow for a particular mold is usually determined mainly by the "flow ratio of the mold", i.e., the ratio of the longest flow path to the section thickness. The higher the flow ratio, the higher the temperature required to obtain fast mold filling and consequently low levels of stress in the IF moldings. The maximum obtainable flow ratio varies for different polymers. Since the obtainable length of flow is also affected by other process variables and conditions, local restrictions, and changes in the flow direction within the mold, precise maximum values for flow ratios cannot be given. But it is common practice to assume the following maximum flow ratios: PS – 200:1, HDPE – 100:1, PP – 100:1, PC – 80:1 [17)]. Particulate and fibrous fillers, and reinforcing agents such as talc, chalk and glass fiber reduce the flow ratio.

Multigating may be used with both unfilled and filled polymer compositions to obtain complete filling of the mold. The number and position of the injection points should be selected so that the flow ratio is within the capabilities of the polymer to be used.

Mold filling in the RIM process is discussed in Chapter 8, while the physico-chemical problems of mold filling are considered in Chapter 13.

3.5.4 Demolding

The smoothness of a mold's inner surface determines the quality of the article surface, the reusability of the mold, and the ease of article demolding. For instance, the ease of demolding is known to increase by 80% if surface roughness is reduced from 1.5 to 0.2 μm and by 70% if the mold temperature rises from 40 to 70 °C [11)]. To make the inner surface of a mold smoother, it is polished or plated with nickel, copper or chrome using galvanoplásty or chemical precipitation.

To facilitate IF part demolding, cone-shaped molds and various reusable antiadherents can be used. The interior of the mold may be coated with teflon, epoxy resin, synthetic rubber, etc. Single-acting antiadherents are also useful, for instance, thin polyester or polyethylene films. Of late, antiadherents based on silicone polymers and standing up to 50 demolding cycles have found wide application. Generally, rigid molds (except those made of silicones) need antiadherents, whereas elastic molds do not.

The experience gathered on ejectors for unfoamed IM articles is also applicable to IF articles. A significant difference, however, lies in the size of the ejectors. These should be large enough so that they cannot deform the molding, even if there is very early mold release. In the case of bow-like moldings with little taper, air ejectors are an essential aid to mold release [13)].

Usually, integral foams require a draft. In the case of articles produced by the gas counter-pressure process, an external draft is always required unless split molds are used. This external draft should be at least 0.5 degree, whilst the internal draft should be about 1 degree [8,14)].

3.5.5 Venting

Prior to injection, the air that has entered the mold in the course of the previons article's demolding must be completely removed, otherwise it may lead to rejects and considerable material losses. Hence, mold venting is necessary for high speed injection. Air vents generally take up to 50% of the mold joint surface and are designed to remove 60–80% of the gas volume quickly. Sometimes, the air is forcevented by a vacuum pump. It is noteworthy, however, that rapid air venting from a mold should not create vacuum as this would lead to cavities in the article structure and foaming gas losses [8,14)].

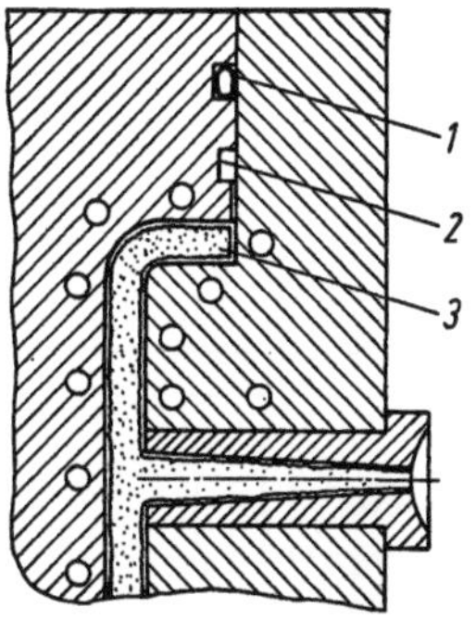

Fig. 3.3. Design of the seal and venting system for the gas counter pressure process; **1**: seal, **2**: venting groove leading to air valve, **3**: mold cavity [8]

Venting is particularly important when processing foamable polymers because the melt is normally injected into the mold within less than a second. The venting system should be designed so that the air trapped in the mold will be expelled by the melt as it is injected at a pressures of 0.2 to 0.4 MPa [13].

If the mold venting system is overdimensioned, the air trapped in the mold will be forced out of the mold with very little pressure. With extremely high injection rates this results in the melt being introduced into the cavity in an unordered way. Weld lines thus occur as the air is trapped between approaching surfaces of the melt.

Inadequate mold venting will lead to material charring at the end of the flow path, similar to that which occurs during molding of solid (unfoamed) polymers. The vents should be about 5–10 mm wide and 0.05 mm deep at the mold cavity end, and 0.2 mm deep (same width) at the venting groove [8,13].

It has proved advisable to provide a large number of narrow vents instead of a small number of broad vents, and to place the narrow vents precisely where they are needed. These comments on venting for the LP process are equally applicable to the gas counter-pressure and the multi-component processes. Care must be taken to make the gas counter pressure mold cavities gas-tight. The seal can be provided by drilling a groove around the edge of the mold cavity and inserting an O-ring (Fig. 3.3) into the groove. This groove may be located between the usual, ordinary seal and the mold cavity and is connected to the mold cavity. The mold cavity can be either subjected to pressure, e.g. by a compressor, or vented from the outside via bores and adjustable valves.

It seems appropriate now to mention negative pressure venting. This is very helpful in molds that cannot be vented in a normal fashion or where there is a high gas build up. This is done by drilling a hole in the area that needs venting, capping the hole at the mold surface with porous metal, and connecting the hole to a water line. This will create a negative pressure when the mold is closed, which in turn will pull gas from the mold cavity through the porous metal and out into the water system [15].

3.5.6 Hot Runner Systems

With much of the interest in IF molding being due to savings in raw material consumption, many molders are interested in hot-runner tooling that will reduce material costs even further. A modular system has been developed by Battenfeld that could ease the problem of mold making (Fig. 3.4). The system is assembled away from the IM machine. The shutoff nozzles with their connecting tubing and the injection element are on the one side of the plate, one of the mold halves is mounted on the other.

These systems are employed either if large articles are being produced which require multiple gating, or several mold cavities have to be filled simultaneously and using two-plate or three-plate molds presents problems. In hot runner systems for IF's, each gate requires a nozzle seal to avoid the premature release of gas from the melt. Systems have been developed which are either a component of the IM machine adjustable to suit the molds, or which are a component of the mold [10].

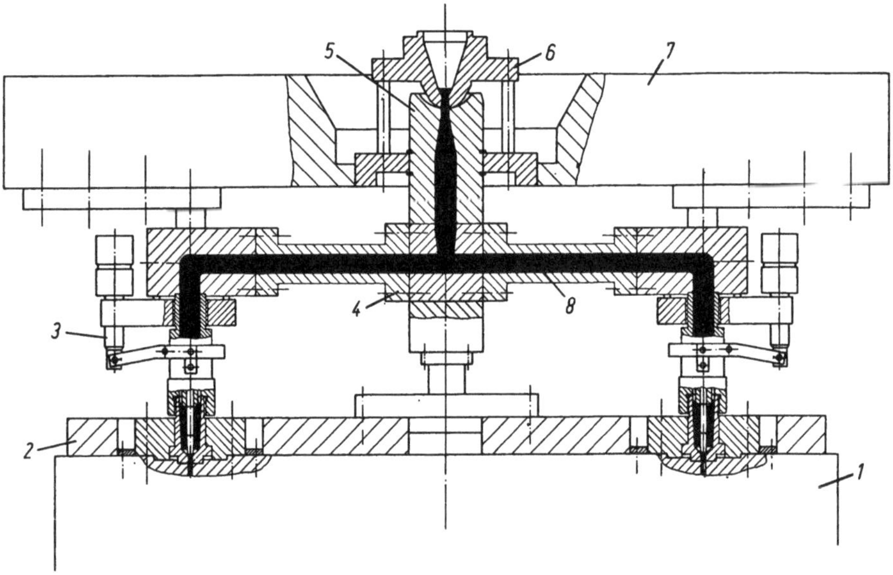

Fig. 3.4. Battenfeld Hot Runner System; *1*: moving platen, *2*: mounting plate, *3*: shut-off nozzle, *4*: hot runner jacket, *5*: sprue heating jacket, *6*: sprue bush, *7*: fixed platen, *8*: polymer melt (Courtesy of Battenfeld Maschinenfabrik GmbH, Meinerzhagen, FRG) [10]

A hot runner system consists of three basic components (Fig. 3.4): the base frame, needle shutoff nozzle with runner network, and controls. The base frame is constructed so as to be adaptable to molds of different sizes. In order to fit several small molds, an intermediate platen, which is clamped to the base frame, is also required. The needle shutoff nozzles are operated separately and can be distributed so that the optimum number and position of gates can be selected. The distance between the gates can be altered by replacing the connection tubes. The hydraulics are located on the base frame, the electrical controls are grouped together in a separate switch cabinet and the control panel can be supplemented by other functions.

3.5.7 Problems

In order to achieve the best results when producing an IF article, the following should be borne in mind before the mold is made[8,14]:

1) Number of articles; if the length of the run and the geometry of the articles are known, the most suitable mold material can be chosen, and a decision can be made as to whether a single-impression or multi-impression mold would be best.

2) Dimensional accuracy requirements; for molds with narrow tolerances, only steel should be used as the mold material; where the tolerances are less tight, less expensive mold materials can be used, provided the other preconditions are met.

3) Selection of a suitable process; the basic design of the mold depends on the design of the article, the surface requirements, the mechanical load, and the raw material selected (see Chapter 12).

It is evident that tooling requirements have become more stringent since the start of mold technology. According to Turner and Koski [15] the following trends towards higher quality tools can be seen: fewer castings, less aluminum, more steel tools, more intricacy of detail, large complicated parts with many side actions, and hot runner manifolds with individual cavity shutoffs.

3.6 Cooling Process

3.6.1 Cooling Time

The most time-consuming step in IF technology is the cooling of the IF article in the mold. Either shortening the time required for this step or combining it with the polymer melt injection step are the main ways of improving the industrial profitability of any IF process.

The thickness of an IF part determines the *cooling time*, τ_c. Note that the cooling cycle for PS and ABS plastics is shorter than that for polyolefins. Thick-walled articles ought to have after-cooling, i.e., the article is demolded hot (but with a sufficiently strong skin) and then submerged in cold water. This can considerably shorten the cooling cycle (Fig. 3.5).

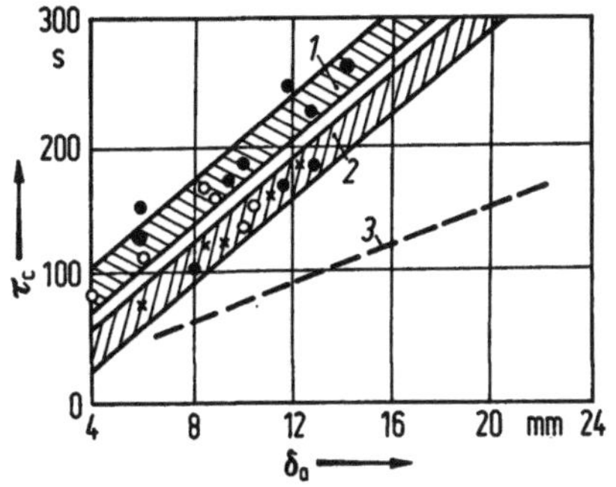

Fig. 3.5. Skin thickness δ_s versus cooling time τ_c for various IF parts produced by the injection molding process; *1*: poor cooling, *2*: fairly good cooling, *3*: after-cooling; +: ABS, ○: PE, ●: PS [11]

Reducing thick sections from 12.7 mm to 6.4 mm can save up to 75% in cooling time, as was demonstrated for some General Electric IF parts [18]:

	Cooling time	
	12.7 mm thick	6.4 mm thick
®Lexan (polycarbonate resin)	3 min	55 sec
®Noryl (poly(phenylene oxide) resin)	3.5 min	1 min
—Valox (thermoplastic polyester resin)	1.75 min	25 sec

Progelhoff and Throne [19] suggested a number of analytical expressions that can be used to calculate the cooling time τ_c for an IF article very accurately, and with due reference to its morphological parameters and process specifications. The authors presented a method for solving the transient heat conduction equation for an IF article having variable density profiles across the part. They carried out the mathematics in detail for slab geometry and verified the accuracy of their solution with a simple transient heating experiment on a slab of polystyrene. The authors showed in particular how the density profile within a rod can affect its cooling rate.

3.6.2 Cooling Units

The cooling time is reduced by using various cooling units and channels. In most cases, these channels are drilled directly into the underside of the cavity plate, or drilled aluminum cooling plates are attached to the mold. The cooling plate technique is particularly useful when designing a family of mold cavities in one mold case [13)].

When designing the cooling layout of an IF mold, some important rules should be followed if the maximum cooling efficiency is to be obtained. Wendle [20)] indicated that the diameter of the cooling lines should be about 12 mm, and the pitch or distance between center lines of two parallel water lines should be three to five times the diameter of the lines. The distance between the center line and the cavity wall should be one to two times the water line diameter. By following these rules as closely as possible an even cooling distribution can be obtained.

Moreover, in order to obtain the best possible heat transfer between mold and coolant, it is necessary to have turbulent flow conditions within the cooling lines. This regime is a function of the pressure drop through out the lines, as well as of the viscosity of the coolant and the cross sectional area of the cooling lines.

Since the injection rate is relatively low (especially at LP process), the shot size is large, and the flow distance long, an initially cold mold surface would probably hinder the mold filling. Some high temperature resistant polymers such as PC may even require that the mold surfaces be heated prior to injection. A 10 °C coolant temperature is recommended by Wendle for high impact PS, linear PS and PP [20)].

The molds for any IF process must be designed so that the cooling passages do not reduce the strength of the mold. The tool must survive the molding and the injection pressure as well as the press cycles without leaking. Because castings frequently leak, Micitano [13)] recommended impregnating them with polyester resin. One method of cooling that can help eliminate leak problems is to incorporate the tubing in the casting itself. Cast tubing costs less than channel-type cooling. This method is especially popular for aluminum molds.

Another way of reducing the τ_c of IF process is to use multi-position molds, either whirling ones or ones involving a rolling table [21)].

3.6.3 Problems

It is true that molding cycle times of IF's are considerably longer than those common for unfoamed plastics. Throne [1,3)] mentioned that, "unfortunately these longer cycles have been erroneously attributed solely to the poorer thermal conductivity of the plastics, thus ignoring the fact that the primary factor governing heat removal in transient conduction is the thermal diffusivity". Though in fact the thermal diffusivities of foamed plastics are approximately equal to those of unfoamed plastics and their contribution to cycle time outweighs in absolute terms that of the conductivity, the latter is nevertheless the cause of the cycle time difference.

The cooling cycle for an IF part involves the cooling of both the plastic and the foaming gas. Before the part can be removed from the mold, it is necessary to cool the plastic to a point where it can withstand the internal pressures of the foaming gas, and to cool the foaming gas to reduce the internal pressure so that a "puff" does not form. High gas pressures require longer cooling cycles [16)].

One further point should be made with regard to the molding cycle. Many customers think that the mold cycle depends on the density of the molded IF part which in fact is not the case. A decrease in density will lower the thermal conductivity of a foamed part containing a *foaming gas which has a lower thermal conductivity than the base polymer* [16)]. At lower densities, there is less polymer and more gas; at higher densities, the conductivity is greater because there is more polymer and less gas,

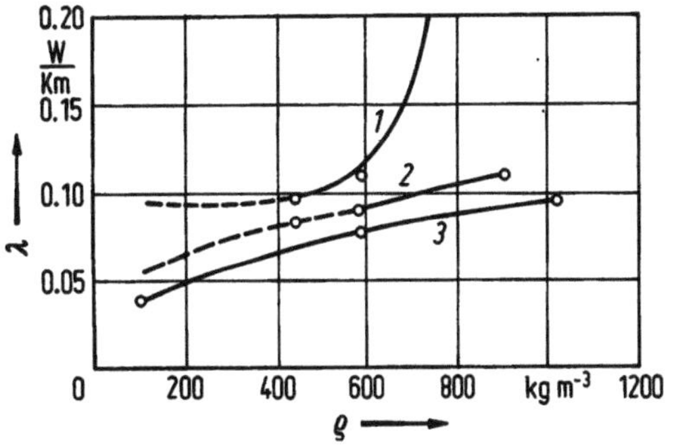

Fig. 3.6. Thermal conductivity λ versus density ϱ of an IF parts; 1: HDPE, 2: PP, 3: HIPS [16]

but more heat must be removed to cool the part to a state of sufficient rigidity before it can be removed from the mold. When molding low or high density IF parts, these competing mechanisms result in a trade-off that yields comparable cooling cycles (see Fig. 3.6). Mold cooling is thus directly proportional to thermal conductivity and inversely proportional to density and specific heat. At normal densities of IF's based on HDPE and HIPS, specific heat and heat of crystallization are the controlling factors.

Since the cellular core structure of an IF article is primarily made up of gas bubbles, heat transfer from the product is very poor, making it difficult to remove the molding heat. If the article is removed before its higher density skin surface is strong enough to contain the internal polymer-gas mixture, the part's wall will continue to expand. As a result, a bulge is formed that is unacceptable for most applications. Therefore, the cycle time is determined by this effect and the heat transfer within the article becomes the limiting factor rather than the heat transfer to the mold.

Wendle [20] pointed out that parts having cured surfaces or a combination of vertical and horizontal planes will not be as susceptible to post-molding expansion due to the built-in structure of the skin surface. In most foam applications the design technique of coring can greatly improve cycle time, at the same time reducing material consumption and providing additional mechanical strength. Another method of improving cycle time is to use high strength polymers, such as PC, which form skins strong enough to contain post-molding expansion. Conversely, polymers with low decomposition temperatures, such as low density PE, are not normally used for making IF's. [23–25]

Proceeding from these general remarks, which are valid for any IM process, we shall make a closer examination of the basic industrial IF processes (Chapters 4–8), and then try to compare the processes in order to formulate objective criteria for choosing a concrete process for a concrete set of specifications of a future integral polymer foam article (Chapter 12).

3.7 References

1. Throne, J. L.: Plastics Process Engineering, New York, Marcel Dekker, 1977
2. Burk, W. R.: The TAF Process. In: Engineering Guide to Structural Foam. Wendle, B. C. (ed.). Westport: Technomic, 1976, pp. 140–150

3. Throne, J. L.: J. Cell. Plast. *12*, 161–175; 264–283 (1976)
4. Chivers, A. V.: Europ. Plast. News *7*, No 10, 50 (1980)
5. N.N.: Plast. Technol. *26*, No 6, 181 (1980)
6. Hara, M., Sano, M.: Japan Plast. Age No 1, 28 (1982)
7. Hazneci, N.: Guidelines for Material and Process Selection for Structural Foam. In: Structural Foam' 78-Expanding Horizons. Westport: Technomic, 1978, pp. 6–13; Plast. Technology *24*, No 10, 75 (1978)
8. Eckardt, H.: Ind. Prod. Eng. No 2, 170 (1980); J. Cell. Plast. *16*, 21 (1979)
9. N.N.: Plast. Technol. *28*, No 7, 172 (1982)
10. N.N.: Mod. Plast. Int. *10*, No 8, 26 (1980); *11*, No 5, 13 (1981)
11. Semerdjiev, S.: Introduction to Structural Foam. Brookfield Center: Society of Plastics Engineers, 1982
12. McBrayer, R.: J. Cell. Plast. *17*, 332 (1980)
12a. N.N.: Canadlans Plast. *40*, No 9, 40 (1982); *40*, No 12, 5 (1982); *41*, No 3, 24 (1983)
13. Misitano, G.: Toolmaker Lokks a SF Tooling. In: Engineering Guide to Structural Foam. Wendle, B. C. (ed.). Westport: Technomic, 1976, pp. 172–177
13a. Components for Plastic Molds and Die Cast Dies. Technical Broshure CE-12 Columbia Engineering, Division of General Defence Corp., Red Lion, PA, USA (1983)
14. Fillmann, W.: Ind. Prod. Eng. No 3, 35 (1978)
15. Turner, N., Koski, G.: J. Cell. Plast. *17*, 89 (1981)
16. Krumm, S., Sauers, M. E.: SPE ANTEC Technic. Papers, *21*, 37 (1975)
17. Norgan, M. R.: Cell. Polymers *1*, No 2, 161 (1982)
18. Brooks, R. L., Colbert, S. J.: J. Cell. Plast. *17*, 94 (1981)
19. Progelhof, R. C., Throne, J. L.: J. Cell. Plast. *11*, 152 (1975)
20. Wendle, B. C.: Mold Design and Constraction. In: Engineering Guide to Structural Foam. Wendle, B. C. (ed.). Westport: Technomic, 1976, pp. 168–171
21. Johnson, P. S.: Elastomerics *115*, No 5, 22 (1983)
22. Eck, E.: Mould technology for RIM and RRIM applications. In: Latest Developments in Reaction Injection Moulding. Meinerzhagen: Battenfeld Maschinenfabriken GmbH, 1982, pp. 6/1–6/d
23. Peach, N.: Plast. Eng. *40*, No 8, 19/1984)
24. Wigotsky, V.: Plast. Eng. *41*, No 8, 17 (1985)
25. Shutov, F. A.: 30th IUPAC Intern. Symp. Macromolecules, Haque Holland, 1985

4 Injection Molding: Low Pressure Process

4.1 Basic Process

Low pressure (LP) injection molding processes for producing IF's are characterized by the following two features:
(a) relatively low molding pressure, from 0.5 to 10 MPa; and
(b) short shots into the mold, amounting to only 65–80% of the mold cavity (Fig. 4.1).
These two features determine the design and operating conditions of the molding machines, as well as the quality of the IF articles.

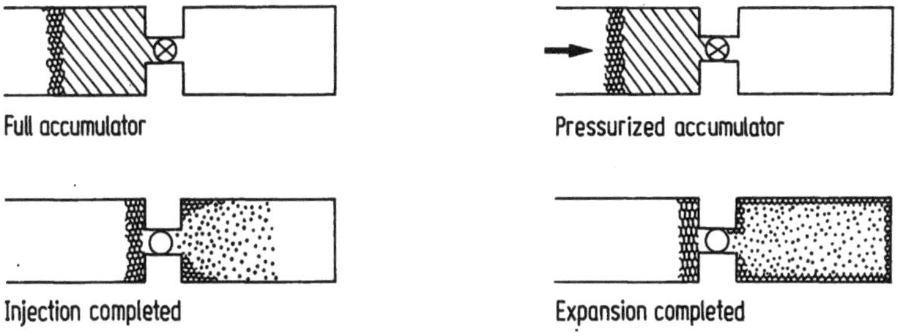

Fig. 4.1. Schematic stages of the low pressure injection molding process; note that the injection and expansion stages are considered to be distinct parts of the polymer foaming process (Courtesy of Technomic Publishing Corporation, Westport, USA) [1]

To realize a LP process in practice, the following conditions have to be satisfied:
1) high injection speeds (up to 100 cm^3/sec) to reduce article density and to improve skin and more uniformity;
2) high injection pressures (30–100 MPa) which require only short shots, permitting thus the use of light and cheap molds;
3) materials and designs of molds capable of withstanding pressures up to the 5 MPa developed by the foaming gas (the mold locking force is hence lower than that required for the production of unfoamed plastics by the IM method);
4) mold wall spacing over 6 mm, otherwise the article density is sharply increased. Walls are generally 7 to 10 mm thick, the melt path in the mold should be at least ten times the wall thickness; thicker walls would lead to longer cooling times of the article. Longer melt paths permit the formation of large-sized articles, as long as several meters, with surface areas 10 to 25 times that of articles produced by the HP-IM method[2–4].

The low pressures in the mold are responsible for a certain roughness of the article surface because of poor contact between the melt and the mold surface, even at very high injection speeds. Hence, the bubbles formed on the surface of the melt's "leading edge" burst and solidify on coming in contact with the cold walls of the mold. Another reason for surface roughness, which is however eliminated by faster injection speeds is that, during injection the leading edge of the melt near the mold walls cools down rapidly, whereas subsequent portions of the melt that touch the film of the curing polymer possess higher viscosities. Since the film is not a continuous one and its thickness is not uniform, the contact between the "fresh" and the "old" melt portions under low pressure conditions is not good, the article's surface ends up with grooves, shrinkage cavities, and sags. Moreover, low pressure molding results in a nonuniform skin density: near the gate it is higher because the pressure in the mold is highest there and decreases with the distance from the gate. This dependence of the IF's skin density on gate position has two undesirable consequences: anisotropy of IF strength properties over the length of the article and irregularities when the article is subsequently dyed, due to different adsorptivities of areas with different densities.

On the other hand, the low molding pressures induce the least stresses in the IF structure during the molding process. Therefore, all integral foams produced by the LP-IM method are characterized by low technological and temperature shrinkages, below 1–1.2%.

4.2 Commercial Processes

4.2.1 The UCC Process

The most widely known method for producing IF's is the one developed by *Union Carbide Corporation* (USA) and referred to as the UCC process. The process uses nitrogen as the BA and can produce articles with various degrees of roughness and texture (imitation wood), a wide range of apparent densities (from 250 to 950 kg/m^3), a clearly defined integral structure with the density increasing continuously from the core towards the skin, a uniform cellular core structure, and large possible size [1-4].

A schematic diagram of a typical UCC installation is given in Figure 4.2. Granulated polymeric material is plasticized in an extruder heated to 205 °C; the middle section of the extruder is supplied with nitrogen which is dispersed in the melt. The resulting composition is fed into the accumulator where it is partially foamed at 165 °C and then on to the water-cooled mold closed by a hydraulic press. Foaming in the mold occurs at a pressure of 1.4–2.4 MPa, to take up the whole volume of the mold. Upon cooling, the article is demolded, the mold closed again and the cycle repeated. Note

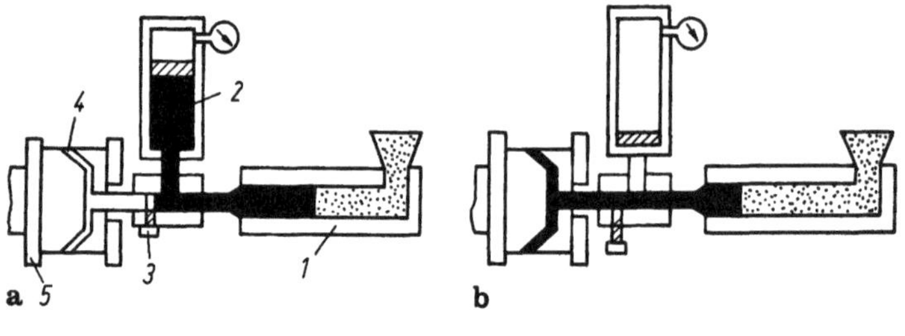

Fig. 4.2a and b. Schematic diagram of integral polymer foam production by the UCC process; **a**: plastification stage, **b**: mold filling stage; *1*: extruder, *2*: accumulator, *3*: valve, *4*: mold, *5*: hydraulic press; ▨ — granulated polymeric composition; ■ — melted composition (Courtesy of Union Carbide Corporation, New York, USA) [2]

that the extruder is operated continuously, filling the accumulator while the article is cooled. The total production cycle for articles with an apparent density of 400 kg/m^3 is about two or three minutes.

Every UCC installation consists of a screw plasticator (extruder), one or more accumulators, and a mold locking block. The extruder plasticizes a polymer mixed with a BA and supplies the melt to the accumulator whence the resulting composition is injected into the mold that is maintained at a temperature of 20–40 °C. Due to the joint effects of accumulator back pressure and the foaming gas pressure, the total pressure in the mold ranges from 0.5 to 10 MPa, depending upon the article's dimensions and density, as well as upon the equipment's capacity and design.

The melt in the accumulator and cylinder is pressurized to prevent premature foaming, the pressures being 15–25 and 2–4 MPa, respectively. The injection speed may range from 1 to 5 kg/sec.

Mold cycle duration (molding time) is determined, other things being equal, by the thickness of the mold walls and ranges from 2 to 10 min.; the cooling times range from 2 to 8 minutes, being about 25% shorter for integral polystyrene foam than for PO.

The low molding pressure makes it possible to utilize lightweight molds made of moderately priced materials, e.g. low-carbon steel, sheet and cast aluminum, and Al—Zn or Be—Cu alloys.

A recent communication has reported the development of a new type of mold — a galvanoplastic multicomponent metal matrix with a support structure that has a low thermal conductivity [5].

As mentioned before, the UCC process permits the production of articles with densities ranging from 250 to 950 kg/m^3. The thinnest articles are 4.5 to 6.4 mm thick, and at these low values the melt paths and injection pressures approach their respective maxima. The best strength-to-mass ratio is achieved at an article thickness of 6 to 6.4 mm and with skins of 0.5–0.8 mm [5].

The main disadvantage of the UCC process, like of most LP-IM methods, is the roughness of the article surface and traces of the melt flow thereon. These surface defects can be eliminated, to some extent, by increasing the temperature of the mold walls up to 120–150 °C and by prolonging the molding time.

The UCC process is largely employed to fabricate integral PO and PS; lately, commercial production of materials based on PC, PPO, and ethylene-vinyl acetate copolymers has been started.

The UCC process was commercialized in 1966 and has been extensively applied all over the world, particularly in the United States where about 80–90% of all IF's are produced by this process [2]

4.2.2 The TCM-Hettinga Process

The *Thermoplastics Cellular Molding (TCM) process* developed by Hettinga Equipment, Inc., USA molds not only IF parts of great quality, but it does so at substantially less cost than the other processes now available. Machinery used in other IF methods is designed with one injection unit intended to serve only one molding station (Fig. 4.3). This results in time-wasting delays for cooling and demolding after each shot. The TCM injection unit (Fig. 4.4) is mounted on a swivel base and immediately after filling one mold, it pivots on its horizontal axis to fill another of the four or six molding stations, that are positioned along a 180° arc facing the nozzle of the injection unit. This method not only makes efficient use of the injection unit's time, it also allows for unhurried, natural expansion and cooling time [6].

Injection pressures are low — the mold needs to withstand an internal pressure from the gases of only 15 MPa. This permits low cost aluminum tooling to replace

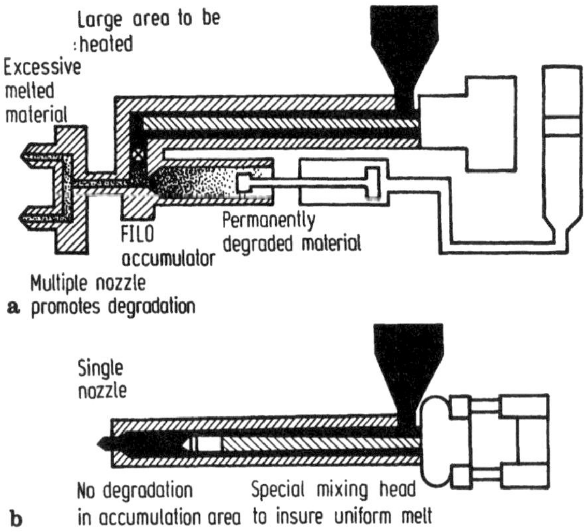

Fig. 4.3. Comparison of two injection unit systems for the production of IF articles by low pressure IM processes; **a**: conventional injector unit results in degraded material and high energy cost, **b**: Hettinga injector unit results in no degradation of material, low energy and superior part quality (Courtesy or Hettinga Equipment, Inc., Des Moines, USA)

Fig. 4.4. Low pressure injection molding Hettinga machine with multiple-station and one injection unit (Courtesy of Hettinga Equipment, Inc., Des Moines, USA)

those made of expensive steel, and savings of up to 70% are possible. Energy savings are substantial too — the TCM low-pressure process saves up to 80% of electricity.

The gentleness of the injection process doesn't force the material to skid across the mold wall fracturing the surface and leaving ripples and swirls that require finishing before painting; TCM IF parts have smooth skins right out of the mold.

In the TCM process the average temperature of the molds is 150 °C, the injection pressure is 4.2 MPa (600 psi), the average clamp pressure 2.3 MPa (325 psi), the cycle time varies from 45 sec to 3 min (depending on the wall thickness, from 3.2 mm to 75 mm). Individual control of all processing parameters for each station is provided by a microcomputer [7].

4.2.3 Other Processes

There are several variants of the UCC process. The "Variotherm" process is characterized by cyclic heating and cooling of the mold, which reduces the molding cycle by about 25%. Another variant involves an additional vessel receiving the melt from the accumulator; as soon as the vessel contains sufficient melt it is injected into the mold [8,9].

Cyclical temperature control (the "Variotherm" process) of the mold has been suggested for some time as a way of improving surface quality and increasing injection speeds. The foam viscosity can be maintained at or near the injection temperature as it flows through the mold channel, thus minimizing resistance to flow.

There is a great deal of attention being paid to rapidly heating and cooling massive mold sections in order to make the cyclical process more economical. It has, however, been shown by Progelhof and Throne [10] that the mold resistance to heat transfer is a negligible part of the overall resistance to heat transfer. For the "Variotherm" process to be economical, the molds must be designed with many, large-diameter coolant lines with unencumbered flow paths. The coolant lines must be heavily insulated and the mold base must also be insulated to prevent energy transfer to the platens and machine frame. Throne states [1] that until these design features are incorporated into all IF molds, the process will not be economically important, and even then the additional time required to cycle the mold temperature may add an additional 20–25% to the machining costs of the article.

Among the other LP-IM methods of producing IF's with a PBA, *Dow Chemical's process* should be mentioned. It uses methylene chloride as the PBA [1,5] and, another peculiarity of the process, uses gas (nitrogen) back pressure in the mold to improve the quality of the article surface.

Over the last few years, commercial LP-IM processes using CBA's, such as sodium bicarbonate, and ACA, have started to evolve. CBA's are known to contribute to smoother surfaces. Examples are the "Variotherm" process mentioned above, and the "Siemag" and "Krauss-Maffei" processes (FRG) [11].

4.3 Equipment Specifications

Low Pressure machines have been classified by Meyer [12] in two distinct areas: injection units and clamping units.

4.3.1 Injection Units

Two different types of injection systems are used for IF molding equipment: two-stage screw systems and in-line reciprocating screw systems. Both system types are economic; the selection of the appropriate system depends largely on the product to be manufactured and the molding operation.

Fig. 4.5. Battenfeld low pressure IF molding machine featuring an in-line reciprocating screw injection system (Model TSG) (Courtesy of the Battenfeld Maschinenfabriken GmbH, Meinerzhagen, FRG)

The following may be taken into consideration as guidelines [12]:

Two-stage screw	*In-line reciprocating screw*
Extremely high injection speeds possible due to low mass of injection plunger. A particular advantage for larger parts when total injection time can be kept within 1 sec for surface quality.	Slower injection speeds due to higher mass of screw.
Fixed screw with constant length to diameter ratio (L/D) to ensure homogeneous melt. Advantageous use of nitrogen blowing agent.	Reciprocating screw with changing L/D, not necessarily disadvantageous because plasticizing capacity is very often much higher than needed.
Successful with engineering resins, except rigid PVC.	Required for rigid PVC.
Exclusively used for producing IF articles.	Preferred for dual purpose machines.
Higher investment.	Lower investment.

A new generation of Battenfeld machines offer IF parts free from sink marks and with low internal stress. Two of these machines have three times the clamp capacity than those of earlier models [13]. One machine range has a screw-plunger injection unit for high shot capacity with standards PS, PO and similar non-heat-sensitive polymers. The other range has reciprocating screw units that are suitable for processing polymers requiring closer control over melt temperature, such as PC, and PPO. The screw-plunger range is available in clamp capacities from 150 to 1,500 tons and shot weights from 2.5 to 50.2 kg (for PS). The reciprocating screw range covers clamp capacities from 70 to 1,800 tons and shot weights from 0.22 to 41.9 kg (for PS). Both are suitable for use with any of the IF molding methods [14].

Figure 4.5 shows an LP molding machine with an in-line reciprocating screw injection system; a two-screw injection unit is shown in Fig. 4.6.

4.3.2 Clamping Units

Cavity pressures in the LP integral foam molding technology do not usually exceed 35 MPa, though under exceptional conditions they might rise as high as 70 MPa. The clamping units therefore have platens and tie bars that are designed for a low

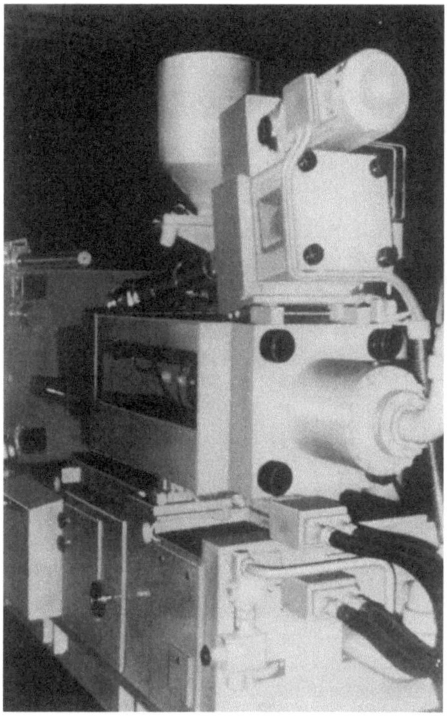

Fig. 4.6. Two-screw injection unit system of a low pressure Battenfeld injection molding machine (Model ST) (Courtesy of the Battenfeld Maschinenfabriken GmbH, Meinerzhagen, FRG)

clamping tonnage. In the past the ratio between clamping force and injection capacity has been 20–40 tons per kg shot capacity [12].

The demand for machines with higher clamping forces has arisen for the following reasons: more complex molded articles, thinner article walls to save weight, increased density to improved surface smoothness, and the move to engineering resins necessitating higher cavity pressures due to the higher viscosities.

Modern IF low-pressure machines use separate modules for the injection and clamping units, and the modules can be adapted very easily for special purposes. Vertical clamps were chosen for easy part removal; multiple-injection set-ups avoid excessive manifolding, which would have been necessary with a single injection unit. For greater efficiency in molding IF pallets, a machine has been designed to run a stack mold (Fig. 4.7). The upper and lower pallet halves are produced in a single cycle, yet the required clamping force is the same as for a single-cavity mold. This machine is designed for automatic part removal [13,14].

For producing relatively small IF articles, rotary table-type LP injection molding machines are particularly suitable. Several clamping units (up to 30 units) are arranged on the rotary table and are served successively by a single injection unit. For producing large IF articles in reasonably large runs, there are special machines which can be assembled from standard sub-units. For example, Battenfeld has designed one such machine for producing two-part pallets weighing 40 kg (see Fig. 4.7).

Fig. 4.7. Low pressure injection molding machine for simultaneously molding two pallet halves using integral PE foam. Clamping force 3,500 kN, maximum shot volume 27,000 cm^3 (Courtesy of the Schloemann-Siemag AG Kunststofftechnik, Hilchenbach, FRG)

4.3.3 Single- and Multi-Nozzle Systems

Low pressure IF molding technology has penetrated markets characterized by small parts requiring precise dimensions, for example parts for business machines, domestic appliances, and computers. These parts demand processing flexibility and control not found, until recently, on conventional low pressure machinery. Some important improvements in design, such as independent extruder back pressures, two-stage and sequential injection, and volumetric gas-control, have done much to broaden the applications of integral foams. However, it took high-speed injection (in conjugation with multiple nozzles) to result in a major breakthrough.

Following the lead of single-nozzle foam machines (comparable to conventional injection molding machines), i.e., improved performance through *higher injection pressure*, manufacturers have boosted the pressures on multi-nozzle units to 80 and even 120 MPa, according to Munns [15]. These higher injection pressures (not to be confused with those used when making high-pressure foam, in which the mold is packed and then cracked open) result in higher injection velocities and faster cycles. Multiple-nozzle systems, from a production point of view, have many advantages. For instance, more intricate parts can be molded, multiple molds can be accommodated more readily, lower cavity pressures and shorter flow paths can be ensured. At the same time, however, multiple nozzles have been a maintenance and set-up nightmare[16].

Munns has pointed out [15] that in old arrangements (Fig. 4.8a) melt is injected by an accumulator, enters a single manifold and travels through spacer blocks to reach the nozzles, and so in order to

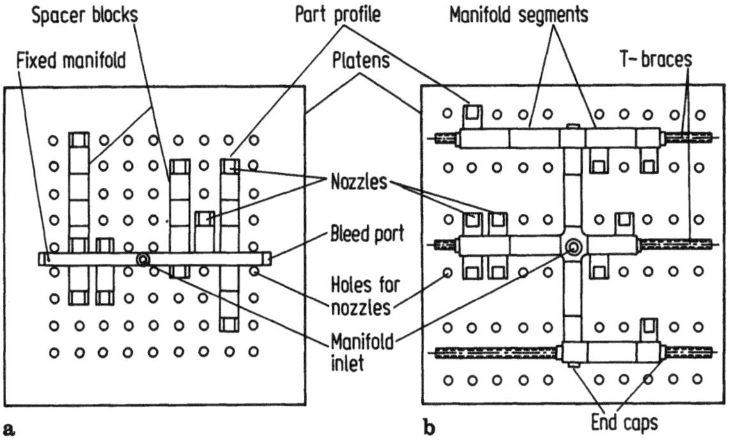

Fig. 4.8a and b. a: single manifold produces pressure drops as polymer melt feeds an array of gates; **b**: multiple manifolds feed gates independently, eliminating individual pressure drops [12)]

Fig. 4.9. Stationary platen (170 ×218 cm) on a Hoover machine (Model HV-450-426), showing 48 nozzle locations on 23 cm centers; note the 4 manifold locating holes (Courtesy of Technomic Publishing Company, Westport, USA) [12)]

change a mold or maintain the nozzles located inside the stack-up, the whole set-up must be dismantled and then reassembled. A more efficient approach uses several manifold blocks positioned on a T-brace (Fig. 4.8b). *The melt manifold* can then be arranged to bring the melt to any area of the platen, contoured end-caps (which replace bleed ports) eliminate corners where plastic could collect and degrade (Fig. 4.9). Since nozzles are bolted to the side of the manifold blocks, they can be moved or removed without disturbing the rest of the set-up. Moreover, coupling the nozzles close to the manifold reduces the pressure drops experienced in the old design, in which melt had to reach an array of nozzles. Using a multiple manifold, 9.9 kg of melt can be shot in 0.5 seconds, as compared to 1.8 kg with the older machines. Thus a high-flow hydraulic system had to be developed that incorporates a servo-operated pressure control valve to regulate the injection pressure and flow rate precisely, the pressures being adjustable from 3 MPa to 100 MPa, and the flow rates from 0 to 20,000 cm^3 per second [15)].

Multi-nozzles machines have many other advantages. The molds are filled before the BA can escape, resulting in finer, more consistent cell structures (preventing weak areas) and more uniform density distributions (reducing stress and warpage); the polymer melt has less time to cool along the fill path, minimizing flow lines; there is less bubble migration and gas entrapment on the surface because the entire melt cross section advances at the same velocity in a plug flow; higher fill rates mean thinner IF skins, and consequently dramatic weight reductions; knit lines are reduced since the melt fronts meet sooner whilst the melt is hottest and the gas is still held in solution; the melt can flow further without solidifying — the quick fill that produces a thinner IF skin also lowers heat loss to the mold and pressure drop [8,17].

4.3.4 Recent Developments

A recent development in IF technology is the separation of extruder back pressure from injection pressure which gives added injection capacity to LP molding. With the addition of a second, *independent hydraulic pump* to charge the gas-over-oil hydraulic accumulators of the injection circuit (Fig. 4.10), a machine can attain the high injection pressure whilst the plastic melt is filling the accumulator [15]. The extruder back pressure (adjustable from 3.5 to 28.0 MPa) becomes completely independent of injection pressure, which latter can be brought independently to its maximum (up to 42 MPa for low velocity machines). The result is a doubled fill rate (up to 77 kg/sec.), improved part consistency, and a smoother IF skin. In the Hoover Model

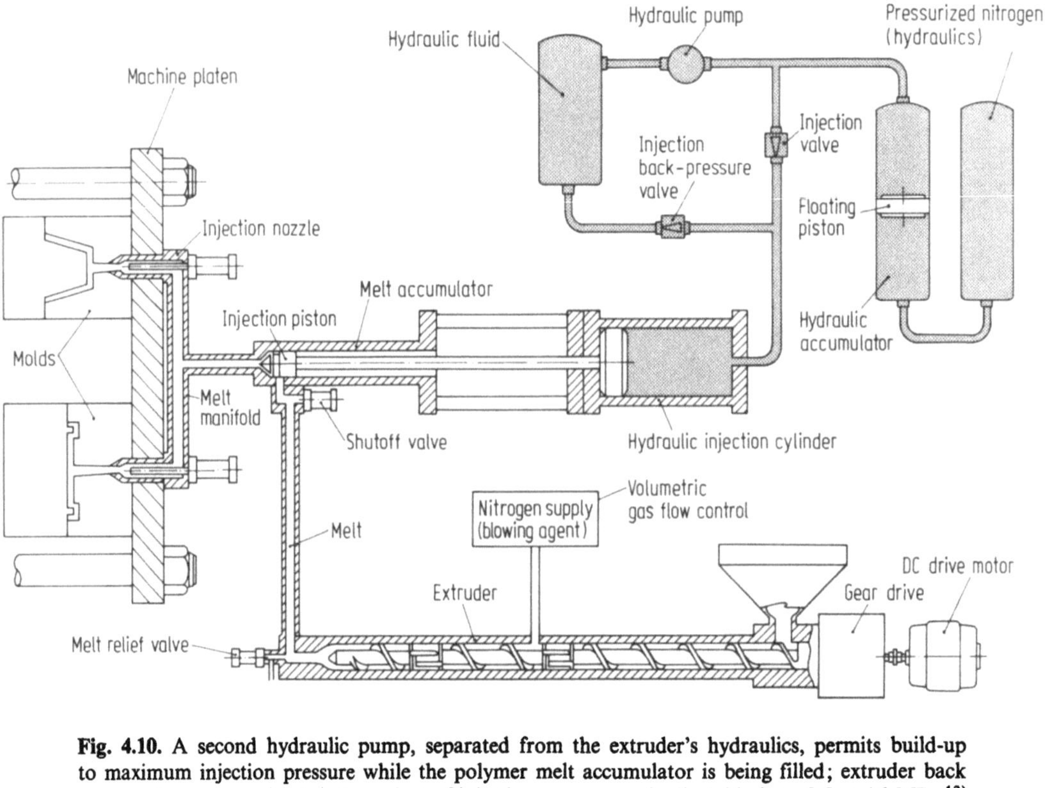

Fig. 4.10. A second hydraulic pump, separated from the extruder's hydraulics, permits build-up to maximum injection pressure while the polymer melt accumulator is being filled; extruder back pressure is now completely independent of injection pressure and adjustable from 3.5 to 4.0 MPa [12]

HV-450, *four injection accumulators* are each capable of a 10 kg shot of PS; each has its own melt manifold for maximum shot control, an independently adjustable shot size, as well as programmable injection rate and injection pressure for optimum distribution of the melt.

The most recent development is *sequential injection.* This method enables portions of the melt in a fully charged accumulator to be shot sequentially into a number of molds of different volumes by programming stops along a linear voltage displacement transducer and controlling the operation of groups of nozzles to coincide with the sequence [15].

It should be mentioned that LP technology for producing IF articles still has room for development. Projects already underway include: close-loop programmable controls to regulate extrusion, injection, and gas flow; a programmable injection profile combined with sequential injection to provide a pressure profile for each segment in the sequence; the Asahi-Dow process (see Chapter 6), which combines IF and blow molding technology, to produce hollow parts; and low-temperature blowing agents (for example freon), in lieu of nitrogen, to shorten the cooling cycle [16–18].

In the case of *thick-walled IF articles* having large areas and low densities, a good surface can be achieved by injecting the polymer melt into a mold cavity that is enlarged after injection, thus allowing the plastic to foam up in the core. A typical feature of this "breathing mold" process is the appearance of parting lines between surfaces which move relative to each other. At present, this process is not covered by patents. For example, using a "breathing mold" process, PS articles can be produced with very low densities (ϱ = 55 kg/m^3), and moreover with wall thickness increased from 6.5 to 11.5 mm [14].

With the advent of new high-flow polyolefins (plus related advances in equipment design) *ultra-thin-wall IF's* have become a dependable and practical way to improve the manufacturing economics without incurring a performance penalty [19,20].

Special-purpose machines with low locking forces, high injection speeds and high-capacity plasticators are used in the "Variotherm", "Siemag" and "Krauss-Maffei" LP processes. Note that high injection speeds are caused by high pressures, and so these special-purpose machines, though still operating in the LP range, develop far higher pressures than those used for producing IF's by the UCC process.

For example, the Structomat machines of Siemag have injection pressures up to 80 MPa, injection speeds up to 35 kg/sec, shots up to 500 kg/h, and locking forces ranging from 75 to 500 tons. They can be used to process practically every thermoplastic except PVC. The TSG machines of Krauss-Maffei are capable of developing even greater injection pressures — up to 100 MPa — and have locking forces of up to 700 tons. They are designed for producing IF's from practically all thermoplastics. Similar equipment intended for large-sized IF articles is also fabricated in Japan by Mitsubishi [5,11]. Based on the Hoover Universal Machine, Horizon Plastics, Canada developed an IM system with nitrogen as the gaseous substance [8]. This system was designed for the use of N_2 from the very beginning and therefore invested in the pumping/metering equipment required for a uniform distribution of the inert gas in the polymer melt.

A new series of molding machines developed by Beloit Corp. converts from LP integral foam molding to high-pressure injection molding with a simple tooling change [20].

A comparison of the economics of the LP processes with that of the other processes will be given in Chapter 12.

4.4 References

1. Throne, J. L.: J. Cell. Plast. *12*, 161, 264 (1976)
2. Structural Foam: Techn. pamphlet of the Union Carbide Corp., F-44669b 11/74-10M, USA, 1974
3. A New World of Strength in Plastics — Structural Foam: Techn. pamphlet of the Horison Plastics Ltd, Cobourg, Canada, 1982
4. Molders and Finishers of Quality Plastic Parts for Industry: Technical pamphlet of the Leon Plastics, Div. U.S. Industries, Inc. Grand Rapids, USA, 1982
5. Semerdjiev, S.: Introduction to Structural Foam. Brookfiled Center: Soc. Plast. Engineers, 1982
6. Hettinga Equipment, Inc.: Des Moines, Iowa, USA (1983)
7. Colangelo, M.: Plastics Technology *29*, No 6, 41 (1983)
8. Law, C.: Canad. Plastics *40*, No 6, 21 (1982)
9. Fillmann, W.: Ind. Prod. Eng. No 3, 35 (1978)
10. Progelhof, R. C., Throne, J. L.: 32nd SPE ANTEC Techn. Papers, *21*, 455 (1975)
11. Berlin, A. A., Shutov, F. A.: Strengthened Gas-Filled Polymers, Moscow, Khimia, 1980 (in Russian)
12. Meyer, W.: J. Cell. Plast. *15*, 292 (1979)
13. N.N.:Europ. Plast. News *7*, No 1, 3; No 2, 31 (1980)
14. Eckardt, H.: Process and Methods for the Manufacture of Molding. Meinerzhagen, Battenfeld, 1979
15. Munns, R. J.: J. Cell. Plast. *16*, 39 (1980); Plast. Eng. *37*, 41 (1981)
16. N.N.: Plast. Rubber Intern. *8*, No 1, 4 (1983)
17. N.N.: Plast. Technol. *27*, No 2, 29 (1981)
18. N.N.: Production (USA) *86*, No 3, 106 (1980)
19. Sneller, J.: Mod. Plast. Int. *10*, No 8, 24 (1980)
20. N.N.: Plast. Eng. *36*, No 1, 14 (1980)

5 Injection Molding: High Pressure Process

5.1 Basic Process

Molding of IF articles at *high pressure* (HP) is characterized by the following:

1. Very rapid and complete filling of the mold cavity with the melt (another term for HP methods is "full-shot", in contrast with "short-shot" for LP methods).
2. Separation in time of the processes of skin formation and core foaming.
3. Foaming due to the expanding volume of the mold cavity.
4. Utilization of CBA's better than PBA's.

The HP process with expansion molds was the first successful method of injection molding IF articles having solid skins free from gas splays marks, and having void-free closed cellular cores.

High molding pressures, up to 150 MPa, yield articles with very smooth surfaces that do not require further finishing; moreover the molding cycle is shorter than that in the LP processes.

As in the case of LP processes, the HP processes also involve back pressure in the mold cavity, allowing an unfoamed surface skin to be formed first; as the pressure is reduced with increasing volume of the mold the core of the article is foamed. High injection and molding pressures, however, do require much higher back pressures than those in LP processes. As a consequence, the physicochemical processes of gas dissolution and saturation in polymer melts exposed to considerable shear stresses have to be taken into account.

It is common knowledge that the saturation gas pressure under such conditions is far greater (over 30 MPa) than that for melts at the same temperature, but not subject to mechanical stresses. Under the latter conditions, the static saturation pressure is generally 1–8 MPa. The back pressure in the mold cavity should therefore be over 30 MPa; otherwise the skin would foam and hence the article would not have the integral structure. This demands considerable expenditures on massive gasproof molds with complicated (extensible) designs and high locking forces. It is for these reasons that IF articles produced by the HP-IM method are generally small-sized and simple in configuration.

A high-speed IF molding process designer must address the three main problem areas:

quality surface finishes,
uniform cellular structures, and
satisfactory molding cycles.

Before turning to a discussion of the advantages of high-pressure IF technology and its solutions for these problems, remember that in the low-pressure IF technology the blowing agent has three functions: distribution of the material, formation of the skin, and formation of the cellular structure. In the HP process the cell structure is also formed by a blowing agent, but the skin is formed and the material is distributed by the action of the injection pressure. Hence, the cellular structure tends to be good because the action of the blowing agent can be manipulated solely to control the formation of the cell structure.

According to Thomas [1], the same mechanism is used to produce a fine surface: CBA decomposition, like other reactions, depends on temperature. A polymer melt meticulously maintained slightly below the activation temperature of the CBA prior to injection, gains a fine surface because *the integral skin is formed on the cold mold surface* in the time it takes for the CBA to generate foam after being triggered by a heat pulse from the nozzle. This system has other advantages. First, it can yield a very low core density because if the BA concentration is carefully controlled, 50% weight reduction is possible in areas of expansion. Secondly, solid and foamed sections can be maintained in the same part simply by controlling the area where the material is allowed to expand. Moreover, very thin sections (up to 0.8 mm) can be molded. Finally, the cycle time is shorter than in LP molding, since no time is lost waiting for the material to pack itself against the cavity surface, and because the thermal conductivity of the material is greatly enhanced by the lack of surface roughness.

An example of an HP IF part is shown in Figure 5.1; it is the back of an access Xerox panel (280 × 482 × 6.4 mm). A "witness line" on the external surface indicates

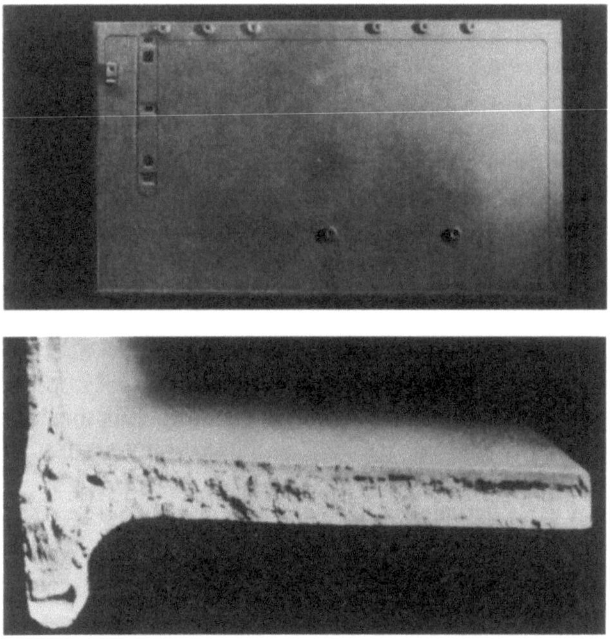

Fig. 5.1. The back side of a Xerox access panel (top) and a cross section (bottom) (Courtesy of the "Plastic Design Forum" Journal) [2]

the moving (expandable) mold section. The fine cell structure can be seen in the lower part of the figure [2].

5.2 Commercial Processes

5.2.1 The USM Process

Of all the one-component injection molding processes currently commercially available throughout the world, the most popular is the one of USM Corporation, USA (sometimes referred to as the Farrel, or Farrel-USM, process). It has been developed for producing any thermoplastic IF, both amorphous and crystalline.

A standard molding machine (based on a single-screw extruder) is used with a special three-plate mold having an "expandable volume", the back plate being movable. After plastication, the injected melt completely fills the closed mold; the initial pressure during skin formation is 100 MPa. The in- moved backwards and the pressure in the mold becomes lower than the saturation gas pressure in the melt, thereby providing the conditions for core foaming. The USM process makes it possible to obtain articles with densities ranging from 200 to 800 kg/m^3 and very precise dimensions (200–500 mm); injection pressure and locking force are about 80% of the values typical for conventional IM methods. Several seconds after injection, when an unfoamed skin has already formed, the back plate is the maximum mass of an article is under 5 kg and its projected area is between 0.3 and 0.5 m^2. As the molds concerned are 5 to 50% more expensive than conventional ones (depending upon the size of the articles), the process is mainly useful for fairly simply configurated articles [3].

The principal advantages of the USM process are:
1) possibility of IF's with very smooth shiny surfaces;
2) the molding cycle is 20 to 25 pe cent shorter than that in the UCC process;
3) possibility of using multi-cavity molds;
4) possibility of fabrication complete sets of components for built-up articles in a single mold and in a single cycle. This facilitates a 15–70% reduction of the unit cost compared to LP moldings.

A large number of IF articles commercially available in the United States are fabricated by the *process developed by Hoover* [2]. This process combines features typical of the LP-IM and HP-IM. On one hand, the pressure in the mold is as low as 2.1 to 3.4 MPa, while on the other hand the injection pressure amounts to 140–210 MPa. A specially designed injection nozzle and mold, as well as the feeding, degassing, and cooling systems yield cycle durations comparable with that in the USM process. ACA is used as BA. The Hoover process can fabricate articles based on PVC, PO, PS, ethylene-vinyl acetate copolymer, and plastics reinforced with glass fiber. The integral foams are produced with densities ranging from 250 to 500 kg/m^3 and a minimum skin thickness of 1.1 mm [4].

5.2.2 The Dow-TAF Process

The TAF, or Dow-TAF, process is designed for making IF parts that have the surface quality of solid injection moldings. The process was developed in Japan in the early 1970's by Asahi-Dow Ltd., and the machinery used to operate the process was developed in cooperation with the Toshiba Machine Company. The TAF process, therefore, takes its name from the co-developers and stands for *T*oshiba *A*sahi-Dow *F*oam [5]. Commercially the TAF concept has been shown to be an economic alternative where good surface and low density are important in the final product. The process mixtures as foaming agents. It can only be used on special-purpose molding machines. was designed to produce IF's based on amorphous polymers, using CBA-PBA The pressure reduction in the mold is not achieved by discharging the gas or by removing some of the composition into a collector, but by increasing the volume of the mold cavity by extending its hold-down plates.

According to Burk [5], the TAF process is most easily understood as an injection molding process from which integral foam can be produced, rather than as an integral foam process. The reason for this is that throughout plastication and *injection the plastic is in an unfoamed state*. The designs of the machine parts and mold parts such as nozzle, sprue bushings, runner systems, and mold gatings are, therefore, the same as for solid (unfoamed) injection molding processes. The IF is formed in the mold by the BA when the mold volume is increased through a mold expansion system. The process has the following process steps: plastication, mold closing and clamping, injection, cooling of solid skin, mold expansion, cooling, and part ejection.

The TAF process requires a mold that can be expanded during the molding cycle. There are many ways in which a mold can be expanded, and any such procedure can be used in conjunction with the TAF process [5]. A common type of expandable mold is the three-plate mold (Fig. 5.2a), in which

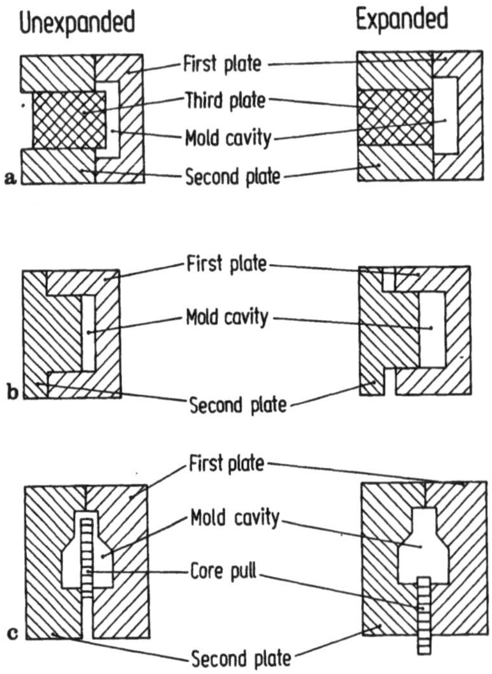

Fig. 5.2 a–c. Different types of expandable molds for the TAF process; **a**: three-plate mold, **b**: two-plate mold, **c**: core pull mold (Courtesy of Technomic Publishing Company, Inc., Westport, Conn., USA) [5]

the third plate is used as the expandable surface. Initially, the plate is in the unexpanded position during the injection sequence, but at the proper time the third plate is moved back to the final desired thickness of the IF article.

Another common type of expandable mold is the two-plate mold (Fig. 5.2b). This mold design is the simplest possible and is expanded by moving the machine platens as if it were partially opening the mold. The cavity is designed so that the actual parting line is recessed in order to prevent the part from bulging at the parting line during the expansion. During the expansion sequence with a two-plate mold (it is often described as a vertical flash mold), the outer edge uncovers an area on the expanded mold's surface where the skin could not have been formed during the initial injection sequence. This area may end up with a poor finish or with swirl marks, but it has been found that by using a proper operating sequence, the surface can be made to have a swirl-free finish.

Finally, a third type of expandable mold uses core pulls to achieve the expansion (Fig. 5.2c). An easy way of visualizing this type of mold is to consider a part with a thick cross section, a blowing pin, for example [5]. The two outer plates of the mold close around a center core pin to form the mold cavity in the unexpanded state. At the proper time, the core pin is pulled out by the amount required by the part design. If the core pin is pulled out so that the top of the pin is flush with the surface of the cavity, an expansion witness line is formed, as is the case with a three-plate mold. If the pin is not retracted completely, a cavity is formed which could be functional and could also hide the witness line.

In general, the practices used in the design of molds for solid injection molding should be applied to the design of molds for the TAF process, that is, the same considerations should be used for choosing the multi-cavity design, runner systems, gating, cooling, and election systems. The only special feature of a TAF mold is the method of mold expansion, and this must be coordinated with the other mold requirements and the design of the finished IF article.

The TAF process is characterized by the development of an integral skin *by controlling the pressure* applied to the melt during foaming. This is in contrast to the uncontrolled foaming that occurs during the low pressure IF processes. One way of controlling the pressure is to gasket the mold to prevent foaming gas from escaping during the filling and foaming operations. In general, this pressure development is not rapid enough to yield good results. Thus, in the TAF process "O" ring gaskets (see Chapter 3) are used and the mold is positively pressurized with N_2 to 0.14 to 0.20 MPa [6]. The pressure is controlled with a relief valve during the foaming process. The mold can either be overfilled initially, as in the cases of the Allied and Bulgarian process or the moving sections can be contracted initially as in the Dow process. Both techniques essentially serve the same purpose of restricting the extent of foam expansion in the early stages of foam formation. The cooling time decreases as the articles become thinner. For example, if the article thickness is reduced from 120 to 14 mm, τ_c decreases from 420 to 180 sec [7].

Although such molds are rather expensive, the cost of articles fabricated by the TAF process is lower than that obtained with the LP-IM method. This is due to the shorter molding cycle and the higher quality surface that does not require any further finishing. A disadvantage of the method is that traces are left by the removable inserts on the surface of the articles.

It has been shown by Throne [6] that during early heat removal from an IF slab, the rate at which the cooling zone moves into the slab is proportional to the square root of the elapsed time. The proportionality constant is the thermal diffusivity of the polymer, which for the TAF process can be assumed to be that of the unfoamed plastic. Throne assumed that the skin formation occurs at or below a certain temperature, such as the molding temperature or the heat distortion temperature at 1.85 MPa. Thus, the compact plastic forms a skin the thickness of which is proportional to the square root of time. The desired thickness of skin can therefore be programmed either by egressing the hot core material or by opening the movable mold sections at an appropriate time. For example, after

20 sec the skin thickness of an IF slab (HDPE, slab thickness 20 mm) increased to 6 mm in the TAF process, while it increased to 5 mm in the Bulgarian process (see Chapter 6).

The process has two drawbacks: the mold opening must be done carefully in order to minimize the collapse of the soft core owing to negative pressure, and the opening requires expensive moving sections.

Egressing an IF part from a mold requires that the material remains relatively soft in the sprue region. Thus the sprue must be exceptionally large, or cyclical temperature control must be maintained in that region. Furthermore, until recently, the egressed material did not have a quality good enough to permit its reinjection into the next shot. Part of this is because of the extensive out-gassing of the blowing agent gas that occurs during egression. A delay in opening the transfer nozzle prior to injection should satisfactorily redissolve the free gas in the hot egressed melt before it is reinjected [6].

Initially, the TAF process was developed for PS resins. At present, formulations are available for ABS, PPO and SAN resins. In principle, the *TAF process can be applied to any thermoplastic resin*. However, before other resin could be used, proper BA systems had to be developed. Parts can be produced with an overall density as low as 50% of the solid resin density. The density of the IF part is controlled by the amount the mold cavity is expanded.

The TAF process can be used to make nearly any part that can be designed in integral foam, but for articles with certain design parameters the process is particularly recommended. Parts with variations in cross-sectional thickness can be considered because the molds are filled by the machine's injection pressure, therefore the polymer melt can flow from thick cross sections to thin ones. In general, hard-to-fill molds can be filled by the TAF system. For best results, the minimum initial thickness should be not less than 3.8 mm, while parts with cross sections greater than 25 mm can easily be made. There are no *limitations on part geometry*, and the part size is not limited by the operation of the process. Up to now however, TAF part weights have commonly been below 2.5 kg [5].

5.3 Equipment Specifications

Most HP process are realized on standard injection molding machines (based on a single-screw extruder) provided with a three-plate "expandable volume" mold and have been achieved, while the possibility to include bosses and ribs results in part molding machine fitted with their proprietary TSG sub-unit, i.e. it is equipped with a device for rapid injection, and with a sealed nozzle (Fig. 5.3).

The main changes to a standard machine for the USM process which was developed by the Farrel-Rochester Division of Emhart Corp., include the installation of two positive-stroke hydraulic punchoff cylinders for expanding the mold (Fig. 5.4), an increase in the L/D screw ratio to 24 to 1 (from 20 to 1) in order to achieve consistent foam quality, and the addition of a positive hydraulic foam shutoff nozzle. Since the expansion is in general in the plane of the platen's movement, the cylinders are mounted so as to bottom on micrometer-adjustable rods on the moving platen. To accommodate larger and multiple-cavity molds, the expansion sequence is closely controlled and a two-stage, infinitely variable, injection-speed capability was added [1].

A TAF molding machine is equipped with a pre-plasticator (a 28-D long screw) in order to control the temperature, pressure, and viscosity of the melt closely, a piston injector to develop high injection speeds, and a mold with inserts featuring three stages, i.e. completely closed (at injection), partially open (at foaming), and completely open (at demolding).

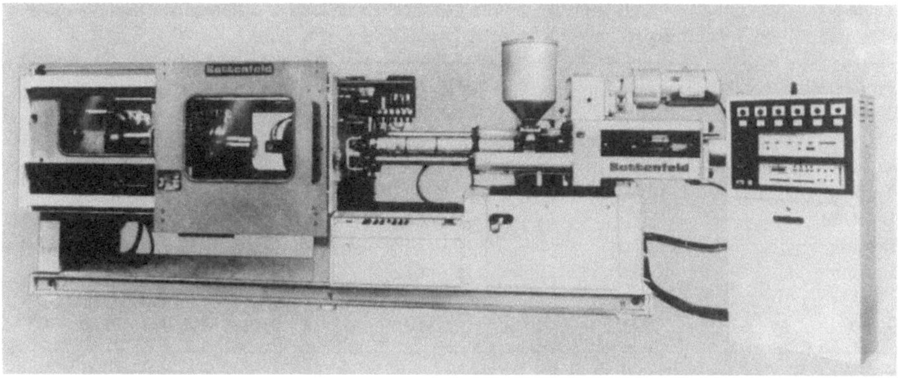

Fig. 5.3. High pressure Battenfeld injection molding machine (series S) with the TSG sub-unit for producing thermoplastic IF parts (Courtesy of Battenfeld Maschinenfabriken GmbH, Meinerzhagen, FRG) [8]

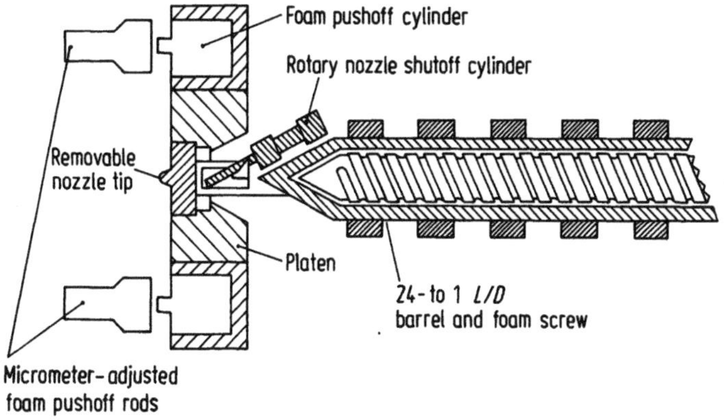

Fig. 5.4. Modification of a standard injection molding machine by adding two positive-stroke pushoff cylinders, increasing the L/D ratio to 24 to 1 (from 20 to 1), and adding a positive hydraulic foam shutoff nozzle (Courtesy of the Society of Plastics Engineers, Inc., Brookfield Center, Conn., USA) [1]

TAF machines were specially designed to maximize mixing and temperature control, and swirl-free IF articles are easily produced with these machines provided correct operating procedures are ensured. It has been shown that most TAF articles could be produced on a standard injection molding machine modified for TAF operation. In general, a TAF machine is an ordinary IM machine where special emphasis has been placed on the control of the plastic melt temperature and the preparation of a uniform plastic melt [7].

5.4 Technical and Economic Analysis

Table 5.1 shows a breakdown of the costs of three differently sized IF articles using favorable-case assumptions. These percentages are probably applicable to almost any low-cost process, and should be carefully studied on this basis [1].

Table 5.1. Cost analysis of high-pressure molding of three IF articles of different size using favorable-case assumption,[a] according to Farrel Rochester Div., USM Corp., USA

Production data	Cost, %		
	Small part[b]	Medium part[c]	Large part[d]
Tooling	9.3	7.9	6.5
Material	32.3	36.5	37.1
Process	21.5	14.3	12.9
Finishing	36.9	41.3	43.5
Total	100.0	100.0	100.0

[a] Favorable-case assumptions: 1) part configuration: 75% of surface in one plane; 2) item weight (sides, bosses, ribs 25%); 3) slightly ribbed structure (6.4 mm thick); 4) material: modified PPO; 5) tooling: steel; 6) tooling ammortization: 2 years; 7) production: 30,000 parts per year;
[b] Size 1,254 cm^2 (196 square inches);
[c] Size 5,760 cm^2 (900 square inches);
[d] Size 10,240 cm^2 (1,600 square inches)

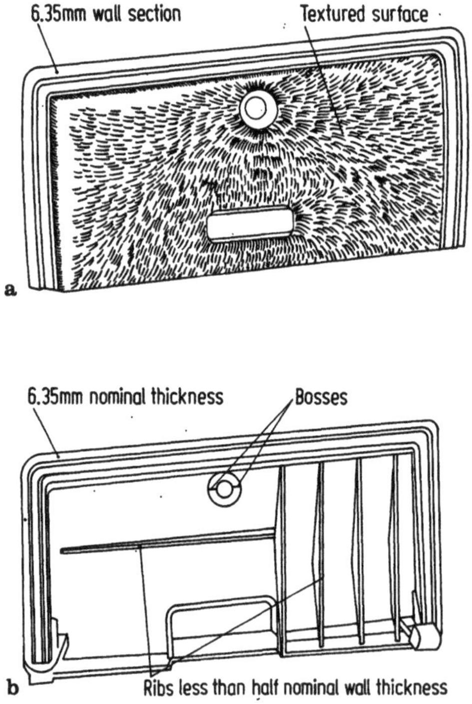

Fig. 5.5a and b. Chrysler glove-box door; **a**: outer surface exhibits fine stitching and leatherette texturing; **b**: underside shows process flexibility in molding thick and thin sections (Courtesy of the Society of Plastics Engineers, Inc., Brookfield Center, Conn., USA) [1)]

It should be mentioned that the two basic disadvantages of the USM process are high mold cost and limited mold geometry. The major use of this process should therefore be to produce *flat panels in large quantities*. Automobile glove-box doors is one such example, and they are in fact produced in very large quantities. The higher tooling cost is offset by the lower unit cost. The parts are produced in multiple cavities

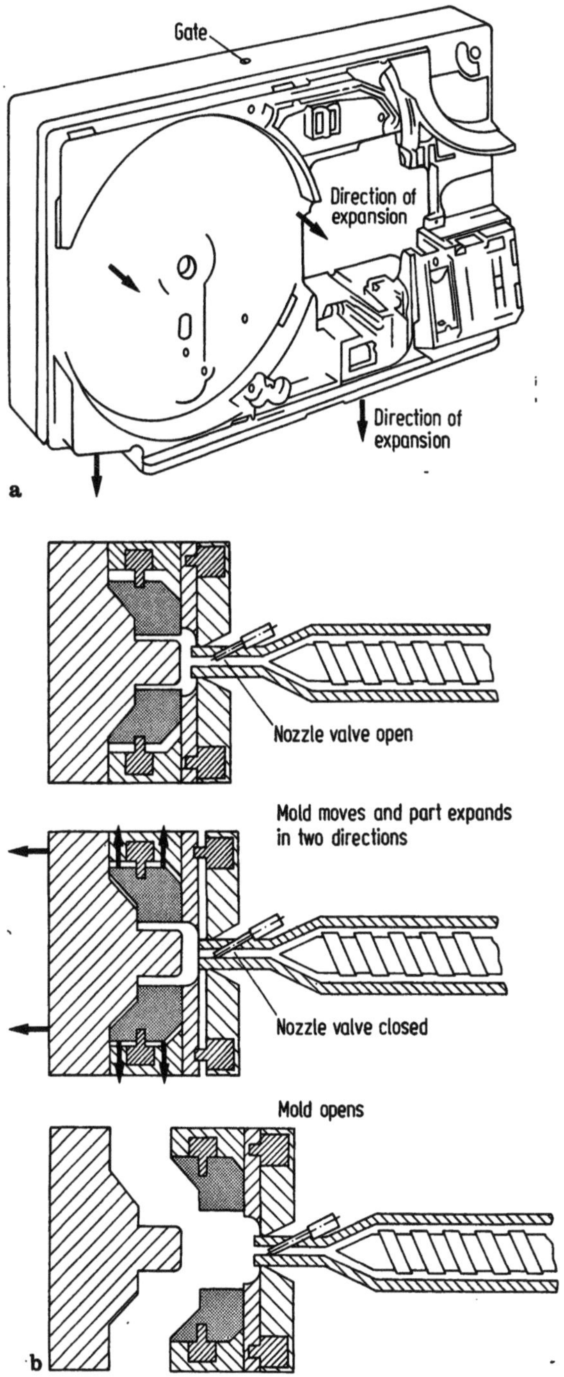

Fig. 5.6a and b. a: a complex part is illustrated by a Bell & Howell movie projector housing, which is foamed in two directions. **b**: mold action for foaming in two directions from filling, to expansion, to release (Courtesy of the Society of Plastics Engineers, Inc., Brockfield Center, Conn., USA) [1)]

in order to achieve the high quantity requirements [8,9]. Recently articles have been developed containing thin, intricate, ribbed sections combined with bosses, and having very good surface definition (Fig. 5.5). The fine stitching and leatherette surface effect on the glove-box door are obtained directly from the dye, while the back side of the article, which is fairly heavily ribbed, has no attendant sinks.

Another recent development is molding a PS article with a very *complex geometry*. It is the housing of a movie projector (Fig. 5.6). Expansion was carried out in the direction of the platen motion and simultaneously at right angles to it.

HP processes initially gained acceptance as lower cost alternatives. However, the capabilities of the process have led to applications that simply cannot be molded by any other means (Fig. 5.7). If an end-user wants to make a part that cannot be molded economically using standard unfoamed injection molding or LP foam molding, or a part with an unusually complex geometry, he should consider HP foam molding. Further cost reductions *via* significant polymer savings (compared to the LP method) have been achieved, while the possibility to include bosses and ribs results in part rigidity and helps to simplify the assembly of large, complicated parts.

To reduce the finishing cost still further, Farrel has recently introduced a modification of its high-pressure foam molding process, called "*Gas sealing*" [1]. In this system, which is similar to the counter-pressure method, but considerably more sophisticated in sequencing and operation, the foam process begins by filling the tool with inert gas, followed by filling it with resin at a reduced volume, releasing the gas, and then expanding the part. The effect on finishing costs is significant, and the procedure appears to provide a no-paint texture. This applies to PC, PPO, ABS, and PS.

Table 5.2 gives a comparison of conventional LP molding using an injection-molding machine, and HP molding using Farrel's special machine. The chart may be used by the end-user to compare the average results for the various processes, as well as to compare the performance of various suppliers [1].

According to LaPlaca [2], high pressure expanding mold IF processes are better than low pressure conventional IF processes in the following ways: 30% cycle time reduction, 30% weight reduction, excellent void-free surface quality, no material

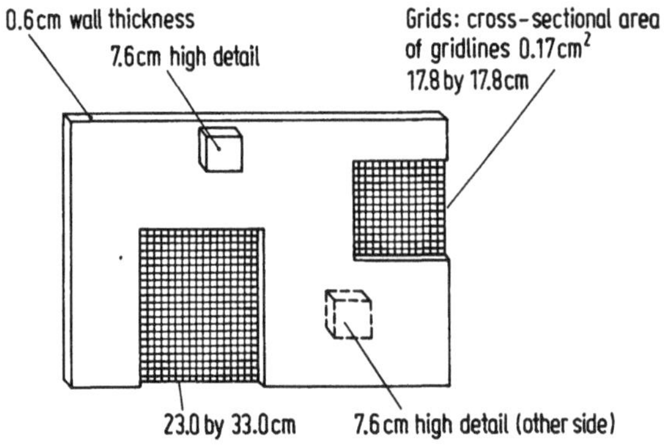

Fig. 5.7. A difficult-to-mold integral ABS part (35% density reduction) with two extensive grid areas produced by the HP injection molding process (Courtesy of the Society of Plastics Engineers, Inc., Brookfield Center, Conn., USA) [1]

Table 5.2. Process comparison chart for an the IF molding of a computer housing panel weighing 2.3–2.7 kg (5–6 pounds)

Production data	Process	
	Low-pressure IF[a]	High-pressure IF[b]
Number of suppliers	Hundreds	20
Tooling		
Metal	Aluminum	Steel
Cost (relative)	100	150
Average lead time, weeks	16	24
Processing		
Cycle time, seconds	150	100
Material	PPO/PS	PPO/PS
Finishing	Sanding, priming, 2-color application, texturing	Textured mold, high-solids polyurethane coating
Part data		
Weight reduction, %	20	40
Wall thickness, mm	6.35	6.00 (3.00 before expansion)
Configuration	Heavy ribs (if required)	Flat parts preferred, multiple density, thin-section ribs
Production cost		
Preferred quantity	10,000	25,000
Unit cost at prefered quantity, %	100	65

[a] conventional injection-molding machine; [b] Farrel machine

degassing cycle required, 15%–20% unit cost reduction. Thus, the advantages of an HP process are: reduced weight, finishing, and molding cycles; void-free surface finish, selective density reduction; commercialized process and versatile molding equipment available.

The limitations of a HP process as compared to a LP conventional IF process are: higher tooling costs (20% above conventional design); patented/licensed process; part size limited to 12 kg by available equipment; limited supplier base; compound configuration limited; increased mold maintenance; steel tooling required [10]. An economical comparison of this process with the other IF processes is presented in Chapter 12.

5.5 References

1. Thomas, J. P.: J. Cell. Plast. *17*, 148 (1981)
2. LaPlaca, J. P.: Plast. Design Forum *5*, No 1, 73 (1980); J. Cell. Plast. *16*, 36 (1980)
3. Menges, G., Schweisig, H.: Chem. Eng. (London) No 330, 194 (1978)
4. N.N.: Canad. Plast. No 2, 47 (1982)
5. Burk, W. R.: Internat. Congr. Cellular Plastics. Montreal Canada, 1976; 6th Structural Foam Conf., Bal Harbour USA, 1978
6. Throns, J. L.: J. Cell. Plast. *12*, 264 (1976)

7. Semerdjiev, S.: Intruction to Structural Foam. Brookfield Center: Soc. Plastics Engineers, 1982
8. Holzschuh, J.: Machinery for processing. Meinerzhagen: Battenfeld Maschinenfabriken GmbH, 1979
9. Berlin, A. A., Shutov, F. A.: Strengthened Gas-Filled Polymers. Moscow: Khimia, 1980 (in Russian)
10. Colangelo, M.: Plastics Technology *29*, No 6, 41 (1983)

6 Injection Molding: Gas Counter Pressure Process

6.1 Basic Process

The gas present in the mold prior to and during injection may prevent the polymer melt from foaming in the mold. In the gas counter pressure process the integral structure and an improved surface quality are achieved by pressurizing the mold to a pressure higher than that of the foaming gas prior to the injection of the melt. If the counter pressure is too high, however, the surface quality is diminished. The mold cavity is filled completely by the melt and the density reduction occurs by volume contraction of the raw material [1)].

The essential parameters affecting IF part quality in this process include the level of the counter pressure and the nature of the foaming gas. The level of the *optimum gas pressure* depends both on the nature and the amount of the BA used. For example, a counter pressure of about 2.7 MPa is suitable for a PS composition containing 0.5% by weight of azodicarbonamide as the CBA. With PBA's, the gas pressure generated in the melt is substantially higher. Accordingly, the counter pressure must be increased, and this can add to the problem of properly sealing the mold [2)].

Using this process the IF parts have smooth homogeneously colored surfaces free from sink-marks, and appear to have only slight frozen-in stresses, compared to one-component processes.

6.2 Commercial Processes

6.2.1 The Allied Processes

The process developed by the Allied Chemical Corporation (USA), referred to as the Allied process, was designed to produce integral thermoplastics with densities ranging from 400 to 800 kg/m^3, skin thicknesses over 3 mm, and smooth shiny or dull surfaces. The IF's are fabricated *in standard molding machines* with an egression-type mold. Prior to injection, the mold cavity is filled with an inert gas (nitrogen) at a pressure of 7–8 MPa. The pressure is determined by the static saturation pressure of the foaming gas at the melt temperature of the given polymer. After melt injection into the heated mold the surface skin forms first (see the Variotherm-process, Chapter 4), because the gas pressure prevents foaming at this stage. Thus, the processes of skin formation and core foaming are separated in time.

After the skin formation, the pressure of the gas is reduced (the gas is discharged via a bleed valve) and the core of the article foams. In doing so, some of the composition

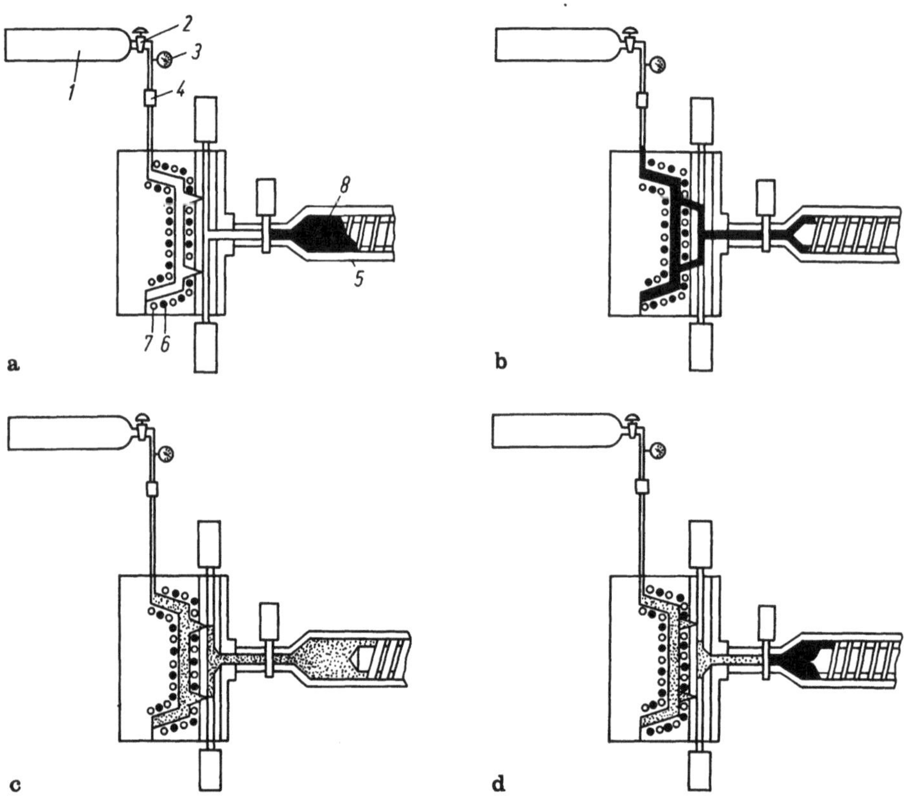

Fig. 6.1. Schematic diagram of IF production by the Allied process; 1: cylinder with inert gas, 2: switch, 3: manometer, 4: discharge valve, 5: extruder, 6: heating pipes, 7: cooling pipes, 8: plastification chamber; description of **a** through **d** see text [4)]

being foamed is forced out through special holes in the mold walls into the collector, whence it is taken again into the process cycle. Another peculiar feature of this process is the cyclic heating and cooling of the mold. At a mold temperature of 150 °C, the surface of the final article is shiny, at 130–140 °C it is dull. Articles are cooled to 3 or 0 °C and depressurized to 1.0 MPa [3)].

A schematic diagram of an Allied installation is given in Fig. 6.1. The inert gas is introduced into the mold cavity via a bleed valve (position a). The screw of the extruder is positioned so that the polymer melt fills the mold cavity (position b). The material is charged into the mold in an unfoamed state (the valve is open). The pressure developed in the mold by the extruder screw creates a hard external layer. During the next stage (position c), the extruder screw retracts and the foaming material partially fills the plastication chamber. Excess material is removed through the vertical gate which has a diameter large enough to prevent premature solidification of the material. Once the composition has foamed to the required extent, the piston moves forward again and closes the gate of the mold cavity, thereby compressing the foamed material somewhat (position d). After the external layer is sufficiently hard and the article is cool, the mold is opened and the article demolded. Large-sized articles are water-cooled in special baths.

A smooth surface and a high-quality internal cellular structure are ensured by the high pressure filling of the mold and by the low pressure foaming of the melt. The thickness of the surface skin is governed by the residence time of the melt under pressure and the cooling time. The apparent density of the core is determined by the amount of the melt removed to the collector.

6.2.2 The TM Process

An important IF technology is the process developed by Bulgarian scientists, and now called the Bulgarian, or TM process [1,2,4]. The process is particularly promising for producing IF's based on compositions containing highly volatile and dissociating components, because the whole process from the melting stage to demolding is carried out *at a controlled gas pressure*. It is used to produce IF's based on polyolefins and polystyrene and is free of the disadvantages typical of the LP-IM method, namely rough surfaces and flow traces at the surface of the end product. The composition containing the CBA and other admixtures is fed trough a charging hopper to the screw plasticator and on to a vertical cylinder where a back pressure is maintained throughout the whole cycle (Fig. 6.2). Then mold 3 is closed and pressurized with a gas (nitrogen), followed by the melt injection. *The foaming process starts after the pressure has been reduced*, and some of the foamed composition is removed through gate channel 2 (at the side of cylinder 4), whence it is taken back into the mold cavity at the beginning of the next cycle. An alternative process injects melt filling only

Skin thickness is in general controlled by several parameters, including residence time of the melt under pressure, temperatures of the melt and the mold, and amount 50–80 per cent of the mold volume, the rest of the volume being filled from the side cylinder; every shot amounts to 1000 cm^3.

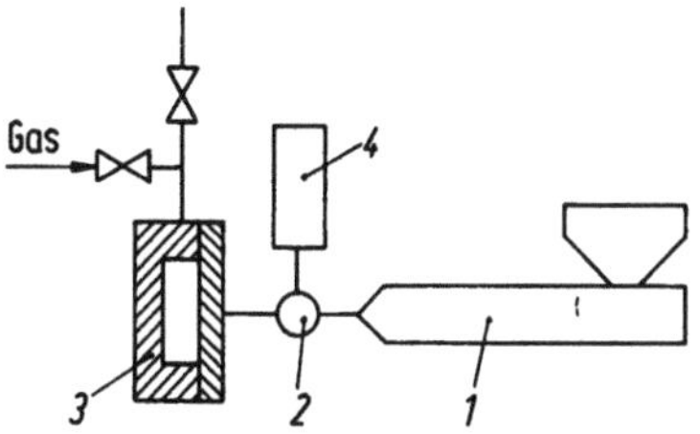

Fig. 6.2. Schematic diagram of IF production by the Bulgarian TM-process; 1: plastification cylinder, 2: distribution gate, 3: pressurized mold, 4: accumulation screw with injection ram [1]

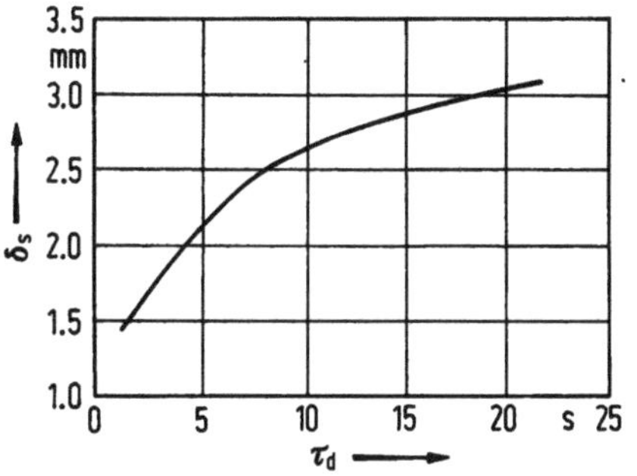

Fig. 6.3. Surface skin thickness δ_s versus dwelling (molding) time τ for integral HDPE foam production by the TM process (IF thickness 20 mm, ACA content 0.5% mass) [1]

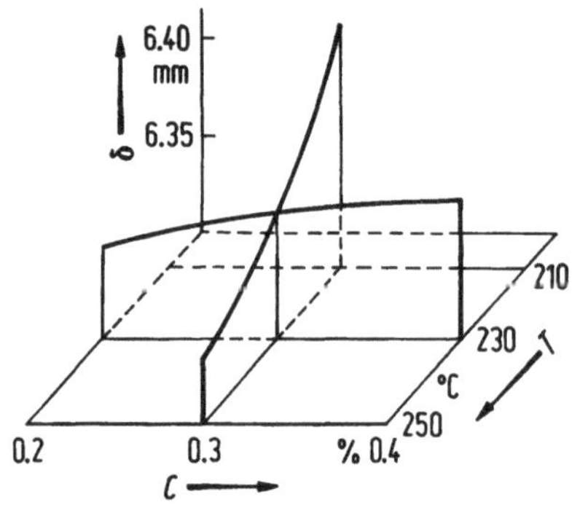

Fig. 6.4. Relationship between skin thickness δ_s, melt temperature T, and CBA (azodicarbonamide) concentration C for an integral PS foam produced by the TM process [11]

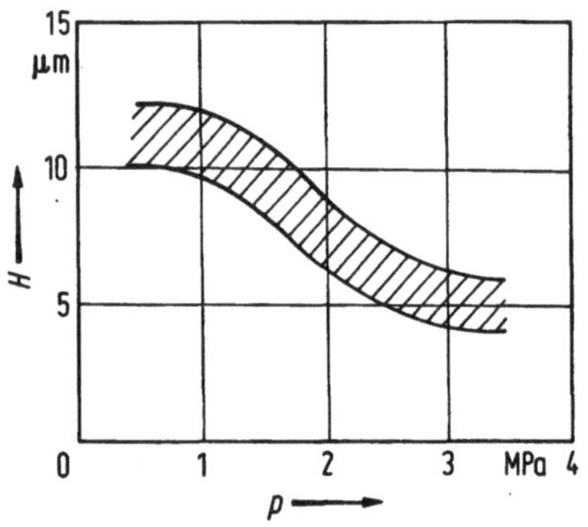

Fig. 6.5. Effect of gas counter pressure P on surface roughness H [1]

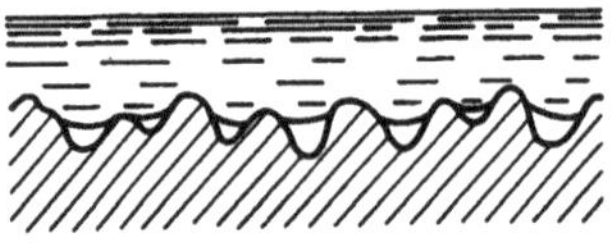

Fig. 6.6. Schematic illustration of the penetration of the polymer melt into the surface grooves of the mold cavity wall [1]

Skin thickness is in general controlled by several parameters, including residence time of the melt under pressure, temperatures of the melt and the mold, and amount of BA (Fig. 6.3). However, the possible increase in the melt temperature is limited by the temperature at which the polymer degrades. In this respect, the Bulgarian method has the advantage of the presence of an inert gas, which means that the polymers can be processed over a wider temperature range. This leads to *shorter molding times* and to better quality articles (Fig. 6.4).

The surface smoothness can be regulated by the gas counter pressure (Fig. 6.5). It is interesting to note that the surface smoothness of the IF part may exceed even that of the mold surface. The mechanism behind this effect is presented schematically in Fig. 6.6. Surface smoothness depends on the ability or inability of the polymer melt to penetrate all the grooves that remain in the cavity walls as traces of machining [1].

The technology developed by Bulgarien scientists considerably simplifies the process of producing IF's and, hence, extends the application of the material.

6.2.3 The Asahi-Dow Process

A new IF molding technology, developed by the Japanese firm Asahi-Dow, produces first class finished articles without any mold expansion or thermal cycling (thus it has nothing in common with the TAF process). The main advantages of the Asahi-Dow process are its *lower tooling and equipment costs* (because conventional molds and injection machines can be used), a reduction in cycle time, and generally lower foam densities [6-8].

The main difference between this process and other gas counter pressure processes is that, instead of using the counter pressure before injection to inhibit foaming at the mold surface, *the gas* (usually nitrogen or compressed air) *is injected through the machine's nozzle together with the foamable polymer* (Fig. 6.7). The high pressure gas entering with the shot (in the case of N_2 at 10–15 MPa) presses the polymer against the mold surface (Fig. 6.8a). At the same time, it forms a hollow (Fig. 6.8b) that accommodates foaming polymer when the gas is released (Fig. 6.8c). This eliminates the need for complicated tooling to expand the mold, and can be used to foam smooth-surface complex IF articles without core pulls. The mold can also be prepressurized with gas or compressed air through an additional vent in the mold to inhibit foaming in the first stage, and thus to improve surface finish.

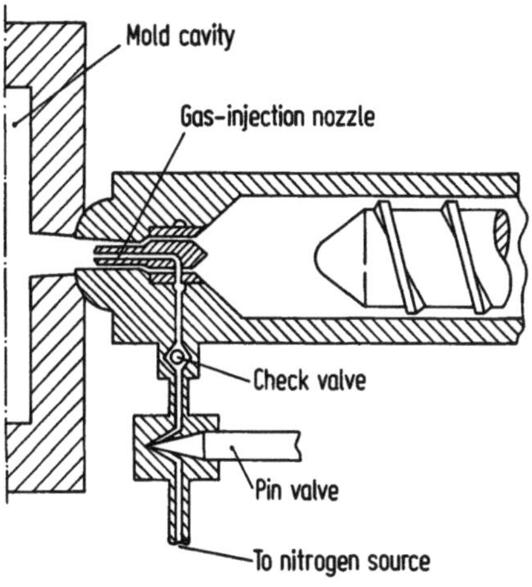

Fig. 6.7. Schematic diagram of the Asahi-Dow process: through-the-nozzle gas-injection head is supplied with high-pressure nitrogen from compressor/gas-recovery unit, and sequence control [6]

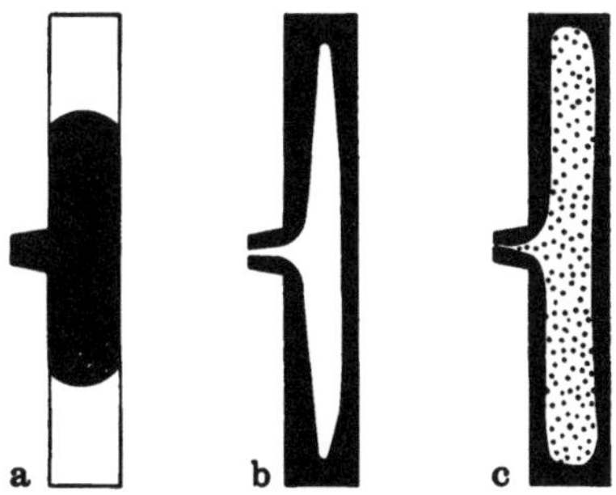

Fig. 6.8a–c. The tree stages of the Asahi-Dow process; **a**: injection of polymer melt into the mold, **b**: formation of a hollow part, **c**: foaming of polymer melt [7]

Almost any thermoplastic can be foamed by the Asahi-Dow process, including glass-filled formulations. The best results have, however, been achieved with high-impact PS, ABS, SAN, PC, and modified PPO.

A typical example of the process, which we shall consider in some detail, is an IF article based on high-impact PS. A 350 g cylindrical part with a specific gravity of 0.57 is made by adding 6% *n*-pentane, 0.2% azo blowing agent, and 1% talc nucleating agent to the plastic. The mold cavity is prepressurized with compressed air at 1 MPa through a separate mold vent, and then 250 g of the melt are injected. Gaseous nitrogen (heated to 200 °C to avoid a mold temperature drop) is injected at 8.4 MPa through the internal nozzle assembly along with the remaining 100 g of the shot. After a 5 sec pressurization cycle, during which a hollow is formed in the part, the nozzle is retracted from the gate to release the nitrogen and allow foaming in the hollow interior (see Fig. 6.8) [7].

6.2.4 Other Processes

Cashiers Structural Foam, Inc., USA introduced its *Smooth-skin process*, an adaptation of the gas-counter pressure process used in Europe [8,10]. Basically, it involves building up pressure inside the mold prior to injection of a predetermined short shot. Upon injection, the counter pressure prevents expansion, allowing a smooth skin to form and set. Then, the back pressure is released and a typical IF foam core is allowed to form. Production savings can reach up to 20% through elimination of sanding and priming operations; savings of up to 31% are possible when parts need no finishing.

Hoover Universal Corp., USA is working on a version of a counter pressure process termed "*thermocontraction*", a modified UCC process (see above, Chapter 4) that takes advantage of the thermal coefficient of expansion of the polymer [10]. At present, introduction of the process is sheduled for 1985.

Applied Molding (USA) is about to enter the thermoplastic foam market with *a gas counter pressure process* of its own. The process will include a proprietary twist designed to get weight reductions of up to 20% using amorphous polymer[10].

One variant of the Asahi-Dow process is the *internal gas-injection method* which provides a solid-skinned hollow part, using conventional injection machines. It yields important weight savings and cycle time reductions for IF parts that do not need the better physical properties imparted by foam cores. A hollow decorative PS goblet (Fig. 6.9) is a typical application, while products that are generally blow molded, such as sport balls, RV fuel tanks, and picnic jugs, could also be added to the applica-

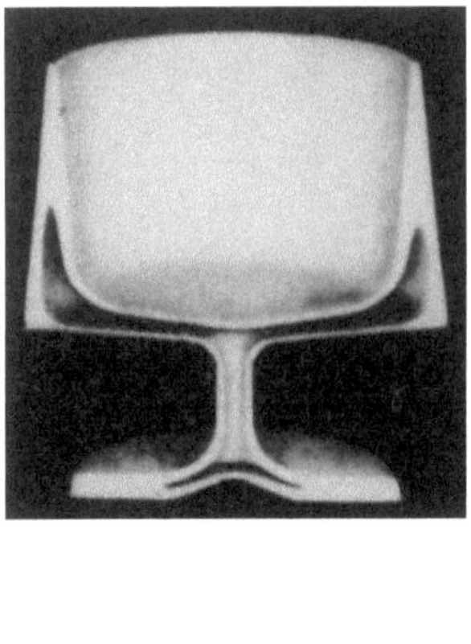

Fig. 6.9. Typical hollow IF part produced by the Asahi-Dow process using nonfoaming materials [6]

Table 6.1. Comparison of structural web (SW) and integral foam (IF) moldings[a] [8)]

Moldings	Density reduction, %	Yield strength, MPa	Flexural modulus, MPa[b]
SW	55	7.48	524.3
IF	55	7.21	460.0
SW	32	11.20	833.0
IF	32	11.48	742.0

[a] based on a 254 × 457 × 9.5 mm HDPE part; [b] ASTM-D-790

tions of the technique. The wall thickness can be varied by controlling the dwelling time of gas pressure within the part being formed.

Another promising offshoot of combining foamable and solid resins is *sandwich molding with two components*. The advantage of the internal gas injection is that the skin thickness can be closely controlled. This eliminates sandwich-molding problems such as exposed core material, unfoamed cores at 90-deg. corners, and variations in skin/core ratios [6)].

Recently a new "*density reduced parts*" *process* was developed by Union Carbide Corporation as a modification of the Asahi-Dow process [8)]. This new gas-injection process results in an entirely different type of parts, with a stiff, strong hollow lattice that could yield an extra 15% resin utilization. Because parts made by this process are not blown in the conventional sense, their solid, noncellular skins allow them to be used or painted immediately after demolding, eliminating the high post-finishing costs. This process, also called "*structural web molding*" adopts a new blowing-agent sequencing to make a non-foamed hollow part that gets its strength from internally formed "webs" or solid polymer "fingers" radiating from the gate to the surface. For identical parts, one with this internal structure has 15% better flexural modulus (at the same density reduction) than one made by conventional IF molding methods (Table 6.1). The process has an impressive set of advantages over conventional methods, including 50% cycle-time reduction, first class finishes out of the mold, and the ability to mold transparent parts. These hollow parts will have a wide range of application areas including construction (doors, pipe fittings, modular building units, translucent ceilings and floors, and solar panels), automotive (bumpers, sun visors, spare-tire covers, hood and trunk lids), material handling (tote boxes, pallets, large containers), and other uses such as food and beverage coolers [9)]. Patented by Union Carbide, the structural web process will be offered under license by Hoover Universal/USA. A 350 ton Springfield machine with a 22.7 kg (50 lb.) shot size is being modified to be capable of running both conventional foam or web[10,10a].

A process *resembling structural web* has been developed by a custom molder, KMMCO Structural Foam, Inc., USA. The difference is that the gas is introduced through a valve located in the tool, most often opposite the sprue. Timing of the introduction of the gas to the solid material is proprietary. The result is the formation of *large cells* of gas rather than webs, within an essentially solid structure, the first such cells forming around the thick ribbed and bossed sections of the part, in order to eliminate sink. Called *Injection Quality* (IQ) it differs from web, too, in that it does not require a modified machine [10)].

6.3 Equipment Specifications

The main features of molds for gas counter-pressure processes are dictated by the injection of melt against mold pressures of 1–4 MPa. The technique is only applicable in special cases, because it is very difficult to seal off reliably the whole mold cavity. This particularly applies to multi-cavity molds, molds with lateral slides, and molds with many ejector pins; these molds are generally made out of steel. Molds with telescoping mold closures are currently used [11–13]. The tooling for the process is generally more expensive due to the gas-tight requirement. However, the process is said to be very suitable for complicated moldings, particularly if substantial wall thicknesses, for example 8–10 mm or more, are required [2,14,15].

The required clamping force depends on many factors and the decision must be made using the technical experience accumulated producing IF's by the molding process in question. The relation between clamping force and shot size, for the IM injection molding machines produced in Europe, is presented in Fig. 6.10.

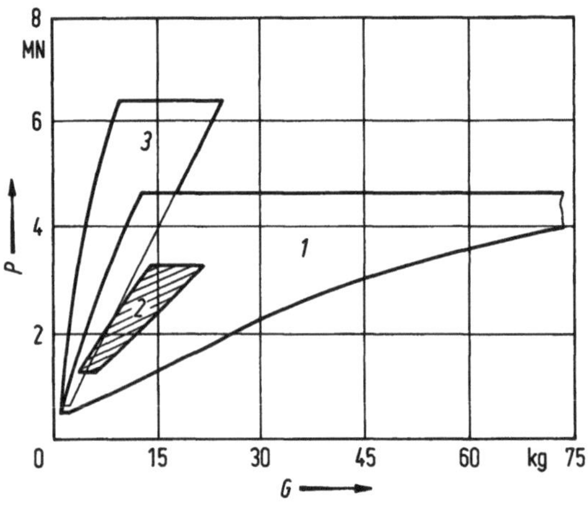

Fig. 6.10. Relationship between clamping force of the mold P and shot size weight G (using PS) for various injection molding machines; 1: "Structumat", Battenfeld, FRG; 2: "Scout", Gebr. Büehler, Switzerland; 3: "TM", Bulgaria [1]

6.4 Technical and Economic Analysis

Hangesbach and Egli [16] carried out a detailed analysis of the impact strength of IF articles produced by different IM processes. Impact testing was done using the falling-weight, or height-of-fall, method with IF discs 200 mm in diameter and 5 mm thick walls. The falling weight was 42 kg and its impact radius was 30 mm. Fig. 6.11 shows the dependence of the impact strength on the density of ABS articles. The "normal" IF discs (i.e. produced by LP injection molding) with a density of 770 kg/m³ have impact heights of only 0.7 cm. Those with a density of 960 kg/m³ have a somewhat larger height of 1.8 cm. However for discs produced by the counter-pressure process and having the same density of 960 kg/m³, the impact height increases by a factor of 19 to 34 cm. This is 36% of the impact height of unfoamed discs produced by CIM techniques. The enormous difference in impact strength of moldings with the same weight and molded using the same ABS, but having different densities, is

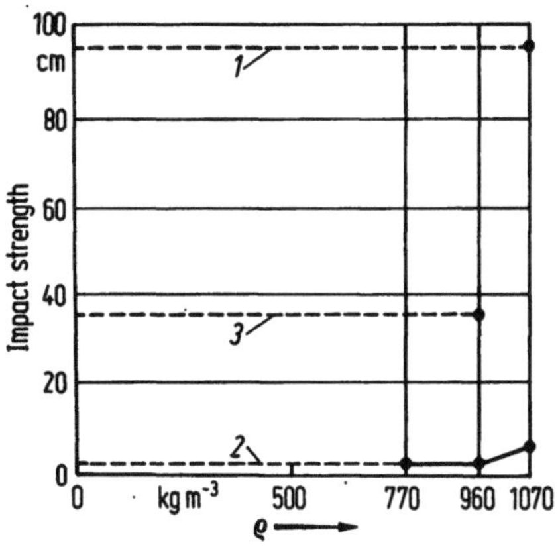

Fig. 6.11. Impact strength versus density for IF parts produced by various injection molding processes; 1: unfoamed ABS parts; 2: integral ABS part made by a "normal" injection molding process; 3: integral ABS made by a gas counter pressure process [13]

remarkable. No voids could be detected throughout the cross-section; the impact height decreased from 95 cm to only 5 cm, most probably due to notches on the surface. The test results for impact PS, modified PPO and PP show similar tendencies.

Compared to LP injection molding process, this process yields IF's with greater impact strengths because parts made of them have better surface smoothnesses, as well as thicker and more dense skins. For example, the impact strengths of IF parts prepared by gas counter-pressure molding from PS and ABS are 5 to 7 times higher

Table 6.2. Comparison of impact strength of IF parts prepared by gas counter-pressure molding or low-pressure molding, for different types of thermoplastics[a] [17]

Material	Density, kg/m³	Drop height cm
ABS-plastic		
Unfoamed	1,070	95
Integral foam		
Counter-pressure	960	35
Low-pressure	960	5
Low-pressure	770	3
High-impact polystyrene		
Unfoamed	1,050	30
Integral foam		
Counter-pressure	950	22
Low-pressure	950	4
Low-pressure	760	1
Polypropylene		
Unfoamed	900	100
Integral foam		
Counter-pressure	800	6
Low-pressure	700	1

[a] Tests were conducted with a 42 kg weight having a 30 mm impact radius on 5-mm-thick samples, measuring 200 mm in diameter. Impact strength is expressed by the drop height at which the sample fails

than those of parts produced by LP injection molding and showing equal densities (Table 6.2).

Density reductions are much less than those achieved in one component LP processes, 5–10% for amorphous polymers and 15–25% for crystalline polymers [2]. However, the increase in material costs due to the lower density reduction (i.e. to higher polymer consumption) is compensated by shorter cycle times and reduced finishing cost.

Gas counter-pressure processes have the following *advantages* over conventional low pressure IF technology: excellent void-free surface quality; polymer materials do not have to be pre-dried; reduced finishing cost; standard processing equipment can be used; reduced cycle time; better physical/mechanical properties. The limitations of the process are: increased part weight/density; limited expertise (being a newly emerging technology); limited part configurations; process is patented/licensed[15,1].

These process are comparable with other IF processes (Chapter 12).

6.5 References

1. Semerdjiev, S.: Introduction to Structural Foam. Brookfield Center: Soc. Plastics Engineers, 1982
2. Semerdjiev, S., Popov, N., Tuleshkov, N., Sgureva, I.: Injection Molding of Structural Foams. Sofia: Technika. 1983 (in Bulgarian)
3. Menges, G., Schweisig, H.: Chem. Eng. (London) No 330, 194 (1978)
4. Throne, J. L.: J. Cell. Plast. *12*, 264 (1976)
5. Semerdjiev, S.: Plaste Kautsch. *27*, No 1; 39 (1980); Kunststoffberater *27*, No 9, (1982)
6. N.N.: Mod. Plast. Int. *10*, No 3, 9 (1980)
7. Kataoka, K.: Plast, Age *24*, No 9, 70 (1978)
8. N. N.: Mod. Plast. Int. *11*, No 5, 13 (1981); *13*, No 6, 82 (1983); *14*, No 7, 14 (1984)
9. Anderson, D., Hunerberg, E.: Structural Foam Conf., Atlanta USA, 1983
10. Colangelo, M.: Plastics Technology *29*, No 6, 41 (1983)
10a. Olabisi, O.: Plast. Eng. *39*, No 10, 25 (1983)
11. Miller, B.: Plast. Des. Process *22*, No 7, 11 (1982)
12. Eckardt, H., Ehritt, H.: Kunststoffe J. *13*, No 3, 6 (1979; Plastic World *38*, No 4, 66 (1980)
13. Meyer, W.: J. Cell. Plast. *15*, 91 (1979)
14. Popov, N., Semerdjiev, S.: Kunststoffberater, *25*, No 7–8, 36 (1980)
15. LaPlaca, J. P.: J. Cell. Plast. *16*, 36 (1980)
16. Hangeshbach, H., Egli, E.: Plast. Rubber Proces. No 6, 56 (1979)
17. N.N.: Mod. Plast. Int. *7*, No 9, 12 (1977)
18. Shutov, F. A.: Intern. Symposium "Plastics in Building", Liège Belgium, 1984

7 Injection Molding: Two-Component Process

7.1 Process in General

7.1.1 Basic Process

It can easily be seen that all the processes so far discussed are based on one component, i.e. the IF's have cores and skins made from the same polymer. At the same time, however, injection molding does enable two-component integral foams to be fabricated in a single mold and in a single molding cycle. The resulting structures are referred to as *sandwich-structures* (not to be confused with the multi-layered sandwich panels mentioned in Chapter 1), *two-component foams (TCF's), or combined IF's*. They are characterized by the fact that their cores and skins are made of different polymers. To fabricate these materials, two compositions based on different polymers are used. One is designed for the surface and has no BA, while the second is compounded for the foamed core with a uniform (non-integral) cellular structure, and contains a BA (Fig. 7.1) [1].

The distinguishing feature and main advantage of these materials is a "linearly" smooth and practically unfoamed surface, due to the fact that the composition used for skin formation does not contain any BA.

In particular cases, both compositions may contain the same polymer or, by contrast, the skins and cores may be made of three different polymers. This is why this technique is sometimes called the *three-component, or multicomponent process*. Other denotations are *co-injection techniques*, or *sandwich-molding techniques*.

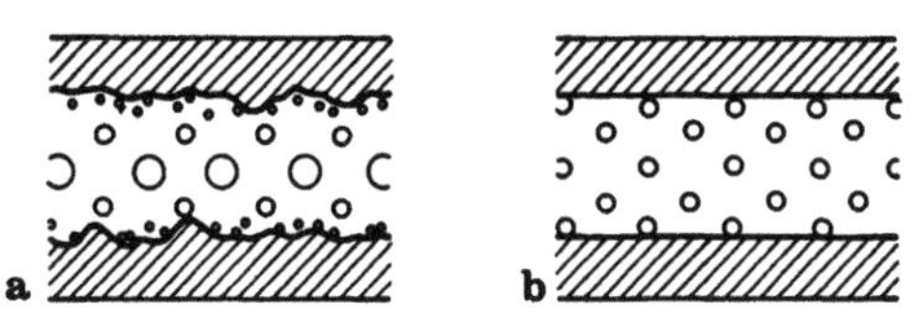

Fig. 7.1a and b. Schematic Sections of (**a**) a one-component IF part produced by a common injection molding process, and (**b**) a two-component IF part produced by a two-component process (Courtesy of Applied Science Publishers Ltd., London, England) [1]

7.1.2 Starting Materials

Two-component methods are suitable for processing thermoplastics, or low-viscosity oligomers, or both at the same time. Commercial hybrid IF's are based on the PO, PS, PVC, ABC polymers and copolymers, polyamides, polybutylene, poly(4-methyl-pentene-1), vinyl acetate, and vinyl butyral, or on polymerizable oligomers derived from urethane, phenol, carbamide, or epoxy resins. Recently, materials with rigid surface skins and elastic cores (PVC — PUR) and vice versa with elastic skins (vulcanized rubber) and rigid cores have started to be produced [2].

The purely polymeric combinations offer the following possibilities:

1) Use of a high-grade raw material for the outer skin in conjuction with a cheap raw material for the core to reduce the overall raw material costs. Waste materials (for example, metal and plastic shreds), and reprocessed and contaminated materials can all be employed for the core. For example, the high cost of flame-retardant engineering resins or UV-resistant formulations can be reduced by making a core of a cheaper resin. As a matter of fact, an 8.1 mm thick sandwich with a 5.2 mm thick skin made of modified PPO on each surface, and a foamed core of PS attains flame-retardant behavior similar to pure modified PPO foam [4].

2) Combination of a filled or reinforced plastic with a nonfilled plastic. It was found in [3] that if the core plastic is reinforced and used in conjunction with a non-reinforced plastic for the outer skin, the mechanical properties of the whole IF article can be improved without any deterioration of the surface quality by the glass fiber. On the other hand, if a reinforced raw plastic is employed for the outer skin in conjunction with a non-reinforced plastic for the core material, the reinforcing is concentrated in the outer skin, and hence the same effect is obtained with reduced raw material costs, or a better effect with the same raw material costs, as compared to a reinforced one-component IF article.

3) The physical and mechanical properties of IF articles can be beneficially modified by the use of different plastics for the skin and the core; for instance, by using a very tough plastic for the skin and a plastic with a high modulus of elasticity as the core material.

The multi-component technique is versatile. Consider, e.g. a tray having a high gloss transparent acrylic skin over a glass reinforced rigid core (the whole tray would be transparent and the glass does not show). Another TCF method uses a mechanical key, obtained by running one material into the other. An item such as a sole and heel unit may be produced so that the sole consists of plasticized PVC and the heel of ABS or rigid PVC [5]. Another IF part consisting of a rigid core (talc filled PP) with a flexible skin (plasticized PVC) has been suggested. This particular combination would be suited to large interior parts of vehicles, aircrafts, boats, furniture and similar applications. Many other combinations are conceivable.

7.1.3 Compatibility Problem

In every case, however, two-component methods have certain requirements of *chemical affinity and compatibility between the polymeric pairs, as well as of rheological similarities* (similar shear moduli and flow rates of the melts concerned), *and physical similarities* (close coefficients of thermal expansion). Examples of compatible systems are PO-PS and PVC-PO. By contrast, the PS-ABS and PP-ABS systems are not compatible, and an IF based on either of them may separate during production or crack in service. However, as Throne has pointed out [6], the mechanical properties of IF's based on poorly compatible thermoplastics are never below the minimum strength values of the least strong polymer of the given pair. This feature distinguishes

Table 7.1. Adhesion between pairs of polymers

Raw polymers	ABS	CA	EVA	Nylon 66	PC	HDPE	LDPE	PMMA	PP	PPO	PS-standard	PS-impact	PBT	PVC-rigid	PVC-plasticized	SAN
ABS	+	+			+	–	–	+	–		–	–	+	+	O	–
CA	+	+	–													
EVA		–	+			+	+		+		+			+	O	
Nylon 66				+	–	–	–		–			–				
PC	+			–	+						–	O				+
HDPE	–		+	–		+	+	–	O							
LDPE	–		+	–		+	+	–	+		–				–	
PMMA	+					–	–	+	–		–	O		+	+	+
PP	–		–	+		O	+	–	+	–	–	–		–	–	
PPO									–	+	+	+				–
PS-standard	–		+		–	–	–	–	–	+	+	+			–	–
PS-impact	–			–	O			O	–	+	+	+			–	–
PBT	+												+			+
PVC-rigid	+		+					+	–					+	+	
PVC-plasticized	O		O			–		+	–		–	–		+	+	+
SAN	+				+			+		–	–	–	+		+	+

+ good adhesion, – medium adhesion, O poor adhesion

the IF in question from a composite unfoamed plastic based on the same polymeric pair.

One solution to polymer incompatibility for a multi-component injection process was developed by the Billion Corporation of France. *This method injects a third material*, selected so that it is compatible with both incompatible polymers, to act as an adhesive. There are, however, obvious machine-cost disadvantages because the runner system is complex and there must be a third injection unit [5].

The constituent polymers should obviously adhere well to one another, particularly if the IF is exposed to mechanical load. If not subjected to load, this point is regarded as of minor importance (Table 7.1).

The shrinkage of the two raw polymers should also be similar. If the shrinkage of the core polymer is substantially greater than that of the skin polymer, the core polymer could easily detach from the skin, particularly if adhesion were poor. If the shrinkage ratio were the converse, high frozen-in stresses would be set up in the molding. In both cases deformation or distortion of the moldings may result.

7.1.4 Density Problem

The density reduction which can be achieved with a TCF process is between 5 and 30%, depending on the configuration of the article and on the polymers used.

The amount of blowing agent in a two-component process not only affects the achievable density and the requisite cooling time, but also affects the ratio of skin to core (Table 7.2). Reducing the concentration of the BA from 0.5 to 0.15% not only results in higher cycle speed but also permits the use of a larger proportion of cheap core material. On the other hand, it increases the weight of the part. Accordingly, the optimum concentration of the BA depends on several factors. In general, however, smaller quantities of BA are used in a multi-component process than in a one-component IF process [3].

Since it is a particular advantage of PBA's that larger amounts can be introduced into the melt compared to CBA's, savings in PBA's are not really important when they are used in conjunction with a multi-component process.

The physical chemistry and rheology of two-component IF's have been reported in detail by Han et al. [7], Young et al. [8] and Kraynik [9].

Table 7.2. Effect of the amount of chemical blowing agent on the weight and cycle time for a seat made from an ABS/SAN integral foam by a multi-component process

CBA concentration	IF part weight,	Number of shoots, per hour	Polymer distribution	
			skin	core
%	g		%	%
0.5	2,100	11	75	25
0.15	2,300	18	62	38

7.2 Commercial Processes

7.2.1 The ICI and BB Processes

Until recently, the most common commercial TCF process was the one developed by Imperial Chemical Industries (ICI) England. While the first developments of TCF processes had clearly sequential injections of core and skin compositions, recent developments allow them to be injected simultaneously (Hanning-process). However, in practice, a small amount of skin polymer has to be injected before the simultaneous injection of the skin and core polymers [4].

The most difficult technical problem encountered in the fabrication of a two-component IF is to prevent the composition containing a BA from becoming mixed with the composition with no BA. In the ICI process this problem is avoided by means of a switching valve situated at the mixing point of the two flows and connecting the mold cavity with only one of the injection nozzles at a time, without interrupting the filling. Later on, two and three-channel nozzles were introduced (Fig. 7.6 below) for mold filling. The Battenfeld-Bayer, or BB, process is a modern version of ICI's process. It is characterized by independent control of the two nozzles, permitting a continuous transition from one polymer composition (for the skin) to the other (for the core), and *vice versa*, without using a distribution valve. This eliminates traces on the IF's surface which are typical for the ICI process and which result from the presence of the switching valve.

We shall now consider the ICI process in more detail, bearing in mind that the BB process is similar. At first, the material (or materials) necessary for injection are plasticized in two different cylinders (1) and (2) (Fig. 7.2, position I). The molding machine has two injection mechanisms and a two-way switching valve which either shuts off the mold cavity or connects it in turn to one of the injection cylinders. Initially, the mold cavity (5) is closed. The mold cavity is then connected with cylinder 2 (position II) and the melted polymer intended for the skin is rapidly (4 sec) injected, using a screw. Later on (positions III and IV), the valve is switched and the melted polymer containing the CBA and compounded for the core is injected from cylinder 1. The second melt "inflates" the covering which is forced towards the walls of the mold. Once the mold is filled, the valve is again switched (position V) and connects the mold cavity to the first cylinder, which again supplies a small amount of the skin material. The composition is held at the final pressure ($1.0–1.3 \cdot 10^5$ MPa) for several seconds, which is followed by a partial expansion of the mold for the core to foam. At this point, some of the composition is driven into the gate (Fig. 7.2). After this the mold is cooled down and the

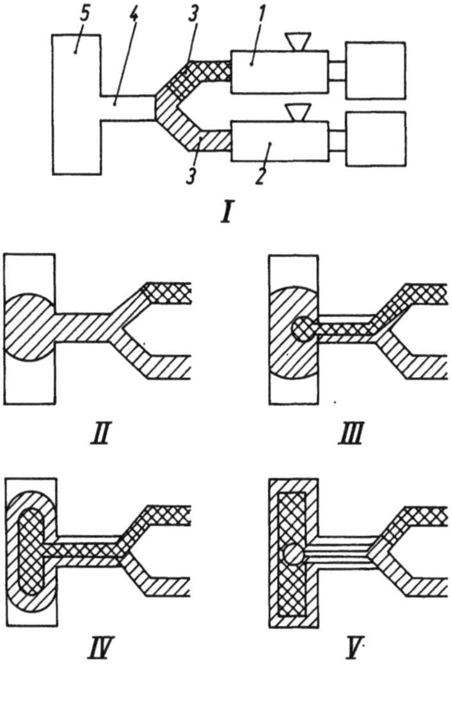

Fig. 7.2. Schematic diagram of a two-component process; I: initial position, II: injection of skin polymer, III: transition phase, IV: injection of core polymer, V: additional injection of skin polymer; 1 and 2: cylinders, 3: hot manifolds, 4: runner channel, 5: injection mold (Courtesy of the Society of Plastics Engineers, Brockfield Center, USA) [2)]

article is demolded. The most critical instant is when the second composition is injected, because it may rupture the covering formed by the first composition. To avoid this, care has to be taken to control the speed and time of injection, and the temperatures of the melts and the mold.

Articles fabricated this way have densities ranging from 200 to 500 kg/m^3, overall thicknesses from 3.2 to 14 mm, skin thicknesses from 0.25 to 0.76 mm. Their skin-to-core ratios are generally 1:2, while their projected areas are 0.37 to 2.2 m^2. In addition to having a smooth surface, the foams produced by the ICI process are very rigid [2,3].

These processes have been used to fabricate IF's based on PO, PS, ABS-plastics, PVC, PMMA, EVA, and polyamides. Either all-reinforced articles or articles with a reinforced surface skin only, can be produced.

The two-component processes do not require expensive molds (they are only 10 per cent more expensive than conventional molds); the molds are made of silicone rubber (for molding temperatures of 26 to 32 °C), epoxy resins (32–38 °C), nickel (40–46 °C), aluminum (38–43 °C), or steel (40–50 °C) [3]. The cost of the necessary molding machines is 20 to 25% higher than that of conventional equipment [4,5], but the difference in equipment costs is compensated for by a short molding cycle of 90 to 100 sec. For example, when producing integral ABS plastics by the conventional HP-IM method (ϱ = 800 kg/m^3), τ_c = 160 sec, whereas when producing the same material, with the same density, by a two-component process, τ_c = 100 sec. As a result, the cost of an IF produced by the ICI process is 60 to 70% lower than that of the corresponding unfoamed plastic. The main limitations of the ICI process is the difficulty of making complicated configurations with deep relief.

7.2.2 The Injection Mixing Process

Another version of the ICI process uses the same polymer for the skin and the core, and uses the same gate to inject the composition with no BA followed by that with a BA, and finally again the composition without the BA (Figure 7.3). Note that the BA is injected into the composition just before it enters the cooled mold. The mold can either be partially filled, or the mold can be filled completely and then have its walls expanded to create the extra volume needed for foaming. The second variant is useful when the molding machine cannot provide high injection speeds [9].

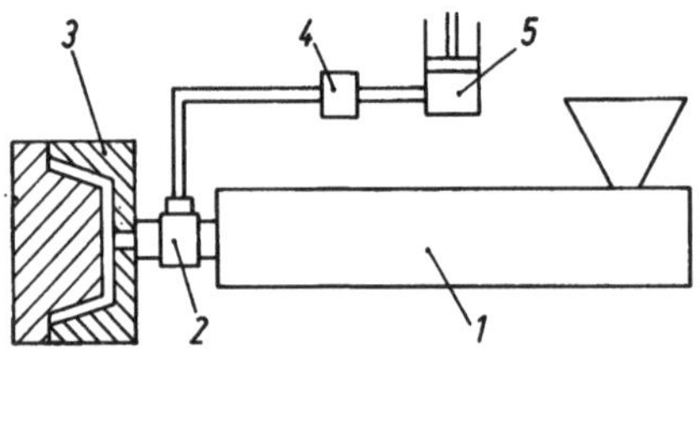

Fig. 7.3. Schematic view of an injection mixing machine; 1: plastification and injection unit; 2: injection nozzle and mixing chamber, 3: mold, 4: pump, 5: reservoir with blowing agent (Courtesy of the Society of Plastics Engineers, Brookfield Center, USA) [2]

In this process the composition does not have to be held under pressure, but it does require that the article being molded is in close contact with the gate while the mold is opened. In order to implement the process on a conventional molding machine e.g. one intended for molding multicolored plastics), only minimum modifications are needed, viz., a device for injecting the BA into the mold must be added. The same device can also be used to supply the plasticators, curing agents, initiators, softeners, etc. Both CBA's and PBA's (freons) can be used.

The advantages of the process are quite evident: excellent quality skin surface, short cycle duration, and low expenditures for equipment modifications. However, in order to obtain a high-quality macrostructure in the IF core, a number of requirements should be satisfied: the blowing agent must be fed in carefully, the precise instant it is injected must be strictly adhered to, and the BA must be thoroughly mixed with the melt using flow mixers.

With these requirements in mind, the designers of the process developed a special mixing chamber (pre-chamber) with a point gate and needle valve [4,9]. The BA is uniformly and precisely fed in using a special piston, whilst the control of the instant and duration of the injection is accomplished electronically. This device facilitates a continuous transition from the first material to the second and at the same time excludes their mutual penetration. The order of operation is similar to that in the ICI process. In fact, the Bayer process can produce an IF not only based on similar polymers but also based on unlike polymers with a wide range of densities — from 250 to 1.000 kg/m^3.

7.2.3 Other Processes

A company in the FRG has recently developed an original method for producing thin IF articles using a PBA in a conventional molding machine [3,9]. The basic principle is the *mechanical foaming* (frothing) of the melt of a plasticized polymer (PVC) by means of *a special foam-generating nozzle*. The nozzle consists of two sections. One (6-mm diameter) lies adjacent to the plasticator outlet, the other (with a considerably larger diameter) is inside the injection device. Owing to the great difference between the cross sections a turbulent flow forms at the nozzle outlet. The turbulence, in turn, contributes to the uniform and rapid distribution of the gas introduced through this nozzle throughout the bulk of the melt. The resulting liquid foam is injected into the cold mold. The structure and quality of the article's surface are determined by the injection pressure; the higher the pressure, the lower the density of the core. Thus, if the pressure is increased from 13 to 20 MPa, the article density is reduced from 800 to 500 kg/m^3.

Two-component integral structures may also be produced using *low-pressure injection molding*. Thus, Siemag of West Germany has modified its process for one-component IF's to produce large-sized multicomponent articles of the following types: (1) an unfoamed skin with a controlled thickness and a foamed core; (2) "reverse" integral structures, i.e., a skin made of foam and an unfoamed core (the polymer is the same); (3) an unfoamed skin made from a high-quality and expensive polymer (ABS-plastic) and a foamed core made from a cheap polymeric composition (PE byproducts or paper).

The trend towards utilizing the LP method for producing multicomponent IF's may become more pronounced in the future.

7.3 Equipment Specifications

7.3.1 Machines

Remember that a machine designed for a multicomponent process is almost universally useful, since it can be employed for virtually any injection molding technology [10-13]. Thus, it could be used for: a one-component process, a gas counter-pressure process (an additional sub-unit needed), unfoamed injection molding, a TCF process with an unfoamed skin and foamed core, a TCF process with a foamed skin and unfoamed core, a TCF process with a foamed outer skin and foamed core (see Chapter 3).

A TCF process requires a short shot to produce a cellular core. Despite the short-shooting, the process requires clamping forces 2.5 to 4.0 times that used in standard IM equipment, i.e. up to 1500 tons. It also needs an injection pressure of up to 1.5×10^5 MPa; the total shot weight may be up to 10 kg.

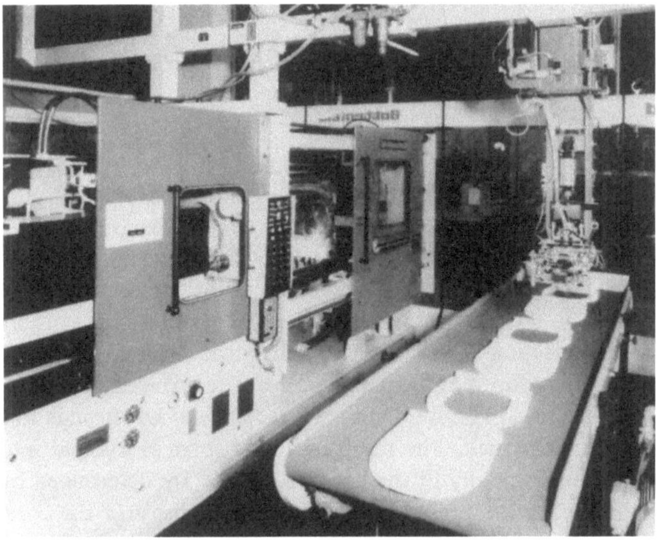

a Battenfeld Machine, and a robot for the automatic take-off of the IF parts from the mold (photo submitted by H. Eckardt, Battenfeld Maschinenfabriken GmbH, Meinerzhagen, FRG)

The latest achievement in IF machinery is a completely *automatic machine* for producing IF toilet seats. It was developed by Battenfeld (Model BM 2 × 2000/400 S-C). This machine has a two-platen mold for producing both the toilet seats and covers in a single shot, and a robot to take the parts off the mold automatically (Fig. 7.4).

The original ICI method was utilized by the Billion Corporation, but it was coupled with the co-extrusion principles [5]. The Billion machines are equipped with nozzle valves and with a variety of different hot runner or standard feedblocks to transport the components to the mold cavities. Injection can be either simultaneous (similar to the laminar flow of a co-extruder), or can be sequential. The choice depends on the design of the item and on the polymer; a simple and low cost simultaneous injection is often sufficient for items such as sun goggles for skiers. But some applications call for the joint between the two components to be visible internally, the outer material or both being transparent. For these applications the rather expensive but commercially established idea of *a moving mold element was developed* [5]. Parts such as car rearlight assemblies can be made by the sequential injection of different colours, the injection forming a perfectly flat interface against the tool element. But such Billion systems increase the tooling costs and require a machine equipped with sophisticated controls and attention to mold temperature.

Presma SpA, Italy, reports an impressive growth of the production of *furniture* made from PP with a foamed core and a solid skin, by multiple carousel-mounted presses, with clamping forces

Fig. 7.5. Five station rotary equipment, developed by Presma SpA, Italy, for two-component molding makes chairs with PP foam core and PP solid skin; each station can be programmed for the shot weight and injection speed of two materials (Courtesy of Presma SpA, Varese, Italy) [13]

down to 27 tons (Fig. 7.5) [13]. Using molds with twin cavities, as many as 10 components can be produced; each station can be programmed for shot weight and injection speeds. Six-component folding chairs are produced at the rate of 8 per hour; total chair weight is 6.6 kg, with components weighing from 0.7 to 1.85 kg. Skin material is 30% calcium carbonate-loaded PP and core material is semi-expanded unloaded PP; core weights are from 30 to 48% of part weight.

7.3.2 Injection Units

Injection velocities have to be set so that the injection of the foam component is completed before the completion of the injection of the unfoamed component. At the end of the injection sequence, a small amount of the unfoamed polymer is injected to seal the gate in order to avoid any open cells after the gate is cut off.

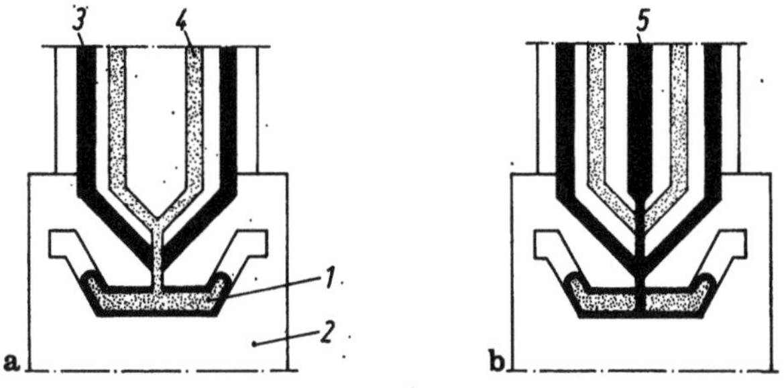

Fig. 7.6a and b. Schematic representations of (a) two-channel and (b) three-channel nozzles; 1: injected material, 2: mold, 3: gate for skin polymer, 4: gate for core polymer, 5: central gate for skin polymer [3]

Compared with a standard IF injection process, a TCF process uses a lower injection rate. In order to reduce the danger of surface imperfections, the injection velocity on both injection units must be programmable. Periodic simultaneous injection was designed to obtain a better distribution of core polymer within the molded part, and was first accomplished with a two-channel nozzle (Fig. 7.6a). However, it was found that with this device the skin thickness opposite the sprue is sometimes thinner than that at the sides of the sprue.

In order to eliminate this disadvantage, a three-channel nozzle was designed (Fig. 7.6b). The additional channel for skin material is located in the center and therefore sends more skin material to the surface opposite the gate. However, a three-channel nozzle is more difficult to control and its advantages are only noticeable when both polymer melts can be injected simultaneously over a long period. A two-channel nozzle is preferred for fan-gated and other parts produced in multicavity molds [4]. A special alternative nozzle was designed for use in two- and three-channel techniques.

7.3.3 Molds

We shall discuss the molds for multi-component processes only insofar as they differ from molds for the one-component IF molding process. Fillmann [10] indicated several peculiarities of these molds. To improve the surface quality of an IF article, a highly polished steel mold should be used. The best cellular structure is most easily achieved with a central sprue gate. Hot runner molds are suitable if separate hot runners are provided for the skin and core materials, which should not be combined until they reach the gating point; but such molds are very expensive. Fan gates have proven suitable for small-sized articles, and three-plate molds for large-sized articles (see also Chapter 3).

7.4 Technical and Economic Analysis

Currently, TCF processes are being used to manufacture outdoor furniture, radio and gramophone cabinets, television screen frames and cabinets, computer cabinets, sanitary and water tanks, etc.[12–14]

Multi-component processes are especially good where IF articles with very specific properties are required, because different polymers can be used for core and skin, or skins. It is possible to produce IF articles which are capable of withstanding greater mechanical loads than those made by conventional LP molding processes.

An example of the advantages of TCF was presented by Mayer [4]. He reported that a dining room chair, molded from non-TCF PS integral foam, failed to meet a dynamic test in which the chair back was pulled back by a 220 kg mass at a frequency of 24 cycles per minute whilst the chair seat was loaded with a 70 kg load. The chair should withstand 50,000–60,000 cycles in order to guarantee a certain lifetime. Only when the chair was molded from a PS/PS foam sandwich by a TCF process were the test specifications met.

The use of off-color materials for cores offers great promise for cutting costs. According to Battenfeld GmbH, one test involved the use of mixed off-color nylon-6 as the core of a handle, with virgin nylon-11 used for the skin [13]. The price per kg

Fig. 7.7. Handle has thick-walled virgin nylon-11 skin with foamed off-color nylon-6 core, produced by two-component process (Courtesy of Battenfeld Maschinenfabriken GmbH, Meinerzhagen, FRG)

of the nylon-6 was one half of that of the nylon-11. As the core material accounts for 50% of the 0.28-kg weight, a material cost saving of 25% is made (Fig. 7.7).

Materials not suited for making even low-quality products on single-component IM machines can be used effectively as core material, according to Battenfeld GmbH [13]. One extreme example, a thick-walled brush, illustrates the possibilities. The skin was made of virgin PP, while the core material consisted of needled felt carpet cuttings composed of PP, adhesives, and PVC coatings. The surface quality was excellent, and mechanical properties adequate.

According to LaPlaca [11] two-component processes are better than low pressure conventional IF processes in the following ways: excellent void-free surfaces, molded to color with texture, low cost core material possible, reduced finishing costs, versatile molding equipment. The limitations of this process are: flow lines visible, minimum wall thickness of 7.6 mm, configuration and size limited, part-weight/density high, processing equipment 30% more expensive, undeveloped supplier base in non-European countries, polymer combinations limited.

Typical process problems are [2, 11]: unfilled sections, core material exposed at outer edges, sporadic solid-filled areas, varying skin-to-core ratios, core material not consistent with flow around 90 deg corners, and separation at skin-core interface. – The process is compared with the other IF processes in Chapter 12.

7.5 References

1. Hilyard, N. C., Young, J.: Introduction. In: Mechanics of Cellular Plastics Hilyard. N.C. (ed.). London: Applied Sci. Publ. 1982, pp. 1–26
2. Semerdjiev, S.: Introduction to Structural Foam. Brookfield Center: Soc. Plastics Engineers, 1982
3. Eckardt, H.: Ind. Prod. Eng. No 2, 170 (1980); Plast. World *38*, 66 (1980)
4. Meyer, W.: J. Cell. Plast. *15*, 91 (1979)
5. N.N.: Brit. Plast. Rubber No 6, 41 (1979)
6. Throne, J. L.: Polym. Eng. Sci. *17*, 862 (1977)
7. Han, C. D., et al.: Polym. Eng. Sci. *18*, 687 (1978); *21*, 69 (1981); *21*, 518 (1981)
8. Young, S. S., et al.: Polym. Eng. Sci. *20*, 798 (1980)
9. Kraynik, A. M.: Polym. Eng. Sci. *21*, 80 (1981)
10. Fillmann, W.: Ind. Prod. Eng. No 3, 35 (1978)
11. LaPlaca, J. P.: J. Cell. Plast. *17*, 36 (1980)
12. Colangelo, M.: Plastics Technology *29*, No 6, 41 (1983)
13. Sternfield, A.: Modern Plast. Intern. *13*, No 6, 25 (1983)
14. Coinjection: Menasha Corp., Molded Protucts Div., Watertown, WI (1984)

8 Reaction Injection Molding Process

8.1 Introduction

About 25% of all polymer products involve polymerization within the final mold. Processes including this stage are monomer casting, rubber molding, thermoset injection and compression molding. Most of these processes use thermosetting polymers, because thermosets must be crosslinked in their final shape and because linear polymers with sufficient molecular weight cannot react rapidly enough. Frequently, the cycle times are longer than those required for melt fabricating thermoplastics. Typically the thermosetting reactions are thermally activated, and thus require time for heat transfer to the material in addition to the time needed for the reaction. Recently, technology has been developed for rapid *in situ* polymerization to form articles directly from monomeric or oligomeric liquids. This process has come to be known as *Reaction Injection Molding (RIM)*, and its modification as the *reinforced RIM (RRIM)* process.

An essential feature of RIM technology is that it combines polymerization and processing.

The first reaction injection molding process was developed by Bayer AG in West Germany. The process has been commercially used for about 15 years in Europe and has been available in the United States for about 10 years.

The RIM process has the following advantages [1]:

- low plant investment,
- low process energy demands,
- large parts manufacture,
- very short demolding times,
- excellent surface quality,
- large variations in part wall thickness without sink marks,
- low densities and thick cross sections,
- wide density variations,
- ability to mold inserts in place.

Only certain features of integral foam RIM technology are covered briefly in this Chapter, excluding those basic factors on unfoamed RIM technology that are well known from the literature (see, for example, the splendid books of Becker [2] and Sweeney [3]).

8.2 Terminology and Definitions

Terminology can sometimes be misleading, especially in such a new field as RIM. The process does not necessarily require that the liquid material be injected directly into a mold, but it can also be combined with an open pour technique [4].

It should be mentioned here that until recently there have been other designations for the RIM process, namely: Liquid Reaction Molding (LRM), Liquid Injection Molding (LIM), Liquid Injection Molding of Elastomer (LIME), Liquid Reaction Injection Molding (LRIM), High Pressure Impingement Mixing (HPIM), Counter-Current Impingement Mixing (CCIN), or (in German language) Reaktions-Schaumguss (RSG) [5]. All these terms refer to processes where reactive components are mixed by impinging streams without mechanical mixers. For the sake of simplicity, we shall refer to all these processes as RIM's.

The RIM process can produce three kinds of parts [6]:

1) Rigid polyurethane foam parts, which can be either structural or decorative, and usually have a dense integral skin surrounding a lower density core. The main applications are furniture and appliance components.

2) Solid polyurethane elastomer parts having relatively thin wall sections, such as ski boots.

3) Microcellular elastomer parts, which are high performance thin-walled cellular products and usually have a dense integral skin surrounding a lower density core.

Ferrari [6] indicated that from the standpoint of physical structure there are many similarities between rigid microcellular parts and the elastomeric microcellular parts made by the RIM process. For example, both have low density cores, although the elastomeric foam may have a less clearly defined skin. From a chemical standpoint, the microcellular elastomers are more similar to unfoamed urethane elastomers. Both are very durable and compliant (damage resistance). While the elastomers generally have better physical properties (such as tensile strength), the microcellular materials do have good physical properties and have better impact absorption or compliance.

Taking into account these considerations, it is useful to redefine the terms skin, core, cell, etc., in relation to RIM. According to Ferrari [6], microcellular structures are, in general, high density elastomeric structures in which the volume fraction of elastomer is higher than the volume fraction of air. This type of foamed polymer is characterized by reactively high densities, and solid or almost solid skin layers. Typically, a RIM fascia may have an overall specific gravity of 1000 kg/m^3, whereby the skin density may be 1100 kg/m^3 and the core density between 800 and 950 kg/m^3. Specific gravities below 500 kg/m^3 are possible with this technology, although the term "integral skin foam" has been used in this case, as well as "microcellular foam".

Some confusion has occurred because the terms "solid (unfoamed)" and "microcellular" have been applied to formulations rather than to the foam part, whereas the structure is often defined by the process conditions rather than by the formulation. It has been suggested that if an elastomeric formulation contains a halogenated hydrocarbon blowing agent (freon), it will produce a microcellular foam, but if the blowing agent is absent, the product will not have a microcellular structure. But at the same time, the microcellular structure is possible with or without the presence of freons, as long as another blowing agent (CO_2, air, etc.) is provided.

It should be remembered that all RIM polyurethane parts are in general integral foamed plastic, even when no blowing agent is added to the starting formulation. The cellular structure of RIM parts is created by the presence of dissolved gases in the liquid reactants. During the polycondensation reaction, considerable heat is generated, and this heat, while contributing to the curing of the polyurethane, also causes the blowing agent, including the dissolved gases, to expand in volume and thus increase the cell size. When the external mold surface is cooled, the gas pressure is reduced faster in the skin zone of a RIM part, generating a high density non-porous skin [7].

8.3 Basic Process

The principle of the RIM process is not new, though when it entered the market in 1969, its advantages were not fully appreciated. The chemistry of the polyurethane products was not completely understood, cure times were excessively long, and suitable processing equipment was lacking. An improved knowledge of the chemistry and the introduction of better equipment came together in the early 70's to ensure development. Most of the early achievements in RIM technology were made in West Germany, primarily with respect to rigid parts. Rigid RIM parts have received a greater acceptance in Europe than in the USA [4].

As mentioned above, RIM is an *in situ* process with in-mold polymerization of highly reactive liquid components to create a polymeric material with a cellular core and a solid non-porous integral skin. In fact, it is a high-speed casting process, and represents an improvement and refinement of the older urethane casting process. It simulates an injection molding process in that it has short cycle times, equivalent to rubber injection molding, however at much lower operating temperatures and pressures.

The RIM process starts, as does a cast-polyurethane process, with two (or possibly more) liquid streams: a polyol premix (monomer A) and an isocyanate or prepolymer

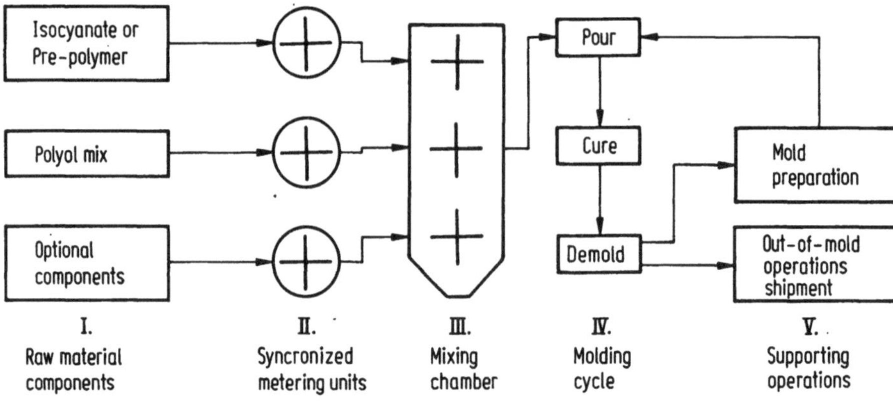

Fig. 8.1. Basic scheme for the liquid casting process: RIM replaces the mechanical mixer (III) with a high pressure impingement mixer [6]

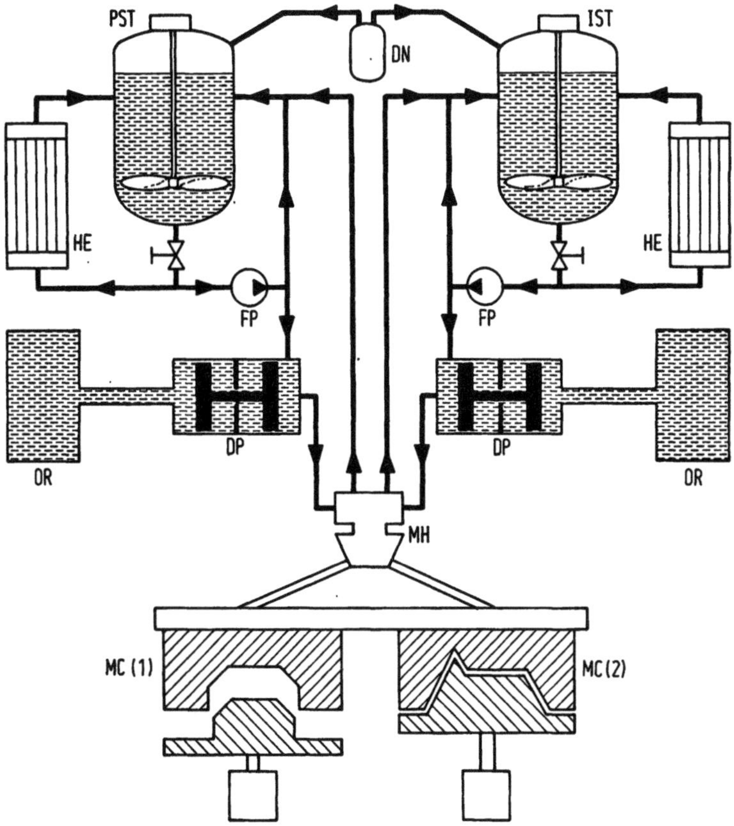

Fig. 8.2. Schematic representation of the RIM process; PST: polyol slurry tank, IST: isocyanate slurry tank, HE: heat exchanger, DP: displacement pump, DN: dry nitrogen, OR: oil reservoir, FP: feed pump, MH: mixing head, MC (1): mold cavity open, MC (2): mold cavity closed (Courtesy of "Process and Chemical Engineering" Journal, Australia) [9]

thereof (monomer B). These streams are metered by a positive displacement to a mixhead and from there they are delivered to the mold or forming conveyor. The flow rate of the two monomers must be accurately controlled to ensure the correct stoichiometry (Fig. 8.1). The mixture takes the form of the mold or restraining fixture, reacts, and gels rapidly into a resilient elastomer or rigid plastic, depending on the formulation (Fig. 8.2). A helpful way of viewing the RIM process is to break it down into several unit operations [8], viz.:

polyol ↘
meter mix — fill – cure – eject – postcure
isocyanate ↗ (mix ↑ mechanical energy; cure ↑↓ heat; postcure ↑ heat)
↑
heat

RIM technology differs from cast-urethane technology in the following key areas: 1) *Formulation*. The formulations used in a RIM process are very fast setting and gelling. In fact it gels so fast, that mechanical mixing is not practical because of polymer buildup in the mixing chamber. Nevertheless, the major ingredients used in a RIM process could be the same as those used for a cast urethane process; thus the chemistry for both process variants is similar (therefore we have not paid special attention to RIM formulations, see also Chapter 13) [1-3,8 10].

According to Frisch [24], competition for RIM urethanes can be expected from other polymers such as nylon-6 copolymers (for example the NYRIM™ system developed by Monsanto, USA [23]), epoxides, and unsaturated polyesters, using RIM techniques. But the great diversity of products which can be made from urethanes or urethane mixtures would appear to have a bright outlook for urethanes in this area.

2) *Metering and Mixing*. Because the formulations react so quickly, high pressure impingement mixing must be used (see Sect. 4.4). The mixhead, unlike those normally used in the casting of urethane, has no rotor, and is usually a simple cylinder with entry port for each or the liquid streams. The formulation is introduced under high pressure and mixing is achieved by turbulence in the chamber. Impingement pressures are usually in the range of 10–15 MPa or higher. As the mixing cycle finishes, *a close-fitting ram* moves forward and clears the mixing chamber. This mixing innovation has made the fast reacting formulations practical in high volume productions.

In addition to these two differences between the RIM and cast urethane processes, the use of tight, close fitting molds has also been described as being a key element. In reality, the use of tight molds is mainly necessary to provide a high quality, dense and paintable surface for certain application areas. This is particularly true for automobile applications (see Sect. 4.4.3).

8.4 Equipment Specifications

RIM installation consists essentially of four modules: a metering unit, a mixing head, a mold head, and a mold clamp. There are many suppliers of the various modules, including a few who offer all parts, for example Admiral, Battenfeld, Cannon-Afros, Cincinnati-Milacron, Desma, Elastogram, Hennecke, etc. All RIM machines have a number of features in common, including:

- high pressure capability,
- high pressure metering,
- non-pulsating flow,
- adjustable output,
- temperature control,
- agitated feed tanks
- recirculating capability.

Most machines today are used as two-component units (polyol and isocyanate); multi-component machines are supplied with third-component units and it is possible to add more components with relative ease[2,3,5,8,11].

8.4.1 Metering Units

A RIM process is usually carried out using high pressure metering which operates in combination with a recycling system. The mechanically or hydraulically synchronized, self-cleaning mixheads are mounted directly onto the molds. Multiple mixheads can be attached to a single high pressure metering unit in a manner that allows individual control of each mixhead. As many as 12 mixheads can be attached to a single metering unit for large volume production. A simplified flow sheet for a RIM process

Fig. 8.3. The electronically controlled high-pressure piston metering unit of the RIMDOMAT machine can be used for processing both RIM and RRIM polyurethane systems (Courtesy of the Maschinenfabrik Hennecke GmbH, St Augustin, FRG)

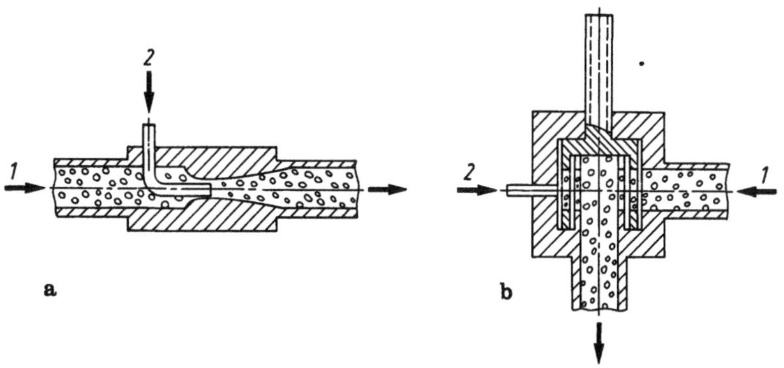

Fig. 8.4. Gas loading units; a: gas loading nozzle, b: gas loading stirrer, 1: polyol, 2: gas (Courtesy of Bayer AG, Leverkusen, FRG) [1]

is shown in Fig. 8.2. The production of high-quality IF parts by a RIM process requires the installation of proper monitoring and safety controls [12].

Most metering devices originally developed for RIM processing utilize a rotating piston pump capable of generating high pressures from 10 to 20 MPa (Fig. 8.3). These pumps are slightly modified hydraulic or diesel injector pumps. Later developments include metering cylinders driven by piston pumps.

The RIM machines are unsuited for RRIM technology as severe erosion occurs in pumps of all types, in the lines and heads. Switching to the metering cylinder principle, increasing the line size to reduce the linear velocity, and use of abrasion resistant materials in critical locations has resulted in machines which appear to meet production needs. Two types of metering cylinders are in use: an accumulator type and a lance cylinder type. The methods of handling the metering, recirculation, and head, vary widely between the producers; provisions can also be made to refit older machines with add-on units to handle fillers.

When larger amounts of gas are required in solution, special devices can be installed to disperse controlled amounts of gas in the polyol and isocyanate compounds (Fig. 8.4).

8.4.2 Mixing Heads

The mixheads on RIM machines are designed to produce mixing by allowing the component streams to collide and so avoid the need for mechanical mixing. The heads are self-cleaning and have adjustable impingement pressures and recirculation capabilities; no solvent flush or air purge is required. Figure 8.5 shows the principles behind the Krauss-Maffei mixhead: the head design features recirculation grooves in the clean-out piston and a self-cleaning mixing chamber, with the piston controlling the flows for mixing and pouring [3,8,12].

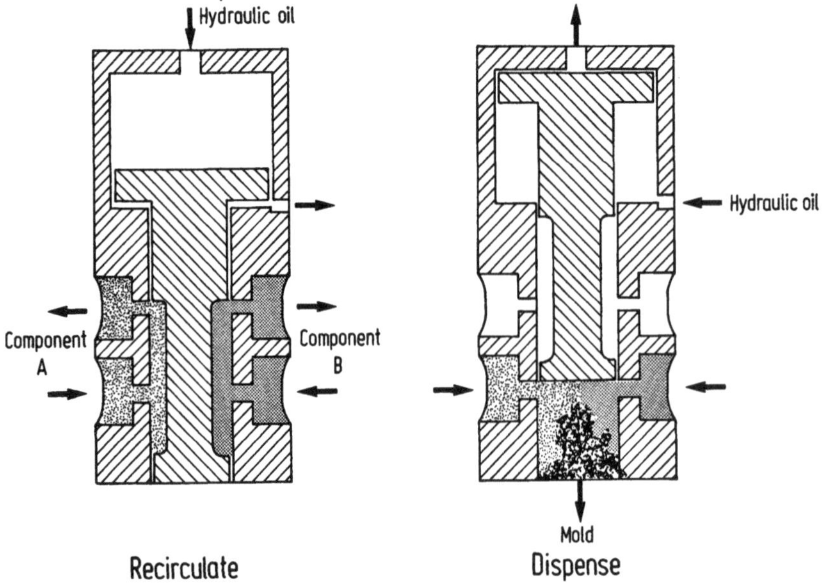

Fig. 8.5. Recirculating, self-cleaning RIM mixing head: closed circuit position (left) and mixing position (right) (Courtesy of the Krauss-Maffei AG, Munich, FRG) [3]

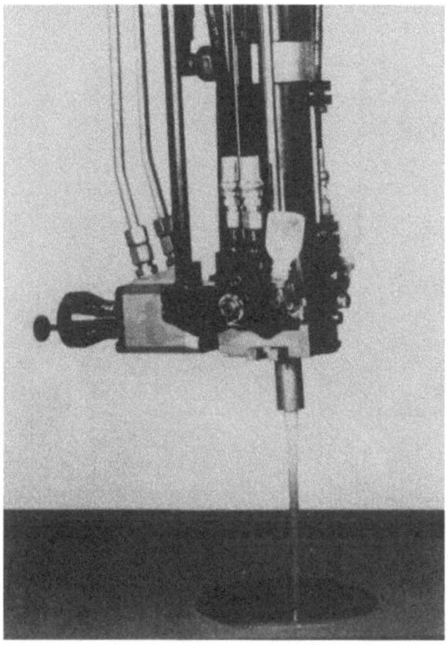

Fig. 8.6. High pressure PUR mixing head (Model MQ) for processing both normal and high viscosity systems, and for formulations having wide variations in mixing ratio (Courtesy of Maschinenfabrik Hennecke GmbH, St Augustin, FRG)

Some suppliers of machinery are retaining the recirculation system, which has stood the test of time in RIM technology and are adjusting their metering and mixing equipment to the properties of the components to be used (Fig. 8.6 and 8.7).

The first European manufacturer to bring to the market equipment for the high-pressure processing of multi-component systems containing fillers was Krauss-Maffei. It used the well-known RIM injection mixing head with piston valve control and the parts subjected to wear were made from abrasion-resistant materials.

Later on, Cannon-Afros, Battenfeld, Desma and Elastogran offered their mixing heads for RRIM technology. Their metering-mixing installations consist of single-throw reciprocating pumps in combination with conventional high-pressure metering devices. Cincinnati-Milacron sells electronically controlled, hydraulically driven installations with submerged plunger pumps and injection mixers for filled and unfilled multi-component systems. Hennecke offers cheap metering equipment in the form of accumulators, as well as positive-drive mechanically friction-coupled and electronically controlled multi-component metering equipment. Another technique that departs from current practice has been developed by Bayer. It is known as the *straight-through system*. In other words, the components are not circulated before and after injection, but the two submerged plunger metering devices are driven by hydraulic multi-throw reciprocating pumps [11-14].

8.4.3 Molds

The choice of mold materials again depends on the anticipated molding conditions, and for a RIM process – for which the temperature and pressures are known – the following are common: steel for more than 100,000 moldings, cast aluminum, forged aluminum, and cast zinc alloys (e.g. "Kirksite") for less than 50,000; electroplated Cu/Ni for less than 100,000; epoxy resin for less than 500; silicone rubber for less than 100 moldings [1].

RIM integral foam is an alternative technology for low-volume programs of 3,000 to 5,000 units. The process is cost effective if low-cost epoxy or aluminum tooling can be utilized.

A typical RIM mold for producing a large-size IF part is shown in Fig. 8.8. Several small RIM parts can be made simultaneously in one filling operation in one mold;

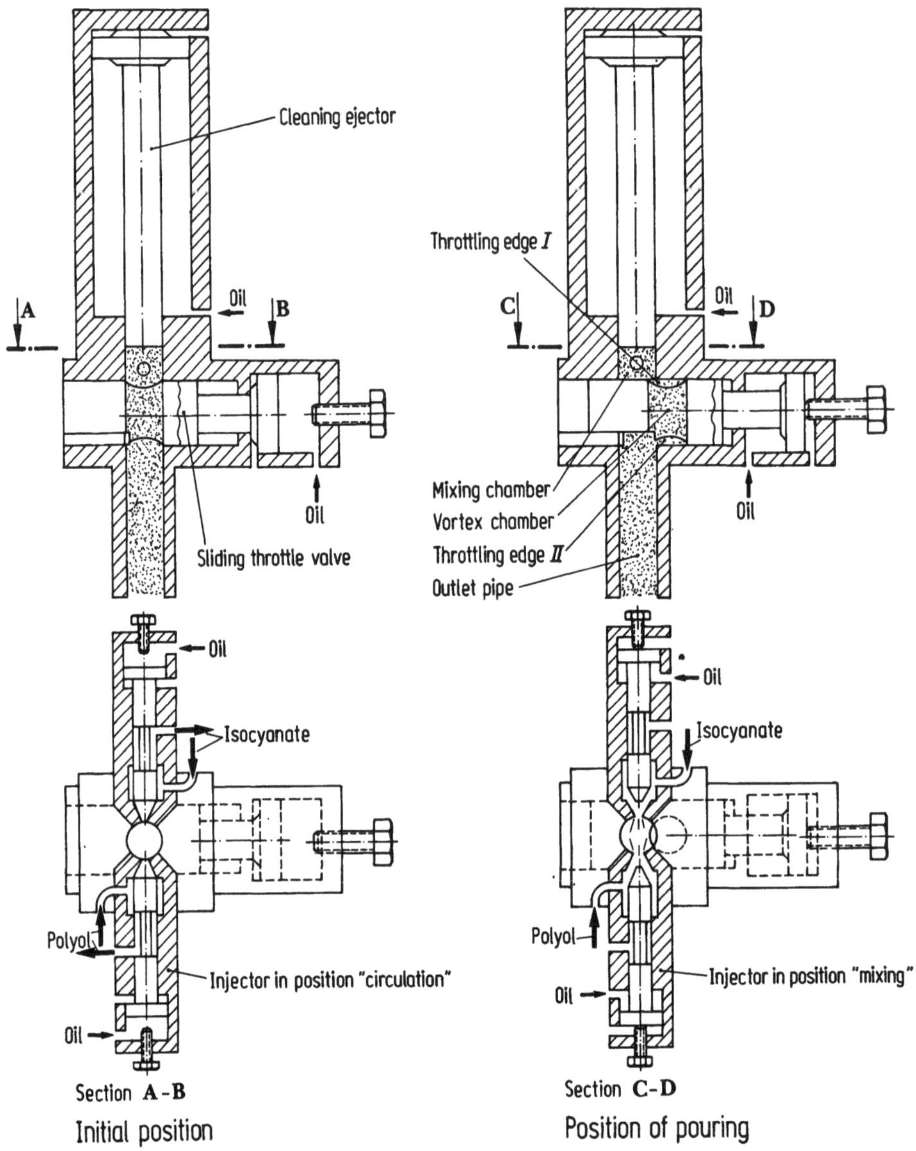

Fig. 8.7. Action of high pressure mixing head, Model MQ (Courtesy of Maschinenfabrik Hennecke GmbH, St Augustin, FRG)

in order to obtain quality IF's, all the cavities should have equal volumes, filling gates and vents, see Fig. 3.2 [13–15].

8.4.4 Mold Filling

RIM articles completely free of entrapped air can only be produced if the reaction mixture is introduced into the mold cavity without entrained air. This is achieved by

Fig. 8.8. Mold fitted into the mold carrier, for the production of a glass fiber reinforced IF polyurethane car-fender by a RRIM process (photo submitted by U. Osinski, Maschinenfabrik Hennecke GmbH, St Augustin, FRG)

feeding it, in a controlled laminar flow, into the lowest part of the mold cavity, allowing the air to be vented from the highest point of the mold parting line. Film gate design for achieving a laminar flow of the reaction mixture is especially critical when producing thick-walled RIM moldings. There are three techniques for filling molds, i.e. direct fill, dam gate, and half-round film gate [1,14a,15].

Direct fill – fountain – head gate

This method yields a scrap free IF part, though it is restricted to thin-walled moldings (5 mm wall sect.). The technique also requires perfect mixing in the mixing chamber, but this is often more difficult to achieve for elastomeric formulations than for rigid PUR formulations.

Log gate – dam gate

The dam gate design is applicable to all PUR systems and for a variety of wall cross sections. The width of the gate is designed to allow the reaction mixture to be introduced directly into the mold cavity with a linear velocity usually less than 1.5 m/sec.

Half round film gate – runner gate

Where thin-walled, high density moldings are to be produced, this design allows a smooth transition from the circular entry into the mold cavity to the rectangular sections of the IF part. In this design, the reaction mixture entrance velocity is normally 2 m/sec.

Gate location

The most favorable position for a film gate is at the lowest part of the mold cavity. This entry point should be selected so that it is not close to an extreme cross-section change, as this would produce great variations in the local flow speeds near the gate and would consequently cause swirling, air entrapment, and other undesirable surface defects [14a].

8.4.5 Mold Tooling

The tooling for polyurethane RIM should be designed to tolerate internal molding pressures of 10 bar (= 1 MPa). This low internal pressure allows prototypes to be produced in plastic molds, but it should be noted that such prototype molds are not capable of producing surface finishes that are representative of the final RIM part. Metal tooling is generally used to obtain excellent surface qualities. Usually RIM tools are temperature controlled by water or other liquid tempering media in a manner similar to that used in the injection molding technique (Chapter 3). Special care and attention must be paid — and this is specific to RIM tooling — to the sealing and vent-

Fig. 8.9. Schloemann-Siemag universal mold clamping press ("W" type) for producing RIM parts requiring a long parallel stroke opener; 1: opening speed is controlled by impulse; 2: five mold clamping units can be operated by one hydraulic unit; 3, 5, and 7: scales for the angles of rotation and tilting; 4: adjustment of mold mounting height; 6: connection possibilities for core-pullers and ejectors (Courtesy of the Schloemann-Siemag AG Kunststoffetechnik, Hilchenbach, FRG)

ing of the tools since the parting lines, cores, and ejectors must be leak-proof to the very low-viscosity reaction mixtures. A useful method is to mill grooves around the cores and ejectors, behind the parting line, and fill them with the PUR reaction mixture. This subsequently forms an elastomeric seal which prevents leakage.

Tool venting is normally placed at the parting line at the highest point of the mold. Whenever a particular tool or part design renders this impossible, an auxiliary parting line should be used [1,15,22].

8.4.6 Clamping Units

Mold clamping units are necessary to seal a mold tightly. These must allow freedom in positioning the tool during molding because of the venting requirements. Most of the clamping units currently on the market satisfy the following requirements: they are normally operable in two axes, and they provide clamping forces of more than 0.5 MPa (5 bar) over the whole clamp area. Specially designed clamping units are offered for the shoe industry, for producing large automobile parts, for window profiles, etc.

The reaction mixture, which only partially fills the mold cavity during the injection cycle, expands to fill the mold and displaces the air present in the cavity. The mold must be positioned to allow the air an escape route to the highest point of the mold. Correct mold position depends on the geometry of the part, the flow pattern and vent arrangement of the mold. The final position of the mold is best determined by trial and error. The geometry of the part also dictates the general filling technique. The sprue (injection part) is normally located on the parting line, and the mold is angled

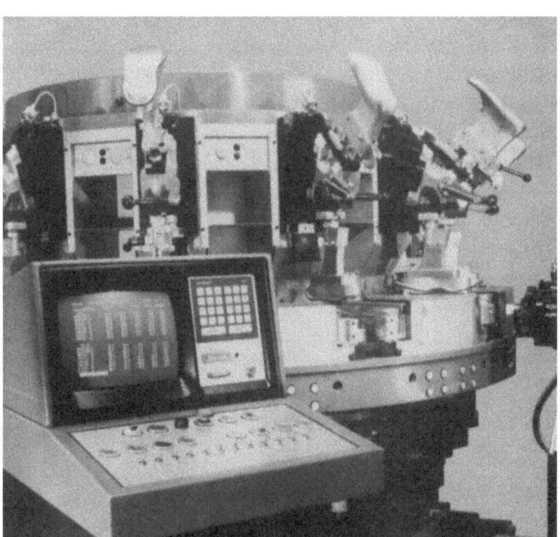

Fig. 8.10. Mold and clamp units in a carrousel RIM machine, with its display-control system, for producing RIM shoe soles, Model Desma 74.16 (photo submitted by A. Malburg, Desma-Werke Hermeskeil GmbH, Hermeskeil, FRG)

so that the reaction mixture is introduced at the lowest point. The air is then removed from the highest point in the mold via a controlled venting.

Mold carriers for producing large and small RIM parts are presented in Fig. 8.9 and 8.10 (see also Fig. 8.8).

8.5 Technical and Economic Analysis

8.5.1 RIM versus Other Polyurethane Processes

Superficially, RIM technology is not different from any other urethane technology since it has all the typical elements of urethane molding. So, what makes the process and its end-products unique? How does its integral foam differ from thermoplastic IF's? And why is its IF structurally superior to high density polyurethane?

Table 8.1. Comparison of the RIM process with other technologies, for polyurethane systems [1)]

	Molding Process											
	RIM for PUR-systems			Injection Molding for thermoplastics						PUR-Resins Prepreg		
				Unfoamed			Foamed					
Mold clamp pressure, MPa	1			100			10			10		
Mold clamp force/ mold furface area, MPa	100			10,000			1,000			1,000		
Achievable molding weight, 10, 40, 70 kg	+	+	+	+	−	−	+	+	−	+	+	−
Flow path limits	Unlimited			Limited			Limited			Limited		
Demold times, min 3 and 10 mm thickness	0.5		1–4	0.7		2–5	1		3–5	1.5		3–4
Surface reproducibility of the tool	Very good			Good			Poor			Moderate		
Wall thickness change without sink marks	Yes			No			Yes			No		
Inserts Small/Large	Yes		Yes	Yes		No	Yes		No	Yes		No
Molding density, kg/m³	300–1,200			900–1,400			650–1,000			1,600		
Pigmenting	Limited			Yes			Yes			Moderate		
Reuse of waste	No			Yes			Yes			No		

To answer these questions, let us consider some general comparisons with the more commonly known polyurethane and thermoplastic IF processes (see below, Sect. 8.5.2).

A RIM IF polyurethane is significantly different from any other IF. A new polymer, formed during the molding cycle, can be designed to fit each particular end-use. Because there are many polyols with a variety of functions, complexities and molecular weights, almost any type of urethane polymer can be produced. Since the components are all liquids, excessive energy is not required for melting as it is the case with thermoplastics. Processing conditions are generally near room temperatures; fluorocarbon blowing agents are used to create the core foam structure. Polyurethanes also require less massive molding equipment and tooling since they generate only 0.3–0.5 MPa in the mold. In contrast to molds for thermoplastics, RIM molds are held at a constant temperature (for example, 60 °C) and are immediately reusable after part removal [16]. The combination of metering and mold carrier is equivalent to — and performs the functions of — an injection molding machine. But because of the low molding pressures and excellent liquid flow inherent to the RIM process, combined with the lower mold weights, the RIM process has three major advantages over the injection molding process [6]:

1) The RIM press or clamp can be one-tenth the weight of that of the equivalent injection molding component.
2) The cost of the RIM metering device can be one-fifth to one-tenth the cost of that of an equivalent injection molding machine.
3) The floor space required for a RIM press or clamp can be one-fifth to one-sixth of that required for an injection molding machine.

A RIM process requires rapid metering and mixing of large quantities of PUR compositions and rapid injection into the mold cavity followed by quick demolding. Very short cycle times and economical production are the main similarities between thermoplastic injection molding and RIM processes for PUR materials; a critical comparison between these two processes is shown in Table 8.1.

Table 8.2 shows a comparison of the nickel and aluminum tooling used in the RIM

Table 8.2. Process Parameter Comparison[a] — reaction injection molding versus injection molding

Parameter	Reaction Injection Molding, Thermoset Urethane	Injection Molding, Thermoplastic Urethane
Tools:		
Construction	Nickel electroform; machined aluminum	Machined steel
Weight, kg	600	6,000
Cost, $	60,000	125,000
Lead time, weeks	20	28
Equipment:		
Molding machine weight, kg	4,500	54,000
Cost of material and mold handling equipment per molding station, $	60,000	300,000
Floor space required, m^2	6.5	37.2

[a] Comparison based on Monza fascia

process and the machined steel tooling for injection molding. A mold used for the RIM process can be one-tenth the weight of a thermoplastic (or thermoset rubber) injection mold. Even if machined steel is used for RIM process, weight savings of 30% are possible. This is a major advantage of the RIM process and translates to weight and cost savings in installation and equipment, as well as lower energy usage [6]. Prices for equivalent RIM equipment from different suppliers can vary quite widely: stationary presses can cost less than $ 50,000 or as much as $ 250,000; the price of multiple station carousels can be as low as $ 100,000 or as high as $ 500,000. In any case, however, RIM facility and equipment costs are higher than those of any injection molding IF process [6].

8.5.2 RIM versus Thermoplastic Processes

Thermoplastic IF processes (see Chapters 4–7) use high molecular weight polymers, the composition of which is predetermined. Moreover, they operate best at pressures of 2–5 MPa and require massive machinery and tooling. The tooling must be temperature controlled because it is thermally cycled, and thermoplastic IF's require one foam machine for every tool. In addition, demolding times for IF's more than 15 mm thick increase significantly, amounting to 5–7 min for parts 18 mm thick (Fig. 8.11). RIM unfinished polyurethane IF parts, on the other hand, have excellent surfaces and very short demolding times, requiring, for example, as little as 30 sec for a thickness of 12 mm (Fig. 8.11). This is approximately 50% the time required by a thermoplastic foam.

Comparisons between a RIM process and a thermoplastics IM process for IF's are given in Tables 8.1, 8.2, and 8.3.

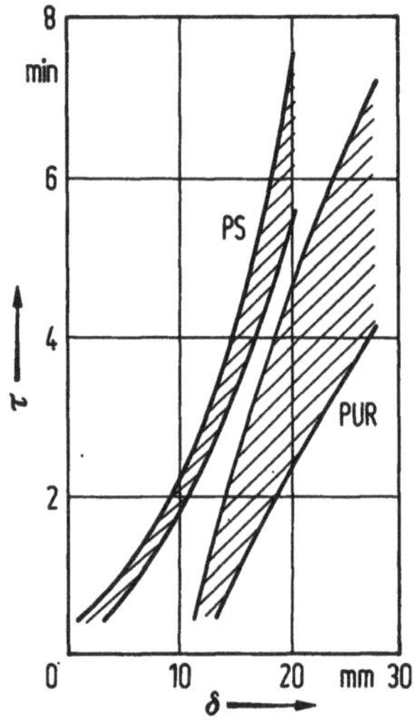

Fig. 8.11. Demolding time versus wall thickness for IF parts based on PS and RIM polyurethane, having the same density [16]

According to LaPlaca [18], RIM processes have the following advantages compared to low pressure injection molding of IF parts (based on ABS and PPO): tooling lead time is reduced by 25% (i.e., 12 versus 16 weeks); lower cost tooling (i.e., epoxy); lower densities (up to 600 kg/m^3), hence reduced

Table 8.3. Comparison between RIM and thermoplastic injection molding [17]

	RIM	Injection Molding
Temperature:		
Reactant, °C	~60	200–300
Mold, °C	~70	ambient
Injection Pressure, MPa	14	70–150
Clamping Force, kg/m²	0.03	6–13
Material Viscosity, Pa · s	0.01–1	10^2–10^5

part weight; and low cost processing equipment. The disadvantages of RIM are: 10–15% higher unit costs; extensive surface preparation prior to finishing; 7.6 mm minimum wall thickness; less complex configurations possible; scrap material cannot be reused; longer cycle time (10 versus 3 minutes); reduced physical-mechanical properties; limited supplier base in non-European countries.

8.5.3 RIM and Energy Saving

RIM technology saves energy in several ways. Because the articles are molded directly from a pre-polymer, which is polymerized as it is being molded, the stages of making and storing the polymer are cut out. There is no need for energy to soften stored polymer, for the pre-polymer is stored as a liquid and the molding is done at low temperatures and pressures. The two components (polyol and isocyanate pre-polymer) are held in separate tanks; when molding is to take place, the two liquids are pumped at controlled rates to a mixing head and into the mold where they react

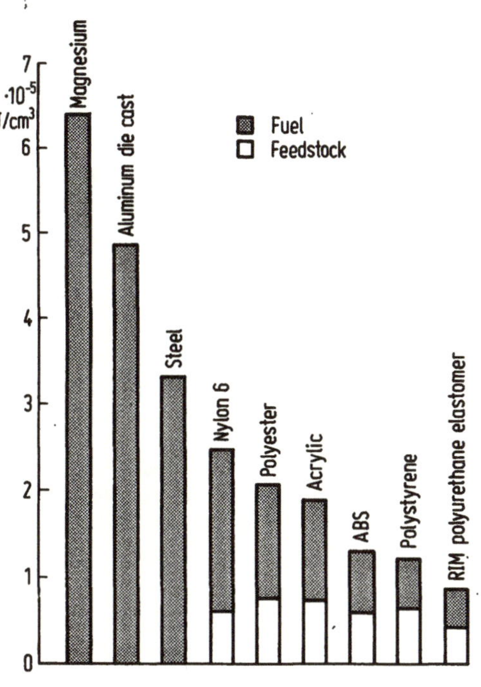

Fig. 8.12. Comparison of the energy requirements per unit volume for producing similar molding items from various materials (Courtesy of Rubber Division, American Chemical Society, Akron, USA)

Table 8.4. Elastomer energy savings (basis: automobile fender liner) [7]

Material	Weight, kg	Energy equivalent of gasoline to make the part, liter (gallons)	Gasoline needed to move item over 5 years, liter (gallons)
Steel	5.90	11.35 (3)	41.64 (11)
Aluminum	2.27	22.70 (6)	15.14 (4)
Plastics	2.72	7.57 (2)	18.92 (5)

to form a molded product after a few seconds. Typical operation temperatures are 55 to 70 °C, and typical pressures about 10–20 MPa. Relatively small equipment means savings in floor space and less capital cost [9].

Fig. 8.12 shows a comparison of the *energy requirements* per unit volume for producing similar molded items from various materials. The energy used in refining and fabricating is quite high for metals. The energy values for polymers include the feedstock.

In the light of recurrent energy crises, the relative economics of steel and RIM polyurethane are very relevant. According to one estimate [19], *the prices* of RIM chemicals have been relatively stable compared to those of other plastics. For the years 1971 to 1979, RIM chemicals, both in USA and in Europe, have increased 40% or less, whilst steel products have increased anywhere from 60 to over 100% over the same period.

An important factor when using polymeric materials in *automobile applications* is their energy efficiency during manufacture. RIM foams are about the most energy efficient materials available today, not only because of their low feedstock requirements (Figure 8.12), but because the fuel costs for processing are very low.

There are a number of economic factors [4,6] promoting the use of RIM parts in automobile applications. There is no question that government legislation on safety and damage resistance has been one key factor that has helped to open the US automobile market for elastomeric RIM parts. The second factor that has pushed RIM polyurethane and other plastics into the USA market is the need for improved gasoline efficiency and thus for reductions in automobile weight and size. Table 8.4 shows the savings which can be obtained by using an elastomer plastic for an automobile fender liner. The comparison includes the energy required to make the part as well as the energy needed to transport it as the automobile is moved (see Chapter 13).

Although government legislation is not so much a factor in the European market, there are common trends in USA and Europe (Ferrary [6]). First, the American reduction in car size ("downsizing") will certainly put American and European automobile manufacturers into more direct competition. Second, gasoline supply and cost problems mean that both Europe and America need to be even more energy efficient than they are today. Third, pedestrian protection has been emphasized more in Europe, though it is now being recognized as important in the USA too. Finally, damage resistance, while not a formal requirement, is becoming more accepted in Europe.

8.5.4 Current Trends

Given a fixed chemical system and machinery, a RIM producer can vary certain conditions to control part performance. A number of detailed studies have been made of selected processing parametern and their influence on part properties[4,7,20–24]. These *processing variables* include: pour rates, impingement nozzle pressures, part density, mold temperature, isocyanate index, post-cure conditions, blowing agent content, mold material, etc. Pour rate and impingement nozzle pressure work together in obtaining good mixing; the best part properties are obtained with high pour rate and/or high impingement nozzle pressure. The density and isocyanate index are more critical to obtaining good properties than is the mold temperature; the latter actually has little or no effect on most properties. Post-cure can significantly affect RIM properties, and this stage can be done either as a separate step or incorporated in the finishing process. The best properties are generally obtained with a post-cure of 30–45 min at 120 °C.

World-wide polyurethane demand continues to grow, with RIM, and more recently RRIM, growing faster than the market as a whole (Table 8.5).

In addition to weight reduction (and proportional cost reduction) available from the more advanced RIM technology, significant process improvements have helped to make RIM parts more competitive:

1) Productivity in terms of cycle time was reduced from over 4 min per part to about 2 or 3 min.
2) Low weight clamps and multiple station lines were developed to take advantage of the low RIM molding pressures, and have resulted in low capital investment requirements.
3) The process ensures low retooling costs.
4) In-house compounding can be used to reduce material costs further.

The short history of RIM technology, which can produce materials from soft polymers to high modulus elastomers and rigid materials, once again illustrates the tremendous versatility of polyurethane, and the future for RIM looks very bright [23,24].

A technical and economical comparison of this process with other IF processes is presented in Chapter 12.

Table 8.5. RIM and RRIM demand in USA (in millions of kg) [4]

	1979	1980	1983	1985
Transportation				
RIM	21.8	23.6	40.9	49.9
RRIM	0.5	0.5	9.1	22.7
Other	0.9	2.3	4.5	7.3
Total	23.2	26.4	54.5	79.9

8.6 References

1. Bayer-Polyurethanes: Handbook. Bayer AG, FRG, 1979
2. Becker, W. E. (ed.): Reaction Injection Molding. New York: Van Nostrand Reinhold Company 1979
3. Sweeney, F. M.: Introduction to Reaction Molding. Westport: Technomic 1979
4. McBrayer, R. L.: J. Cell. Plast. *16*, 331 (1980); SAE Techn. Papers No 800515 (1980)
5. Börger, H.: Kunststoffe *69*, 863 (1979); *72*, 359 (1982)

6. Ferrari, R. J.: Manufacturing Technology Using the RIM Process. In: Injection Molding. Becker, W. E. (ed.). New York: Van Nostrand Reinhold Company, 1979, pp. 56–106, 107–179
7. McBrayer, R. L., Carver, T. G.: Elastomerics, *113*, No 6, 30 (1981)
8. Lee, L. J.: Rubber Chem. Technol. *153*, 542 (1980); Polym. Eng. Sci. *20*, 868 (1980)
9. Webb, S.: Proc. Chem. Eng. *33*, No 12, 73 (1980)
10. Camargro, R. E. et al.: Polym. Eng. Sci. *22*, 719 (1982)
11. Fiorentini, C.: 5th Internat. Conf. Cellular and Non-Cellular Polyurethanes, Strasbourg France, 1980
12. Schlotterbeck, D.: Kunststoffe *71*, 775 (1981)
13. Botsch, H.: Europ. J. Cell. Plast. *3*, No 5, 115 (1980)
14. Thiele, H., Zettler, H. D.: Ind. Prod. Eng. No 3 94 (1980)
14a. Misitano, G.: Plast. Eng. *35*, No 2, 27 (1979)
15. Knipp, U., Becker, W. E.: Mold Design. In: Reaction Molding. New York: Van Nostrand Reinhold Company , 1979, pp. 215–240, 241–300
16. Horvath, M.: J. Cell. Plast. *12*, 289 (1976)
17. Gerkin, R. M., Critchfield, F. E.: Meet. Plastics Inst. of Amer., Acron USA, 1976
18. LaPlaca, J. P.: J. Cell. Plast. *16*, 36 (1980)
19. N.N.: Ind. Week *1979*, Sept. 17, 140
20. Zahn, E., et al.: 5th Internat. Conf. Cellular and Non-Cellular Polyurethanes, Strasbourg France, 1980
21. N.N.: Plast. World *38*, No 1, 11; No 5, 28 (1980); *40*, No 12, 12 (1982)
22. Von Hassel, A.: Plast. Technology, *29*, No 1, 21 (1983)
23. Hölzl, E.: Kunststoffmaschinen No 14, 3 (1983)
24. Frisch, K. C.: Plast. Rubber Intern. *8*, No 1, 17 (1983)

9 Extrusion

Extrusion methods of producing integral foam structures have recently been widely applied in many countries. The main advantages of these methods are: continuous and high productivity, simple equipment, and precisely sized articles. Solid and hollow profiles made of IF successfully substitute for wooden articles (without any further machining) and profiles made of unfoamed plastics, thereby reducing feedstock consumption by 15 to 40 per cent. Extrusion methods are particularly economic for fabricating finished articles that do not require any secondary processing (furniture parts, complicated cross sections, etc.), and where new sorts of articles are to be produced, like, for example, "integrated" pipes (see Chapters 15 and 17).

9.1 Basic Process

9.1.1 Free and Control Extrusion Processes

Until recently, there was considerable industrial activity involving conventional free-foaming extrusion techniques (FFE). This process is similar to normal unfoamed plastic extrusion except that the polymer melt contains blowing agents which cause foaming at the extruder exit (or die). In FFE, dies are cut so as to deliver the foaming material with only an approximation of the shape and size of the desired profile. Thus, the polymer expands freely upon leaving the die. The hot-melt strength of the polymer is a critical factor as is the significant foaming pressure which usually limits the density reduction obtainable with this technique; densities are generally above 500 kg/m^3. Profiles produced by FFE frequently have very fine cell sizes and uniform distribution but there is little or no solid skin on the surface [1].

In order to obtain an IF profile with a solid skin and a cellular core, a controlled foam extrusion (CFE) process, particularly the Celuka process, has to be used at present.

The main advantages of a CFE process over an FFE process are:

1) The skin of a CFE profile may be from 0.2 to 3.0 mm thick.

2) The reinforcing effect of the unfoamed skin allows a significant reduction in density for equivalent properties. Hence, to have the same rigidity as a CFE profile, an FFE article of the same configuration would require a density 20 to 25% higher, i.e. 500 vs 400 kg/m^3.

3) In a CFE process, an extruded IF article is mechanically constrained in at least two dimensions to the size of a shaper until it becomes stable; thus, complex and intricate cross-sections can be extruded to close tolerances with dimensional control to ± 0.1 mm.

4) CFE profiles can be extruded over a broad density range, viz., PS from 150–800 kg per m^3, rigid PVC from 300–900 kg/m^3, and PP and PE from 350–600 kg/m^3. In addition, good control of density to ±20 kg/m^3 can be achieved.
5) Theoretically, there is no limit to the maximum size of a CFE profile. The relative economics of tooling and equipment costs vs the market potential will be the main limiting factor. Commercially the shapes so far produced have been as wide as 750 mm with thicknesses down to 5 mm. Obviously, extrusion of profiles with thicknesses below 5 mm may be achieved by increasing the foam density [1].

9.1.2 Practical Realization

The general principles of producing an extrusion IF are the following: The components of the formulation are mixed, as a rule, in a two-stage turbospeed mixer that is charged in steps. The first to be charged into the heated first mixer are the solid components; they are heated to 50 °C, with a first speed setting. Then, the liquid components are charged and heated to 110 °C, with a second speed setting. The resulting composition is fed into the second mixer, cooled to 50 °C, mixed with a CBA or a PBA (hydrocarbons, freons), nucleating agents, pigments, stabilizers, etc., and shut into the extruder. The melt temperature in the extruder is increased by 30–40 °C and the pressure is increased to 10–15 MPa.

In order to obtain high-quality articles with a predetermined degree of foaming and controlled skin thickness, the physical processes in the extruder itself and at its outlet must be understood (Fig. 9.1). Initially, as the BA decomposes (or boils) the back of the screw should be completely closed to avoid losses of foaming gas. The viscosity of the melt should not be too low at this stage, so that the mixture does not flow out through the die nozzle. The pressure in the cylinder should be chosen such that the total melt (along the extruder length) is saturated with gas, on the one hand, and such that the foaming process can only start at the exit of the die, on the other. As a rule, the pressure in the melt should be about 10 MPa at the instant of blowing [2].

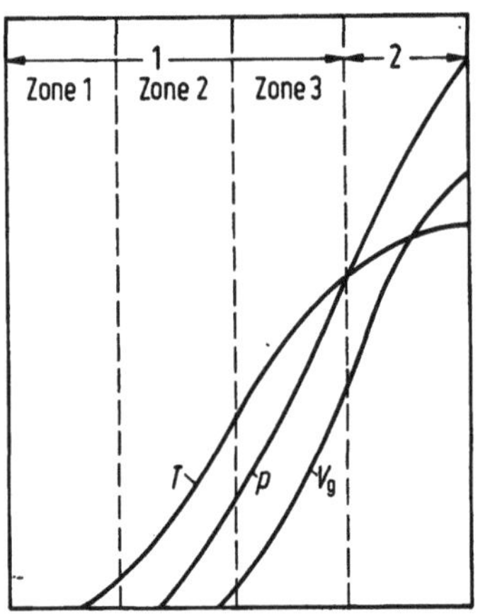

Fig. 9.1. "Ideal" curves for IF extruder process: variations of the temperature T, pressure P, and volume of the foaming gas V_g, in the cylinder (1) and in the die (2) of the extruder [6]

The material plastication in the extruder should start at a moderate temperature and should be finished before the chemical blowing agent decomposes; the temperature profile within the extruder is dictated by the decomposition temperature of the BA, and the residence time of the melt in the extruder is dictated by the kinetics of the BA decomposition. The temperature at the exit of the extruder head is generally 180 to 190 °C. A smooth and thick surface skin is formed if the temperature of the head is 140–150 °C. If it is 150 to 170 °C, the skin is thin and not very strong. In the first zone of the mandrel the temperature is 50–90 °C, the core being foamed at 180 °C. To control the temperature profiles in the extruder itself and at its outlet, computer-controlled devices are used [3].

The integral structure forms due to the suppression of the foaming process by the rapid (less than 10 sec) cooling of the inner and outer surfaces of the extrudate as it exits the extruder head.

9.2 Commercial Processes

Commercial processes for producing extrusion IF involve practically all types of thermoplastic polymers. CBA's, particularly ACA and potassium bicarbonate, are the main BA's used; PBA's are not widely used at present.

9.2.1 The Celuka Process

Basic Process

The most common extrusion IF method is the "Celuka" process. It was developed by Ugine Kuhlmann in France about ten years ago and was intended for processing any thermoplastic polymer [4,5]. The process can be described as follows. After plastication in the extruder cylinder, the polymeric melt containing a CBA or PBA is passed to the extruder head which is partially plugged by a "torpedo" (Fig. 9.2). The melt flows around the torpedo and is forced towards the walls of the head. In the shaper (which has the same size as the profiling slit), the outer layer of the extrudate is cooled intensely by the walls and forms the skin of the article, the core foaming proceeding "inwards", i.e. from the skin towards the center of the article.

The article density, which is generally 20 to 40% of that of the starting polymer, is controlled by the rate of extrusion and the torpedo cross-section. The thickness of the surface skin may vary from several microns to 2 mm and is controlled by the rate of cooling, melt temperature, rate of extrusion and speed of screw rotation. The "inward" foaming results in good size precision (±0.1%) on the one hand, and in

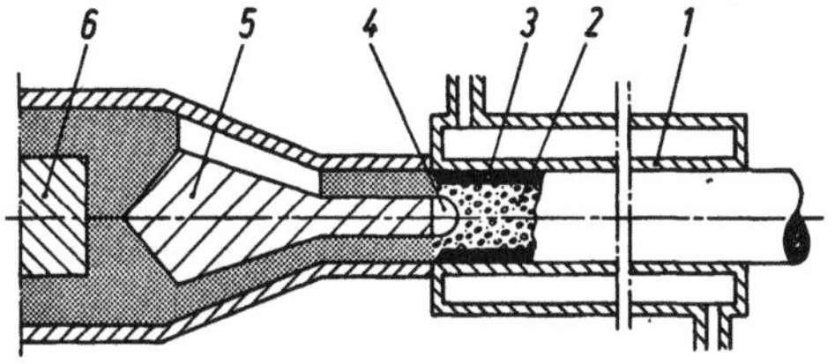

Fig. 9.2. Schematic of IF extrusion by the Celuka process which prevents the outer skin of the IF profile from foaming; 1: cooled shaper, 2: foamed core, 3: unfoamed skin, 4: void, 5: fixed torpedo, 6: screw [8]

Table 9.1. Controlled IF extrusion process: typical extrusion and weight rates (Celuka PVC profiles)

Profile size Shape Example	Small Simple Moldings	Medium Simple Board sidings	Large Complex Door and window frames	Large Complex Hollow panels	Large Complex Special[a] hollow panels
Cross section, mm²	160–580	1,290–2,580	1,290–6,450	3,225–7,740	3,225–7,740
Linear rate, m/min	4–2	1.8–1.2	1.2–0.9	1.0–0.8	0.7–0.6
Density, kg/m³	400	400	550	500	500
Weight rate, kg/hr.	14–30	57–75	52–195	103–177	73–142

[a] With high quality appearance and intricate design specifications

an unfoamed surface skin and hence a rugged and strong profile, on the other. However, the larger the articles, the lower the size precision.

It is possible to modify the skin thickness by changing the formulation or the extrusion conditions, but in general terms, the quicker the material is allowed to foam, the thinner the skin. The reverse is also true: thick skins are obtained by delaying, as far as possible, the start of foaming [6]. Table 9.1 gives some typical production rates [1]. The productivity of the process (for example, 18 kg/min at an extruder diameter of 400 mm and 60 kg/min at 60 mm) is similar to that of unfoamed plastic extrusion, whereas linear productivity is twice as high for foams (2–3 m/min). The rate of extrusion is limited by the cooling time. To improve productivity, several profiles can be extruded in a single head.

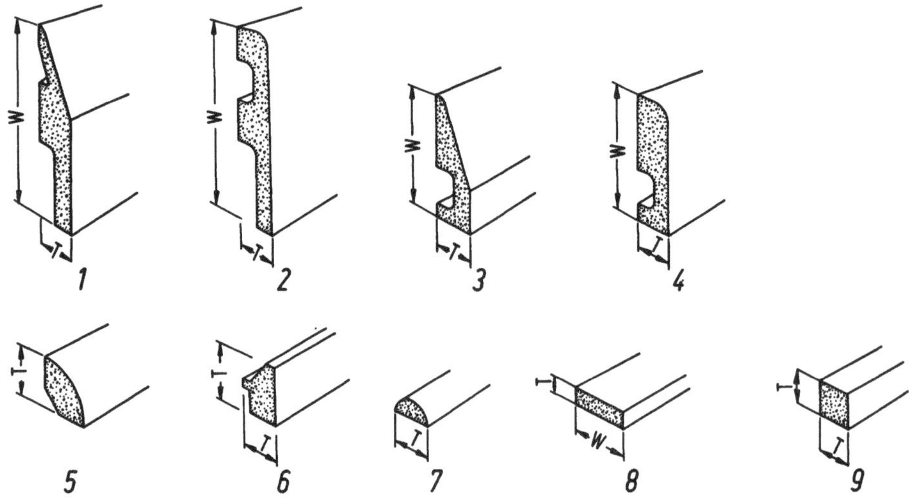

Fig. 9.3. Some typical PVC integral foam profiles produced by the Celuka process, and their dimensions (T (height) × w (width), mm); 1: 14 × 70, 14 × 95; 2: 14 × 70, 14 × 95; 3: 17 × 45; 4: 14 × 45; 5: 12 × 12, 19 × 19; 6: 15 × 21; 7: 12; 8: 6 × 22, 9: 12 × 12, 19 × 19 (Courtesy of Ugine Kuhlmann, Paris, France) [4]

It should be mentioned that the technological features of the Celuka process are common to most IF extrusion processes.

Types of Articles

Very complex shapes can be made by the Celuka process, mainly because the die is cut to a shape and size very close to that of the profile required, and the material is never allowed to foam freely, outside the constraints of the tooling. In fact, the foaming of the polymer melt may be viewed as the reverse of a FFE process, where the polymer melt expands from the inside to the outside, the skin being formed merely of crushed cells. In the Celuka process the skin is formed first and the remaining material foams at ever decreasing density towards the centre of the profile, i.e. from the outside to the inside [1,6]. This is a fundamental difference that enables the Celuka process to make low density complex profiles (Fig. 9.3). As Figure 9.4 shows, the skin of a Celuka profile is unfoamed (see also Fig. 1.2) and consistent in density and there is a very large drop in density between the skin and the cellular core. This is why a Celuka profile might be called a sandwich material.

The *Celuka profile* has the following advantages:

1) Low thermal conductivity, and consequently high insulation, which is particularly favorable when compared to metal frames; Celuka profiles have the practical advantage of not attracting condensation and the subjective advantage of feeling warm to the touch.

2) Freedom from maintenance. Celuka frames will not rot as timber frames do, nor will they corrode as metal frames do, in the long term.

3) Possibility of assembling profiles combining Celuka PVC and metallic reinforcement; less expensive than when using unfoamed PVC profiles (Fig. 9.5).

4) Possibility of repairing Celuka frames damaged during installation. This possibility does not exist with metal frames where a complete replacement is nearly always the only answer.

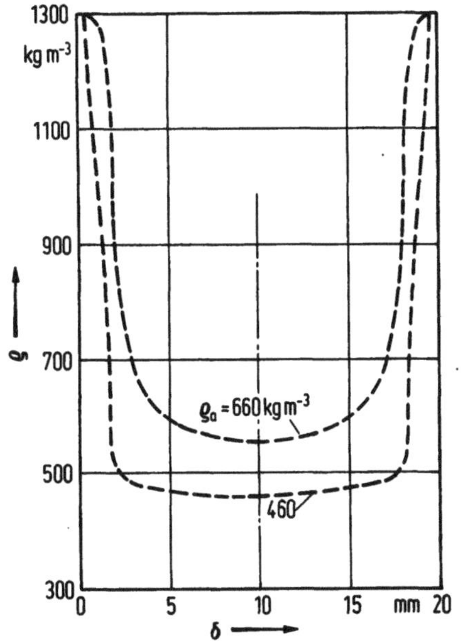

Fig. 9.4. Density profile for two Celuka PVC integral foams [6]

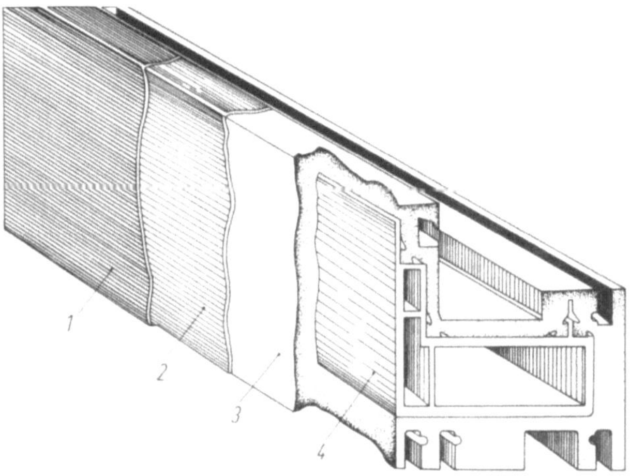

Fig. 9.5. Complex multi-component profiles of the ®Combidur Decor panels for construction applications; 1: weather-proof acrylate covering, 2: one-color painting, 3: PVC Celuka integral foam, 4: aluminum profile (Courtesy of Gebrüder Kömmerling Kunststoffe Werke GmbH, Pirmasens, FRG)

5) Good compatibility with concrete and other building materials; Celuka parts can be fixed directly into masonry without any protection; moreover, any plaster or concrete splashes can easily be cleaned off.

Recently, a process has been developed to produce light ($\varrho = 200\ kg/m^3$) hybrid profiles, whose skins and cores are made from different polymers [4,6].

9.2.2 Coextrusion Process

A special kind of extrusion process for foam-core articles is the coextrusion for making sandwich-like sheets and pipes. Strictly speaking, this process produces articles whose structures are closer to sandwich structures than to integral foams. But the importance of these foam-core materials in modern industry is so great and they promise to be so useful that we must discuss them here.

Over the last five years a number of successful methods of coextruding foam-cored, plastic pipes have been developed. Coextruded PS, ABS and PE foam-core pipes have been made on a commercial scale. Because of the encouraging technical success and favorable economics (see Chapter 17), ABS foam-core pipe promises to become an important material for drain, waste and vent (DWV) pipes [7].

Let us consider the main features of the process. Two conventional extruders are set at right angles, one for the foam core and one for the solid skin (Fig. 9.6). The successful extrusion of the foamed core depends on coordinating the temperature in the extruders with the decomposition rate of the CBA. A vented extruder can be used to produce the solid skin. However, the need to hold the foamed polymers under pressure until the time for expansion rules out the use of vented extruders for the foam core. The solid-skin extruder should be of the single stage variety, and have an L/D ratio of 30:1. The foam-core extruder, which is responsible for only 40% of the total output, can be a smaller unit, and have an L/D ratio of 24:1. The main parameters of the process for an ABS pipe are shown in Table 9.2 [7].

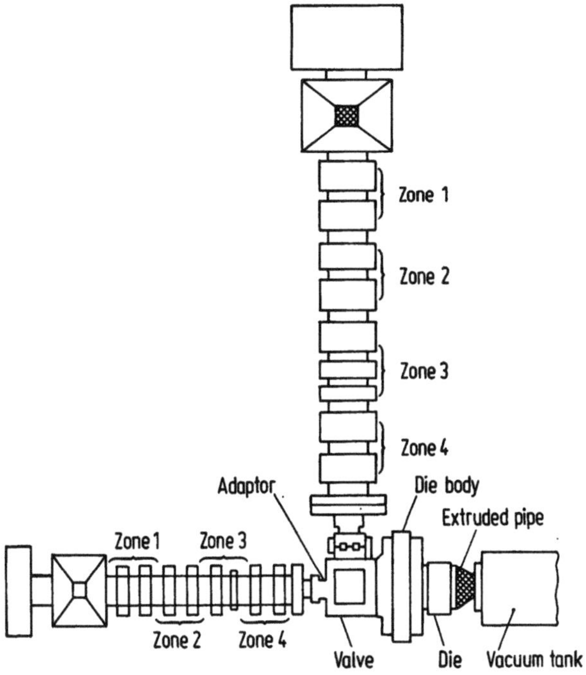

Fig. 9.6. Layout for a foam-core pipe coextrusion process [7]

Table 9.2. Equipment and conditions for an ABS foam-core drain, waste and vent pipe

Parameter	Foam-core extruder	Solid-skin extruder
Size, mm	50.8 mm	76.2
L/D ratio	24:1	30:1
Compression ratio	2.7:1	3.0:1
Screw	Single-stage, constant taper	Two-stage, unvented
Temperatures, °C:		
Zone 1	166	160
2	168	160
3	165	182
4	207	188
5	202	193
Gate	190	204
Die 1	182	204
2	171	204
3	185	204
Melt temperature, °C	204	204
Head pressure, MPa	19.25	31.5
Screw speed, rpm	16	16
Drive, MPa	10	14
Line speed, kg/min	3.6	3.6

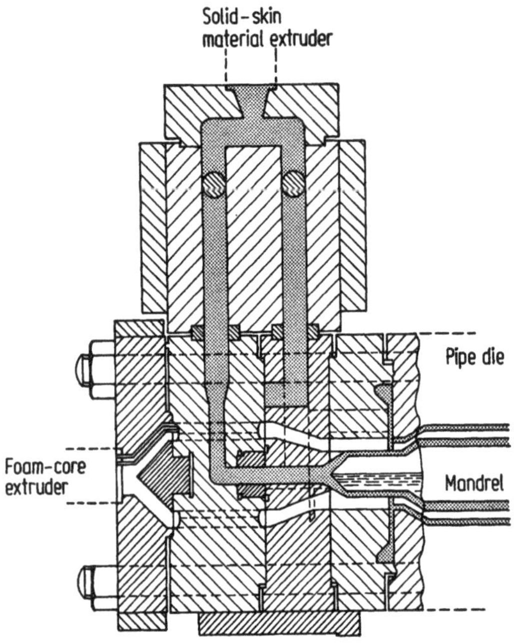

Fig. 9.7. Pipe coextrusion feedblock adaptor [7]

Foam-core pipe can be coextruded by two types of tooling. One uses conventional pipe dies in conjunction with a combining feed-block adaptor fed by the two extruders. The other uses three extruders with a multi-manifold die design. The adaptor (Fig. 9.7) layers the foam core between the solid walls before they enter a conventional pipe die. The pipe produced with this tooling has knit lines where the extrudate recombines after passing around the legs supporting the mandrel.

9.2.3 Other Processes

In the "*Armocel*" *process* developed by Armosig, France, the extrudate coming from the head gets to a special additional device with an increasing gap (Fig. 9.8) which then turns into a constant-diameter gap [2]. It can easily be seen that in contrast to the "Celuka" process the Armocel process forms a core due to "outward" foaming. The process is intended for producing hollow articles and pipes made from PVC, PS, and PO with densities 2–3 times lower than those of the initial polymers.

An original way of producing integral PS foam with a smooth surface that imitates the structure of wood underlies the *Woodlike process* of the Japanese company "Sekusui-Kogyo" (Fig. 9.9). A perforated plate is placed at the exit of the extruder head so that the extrudate passing through the plate is divided into a series of jets (filaments) which foam separately and then fuse. The density of the articles is determined both by the BA concentration and by the number and diameter of the holes in the perforated plate. The device may be implemented on standard equipment as well as on a special-purpose equipment [2, 3].

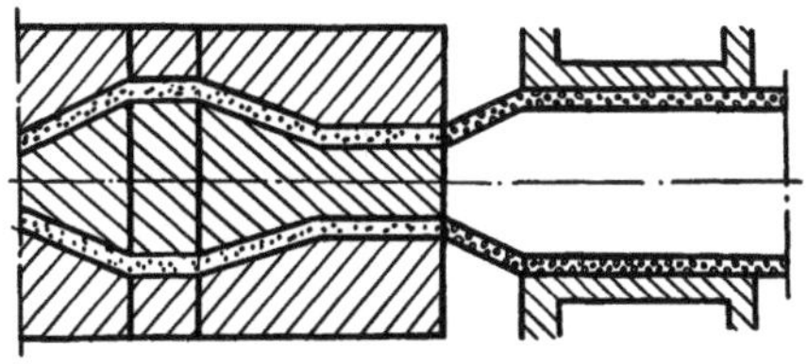

Fig. 9.8. Installation for producing IF pipes by the Armosig extrusion process [2]

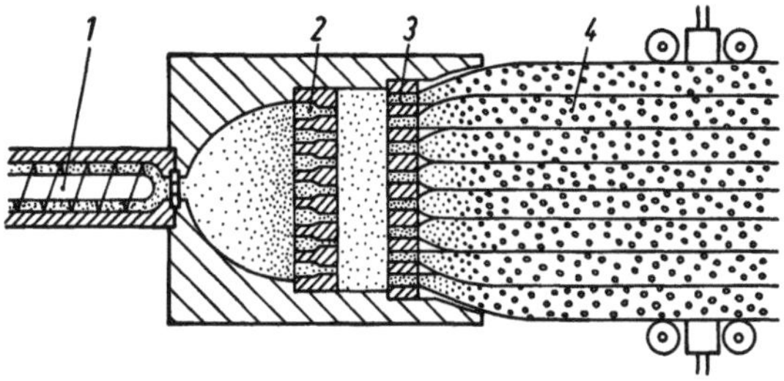

Fig. 9.9. Scheme of the Woodlike process for PS integral foam boards; *1*: extruder screw, *2*: inner plate with holes, *3*: outlet with holes, *4*: foamed strands [8]

A similar principle — division of the melt into several streams — is used in the *Sherer & Trier process* (FRG). Here, the innermost stream is heated to a higher temperature than the outer streams [2]. In another process [3,8] the extrudate is passed through a head provided with projections. These leave strips on the outer layer of the extrudate and after passing through the die, the strips fuse, but are partially rather than completely smoothed out.

In order to improve the structure of core and skin, *Dow Chemicals* has developed a process for producing an IF based on PS in a vertical extruder [2,3]. This process is carried out on conventional equipment and has the following distinctive features:

— an elevated pressure (0.35–1 MPa) is developed in the extruder head which prevents the melt from foaming;

— a PBA-CBA mixture is used, the PBA content determining the density of the articles, the CBA content determining the size of the cells.

BASF in West Germany has developed a process for extruding IF profiles where the composition is first supplied to an accumulator and then extruded via the mandrel nozzle to produce either a solid or hollow profile. Integrated profiles with different cross-sections are produced by the extrusion processes of Montedison (Italy), CdF Chemie (France), and B. F. Goodrich Chemicals (USA) [2,3].

The extrusion process of *Cedal of France* can fabricate hybrid integrated profiles on conventional equipment based on PS (ϱ = 400 to 800 kg/m^3) and PVC (ϱ = 500 to 1,000 kg/m^3). The main difficulty involved in producing foamed thermoplastics is the considerable frictional force suffered by the profile exiting from the die head. This is avoided in this process by a careful choice of the relationships between the diameter and the length of the screw (the L/D ratio is 26:1), and between the surface area of the profile and the rate of extrusion [2,3]. Thus, a 300-mm diameter extruder is used for profiles with cross-sections under 5 cm^2, whereas for 10-cm^2 cross-sections a 45-mm diameter die is used. Small cross sectioned profiles (under 4.5 cm^2) are extruded at a linear speed of 3.75 m/min. with a composition consumption of 50 kg/hr. Profiles having 24 cm^2 cross sections (ϱ = 500 kg/m^3) can be extruded at 0.83 m/min, with a consumption of 60 kg/hr.

Employing the principle of joint extrusion, hybrid IF's consisting of either two ("one-sided" IF's) or three ("two-sided" IF's) layers can be made. *Raifenhauser* (FRG) has developed a technology for fabricating profiles by co-extruding the core and skin from like or unlike polymers, their plastication taking place in different extruders. The thickness of these hybrid IF's varies from 0.5 mm to several millimeters, their density being 400 to 900 kg/m^3 for PVC, 300 to 700 kg/m^3 for PS, and 300 to 500 kg per m^3 for PO [8].

In 1965 Phillips Petroleum Corporation (USA) brought out a process for fabricating thermoplastic IF's which is now known as the *Engelit process*[3,8]. This process has a rather unusual plastication procedure — on a turn-table (disc). The granulated polymer (PS or PE) goes from the feeding channel to a table that is heated to 100 °C and is rotating at 6 r.p.m. There it is distributed into a thin layer, plasticized and then fed into the extruder cylinder, receiving at the same time either a foaming concentrate or a PBA mixture, pigments, and other admixtures. In order to prevent premature foaming, the pressure in the cylinder is maintained at 0.5–0.6 MPa. The melted composition from the cylinder is injected at high speed into a metal mold kept at a pressure of 0.84 MPa where the foaming occurs. Articles produced this way have smooth surfaces and minimum densities, down to 200 kg/m^3. One recent achievement of the Engelit process is the production of thin integrated films based on PE and used as paper and cardboard substitutes. Multiposition molds (up to 8 positions) provide up to 120–150 parts per minute, the overall installation capacity being 10 to 100 kg/hr. The advantage of the process is its favorable economics due to the low cost of the equipment, only 20% of that of LP machine with the same capacity. To improve the quality of the surface, high molding pressures (up to 120–180 MPa) are used and the machines themselves are equipped with multiposition (16–20 positions) molds.

9.3 Equipment Specifications

Extrusion methods can fabricate integral foams with various types of external surface — from smooth and shiny to dull and rough. To obtain a dull surface, the extrudate should be passed through a mandrel at a certain distance (25–250 mm) away from the head. In this case, the critical parameters of the process are the shear strength of the melt, the foaming temperature and the pressure. At elevated pressure, very dense (over 500 kg/m^3) IF's cannot be produced. By contrast, in order to produce smooth surfaces, the mandrel should be close to the head. Depending upon the intensity of mandrel cooling and the extrusion rate, the skin thickness may vary from 0.2 mm to 5 mm and the article density may vary from 150 to 450 kg/m^3 [8,9].

IF articles are fabricated in standard single- or double-screw extruders. The former (screw diameter 45 to 60 mm, linear rate of extrusion 12 m/min) are used to produce small cross-section profiles, the latter (screw diameter 60–85 mm, rate of extrusion 1.5 m/min) are used for large cross-section profiles. By varying the speed of screw

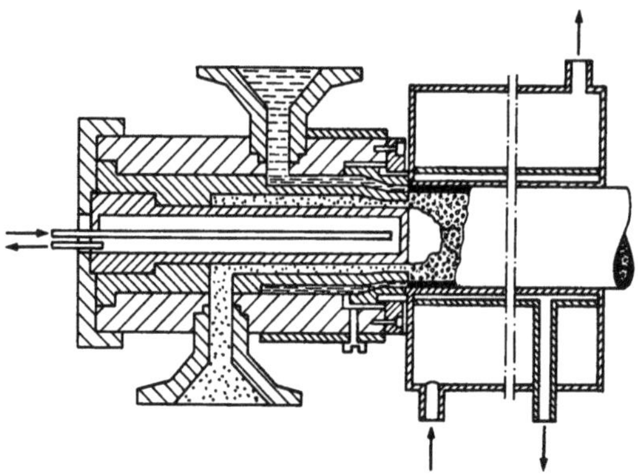

Fig. 9.10. Die for the coextrusion of foam-cored pipes: the top extruder feeds the skin layer and the bottom extruder feeds the core [8]

rotation, the apparent density of the articles can be varied by up to 200 kg/m^3. The extrusion rate determines the density of the articles and the thickness of the skin. The average density of an extrusion IF is generally 350 to 700 kg/m^3, and its skin thickness is from 1 to 2 mm, while the article thickness is from 10 to 300 mm; the production cycle lasts from 3 to 30 minutes [2,5,8].

Two different types of head designs are used: a conventional extruder head (with no mandrel) with a diameter smaller than that of a mandrel, and a head (with a mandrel) whose opening diameter is equal to that of the mandrel. The first design is used for producing pipes by the Armocel process; the second design is used in the Celuka process.

Different head designs can produce solid (Fig. 9.10) and hollow foam-core profiles and pipes, as well as symmetrical and asymmetrical profiles, alternating sizes and different colors and textures (Fig. 9.5) [9,10].

Since the design of a head with a torpedo is relatively simple and the process can be used on conventional extruders, the cost of articles using the Celuka process is only 15–20% higher than the production cost of the unfoamed plastic. The properties of extruded IF's based on different polymers will be discussed further in Chapter 15 (Table 15.4).

9.4 Technical and Economic Analyses

It should by now be clear that extruded integral foam and foam-core parts have the following advantages:
- possibility of producing intricately designed and thus multifunctional parts;
- good surface appearance;
- traditional and cheap assembly possible, similar to wood;
- good sound and thermal insulation.

Injection molding as well as extrusion of IF's permit one-step production of complex shapes that, if machined or assembled from metal and wood, would require a series of costly operations. A detailed comparison of the two processes by Kiessling [1] indicates that extrusion has the following advantages over injection molding: lower equipment investment, lower tooling costs, better surface appearance, broader density ranges, more material versatility. Let us consider these points in more detail.

1) In terms of *capital investment*, an extrusion process requires approximately one half the capital needed for injection molding, for the same production capacity. The tooling costs for an extrusion process are usually somewhat lower than those required for injection molding.

2) *Manufacturing costs* are generally lower for the extrusion process; labor costs for an injection process are 20–30%, and utility costs about 100% higher (Table 9.3) [1].

3) Although improvements are being made in the surface appearance of injection molded IF's, extruded IF's have long offered a wide variety of *surface finishes*, from a high gloss to matt finish, as an integral part of the process.

4) Generally, *lower densities* can be obtained with the controlled foam extrusion than with the injection molding.

5) Extrusion of IF's has greater *material versatility*, particularly in view of the ability

Table 9.3. Comparison of IF extrusion and injection molding processes

Component of cost	Extrusion, %	Injection molding, %
Material	60	57
Labor	8	11
Utilities	1	2
Indirect costs	31	30
	100%	100%

of extruding rigid PVC. It is possible to manufacture heat-sensitive polymers (such as PVC) more easily than with injection molding [11].

On the other hand, it should be recognized that the extrusion process is limited to linear shapes that can be produced continuously, and it certainly cannot produce the intricate, three-dimensional shapes that can only be injection molded. However, for continuous hollow shapes such as hollow panels, pipes, etc., and for many complex two-dimensional profiles, the extrusion process is certainly very advantageous.

A technical and economical comparison of this process with the other IF processes is presented in Chapter 12. Detailed cost information concerning extruded IF parts will be presented in Chapters 15 and 17.

9.5 References

1. Kiessling, G. C.: 6th Structural Foam Conf., Bal Harbour USA, 1978
2. Domininghaus, H.: Maschinenmarkt *81*, No 47, 850 (1975)
3. Berlin, A. A., Shutov, F. A.: Strengthened Gas-Filled Polymers. Moscow: Khimia, 1980 (in Russian)
4. Celuka: Technical pamphlet of Ugine Kuhlmann Produits Chimiques, Direction de la Technoloqie, Paris, France, 1976
5. Brard, F.: Offic. Plast. Caoutch. *26*, No 273, 803 (1979)
6. Jentet, P.: J. Cell. Plast. *15*, 151 (1979)
7. Ahnemiller, J.: Plast. World *38*, No 6, 82 (1980)
8. Barth, H.-J.: J. Cell. Plast. *15*, 103 (1979)
9. DelMonte, E., Pessina, V.: Mater. Plast. Elastom. No 9, 479 (1981)
10. N.N.: Mod. Plast. Int. *10*, No 2, 16 (1980)
11. N.N.: Mod. Plast. Int. *13*, No 6, 34 (1983)

10 Rotational Molding and Other Processes

10.1 Rotational Molding Process

The most universal and efficient method for creating a uniformly thick and equally dense skin is the *rotational molding*, or *rotomolding process*. Commercial production of IF's by this method was first initiated in Japan in 1966, and today it is common in many countries [11].

10.1.1 Basic Process

The principle underlying the rotational molding method helps avoid the main drawback of integral foams, namely a nonuniform thickness and density of the skin. The nonuniformity is attributed to the fact that the skin formation mechanism in different parts of the mold may differ and depends on whether the surface of the stationary mold is wetted by the initial composition or not. For instance, when a PBA is used, the bottom skin is thicker because the temperature of the melt touching the cold walls of the mold cavity decreases rapidly, and because the PBA vaporization is suppressed. By contrast, the top skin is formed in a part of the mold that is not wetted by the melt (or solution), therefore it is thinner because the skin formation obeys quite another mechanism: condensation of the PBA vapour released as the composition foams.

Hence, equal thickness and density of the skin can be achieved if at the first stage of producing the integral structure, i.e. at skin formation, the surface of the mold cavity is covered with a layer of the starting composition. This, in turn, can be achieved either by rotating the mold or by spraying the initial composition onto the inner surface of the mold cavity.

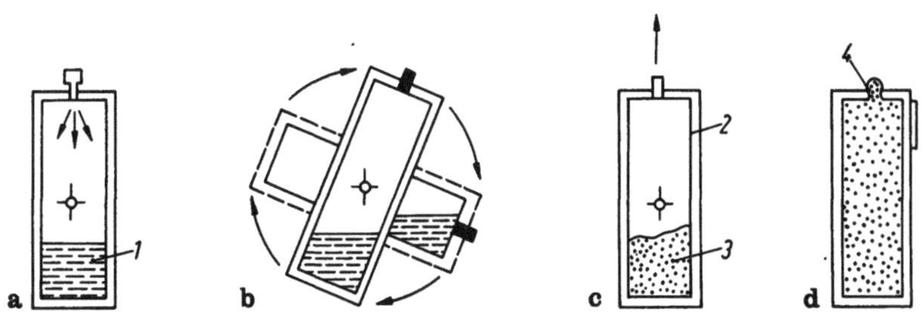

Fig. 10.1 a–d. Schematic diagram of IF production by the rotational molding process; **a**: mold filling, **b**: mold rotation, **c**: foaming and degassing, **d**: curing; 1: liquid composition, 2: surface skin, 3: core of an IF part, 4: excess foam [2]

The main idea of rotational molding is to rotate the mold around one axis, or two mutually perpendicular axes, for a short time, just after the starting composition has been injected (Fig. 10.1) [2]. Core foaming takes place when the rotation is discontinued.

One important aspect is that the method considerably reduces the molding pressure needed – down to 0.02–0.1 MPa – due to a lower required PBA concentration. Since the skin formed during rotation prevents the composition from flowing out of the mold, it does not need any additional fittings such as sealers or locking devices. Despite the higher costs of rotational machines (by a factor of 1.5 to 2) compared to those for molding, the cost of an article from a rotational machine is about 30% lower than a conventionally molded one [3]. This is due to the lower molding pressure, feedstocks, and labor costs, as well as to the far shorter molding times and higher machine capacities. For example, an integral foam 75 × 40 × 2 cm panel is produced on a rotational six-mold machine in 45 seconds, whereas the same panel would take 232 seconds (with an equal cooling time of 210 sec) on a low-pressure molding machine. The working capacity of a low-pressure machine is far lower than that of a rotational machine – 15.5 cycles (or 31 articles per hour) compared to 80 cycles (or 160 articles per hour) [4–7].

The advantage of the rotational molding method is particularly evident when producing articles with low or moderate average densities (up to 500 kg/m^3). For heavier articles, the stationary methods, giving high quality surfaces by high-pressure methods, are better.

10.2 Commercial Rotational Molding Processes

10.2.1 The Uniroyal Process

Escalating resin and freon costs and tight supplies are making IF rotomolding with chemical blowing agents attractive for large, lightweight parts. However, according to Kravitz and Heck [5,7] this process requires a level of technological sophistication not always demanded by traditional rotational molding techniques. Far more testing and attention to mold design, along with careful attention to CBA dispersion are required.

Table 10.1. Rotational molding process: effect of chemical blowing agent on density reduction and wall thickness of a PE part[a]

CBA level, %	Wall thickness, mm	Specific gravity, kg/m^3	Density reduction, %	Increase in wall thickness, %
None (unfoamed)	3.5	931	–	–
0.2	6.0	639	32	42
0.5	7.8	451	52	56
0.8	10.8	373	60	68
1.0	13.0	310	68	73

[a] 305 × 305 × 152 mm box, resin: MDPE, CBA: Azodicarbonamide

By carefully matching the CBA and polymer (PE or PVC), and by using optimum processing conditions, the properties of a rotomolded part may be expected to be comparable to those obtained by other IF processes. Since CBA's are usually finely powdered, they can be easily blended with the powdered resins commonly used in rotomolding. The most typical method of incorporating them is by tumbling the resin and CBA together in dry. Generally, a minimum tumbling time of 15 min ensures a good, consistent mix. The amount of CBA used should be within the 0.5–1 % range, and 1 % has been found to provide density reductions of as much as 68 % (Table 10.1) [5].

This principle is realized in two Uniroyal processes: the solid outer wall/foamed inner wall, or *two-drop*, process; and the solid/foam/solid, or *three-drop*, process. The first process produces "one-sided" (one skin, one core) IF articles, whereas the second produces classical "*two-sided*" (*two skins, one core*) IF's.

The *Uniroyal process* is shown schematically in Fig. 10.2. The mixture of the CBA (Celogen OT, or Celogen AZ, see Table 2.1) and the polymer (HDPE, MDPE) is melted and fused to the mold during the oven cycle. Resin foaming takes place during the air cycle (Fig. 10.2b) followed by water cooling (Fig. 10.2c) to arrest the expansion process. The air cooling time is important for rotationally molded IF parts. Foaming of the polymer takes place at this stage and by lengthening or shortening the air cycle, the degree of foaming (wall thickness, density reduction) may be controlled. Too short an air cycle may lead to little or no foam being formed, while extreme overblow and possibly "burning" of the polymer will occur if the cycle

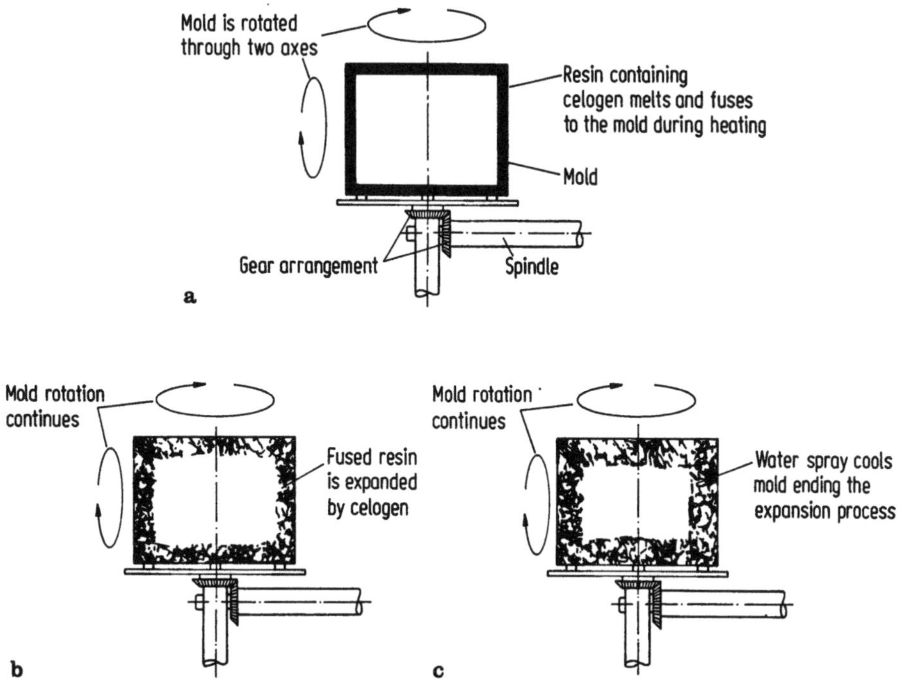

Fig. 10.2. Schematic diagram of the rotamolding Uniroyal process (Courtesy of Uniroyal Chemicals, Uniroyal Inc., Naugatuck, USA) [6]

Table 10.2. Time/Temperature effects on the foam core of a HDPE part [5)]

CBA level	Oven temperature, °C	Oven time, min	Results
1% OBSH[a]	246	10	Good inside skin, limited foaming
1% OBSH	246	12	Good inside skin, good foam
1% OBSH	246	14	Fair inside skin, good foam
1% ACA[b]	260	10	Good inside skin, little foam
1% ACA	260	12	Good inside skin, good foam
1% ACA	260	14	Poor inside skin, overblown foam with large cells

[a] p,p'-oxybis(benzenesulfonyl) hydrazide, [b] azodicarbonamide

is too long. Other factors such as a bad choice of the CBA, or extremes in oven residence time and temperature may also contribute to these defects (Table 10.2) [5)].

It is recommended that the conditions for rotomolding a solid article be used as the starting point when attempting foam molding, although some adjustments in residence time and/or temperature may be necessary.

A water cooling cycle which is too short will give a product which is difficult to remove from the mold. Some loss in density reduction may also occur since improperly cooled parts may collapse when removed from the mold. Experience has shown that rotomolded foams require larger amounts of mold release agents to make removal from the mold easier. The mold release agents normally used have been found to be effective, but generally 0.5 to 2 times the amount used for unfoamed pieces is needed [6)].

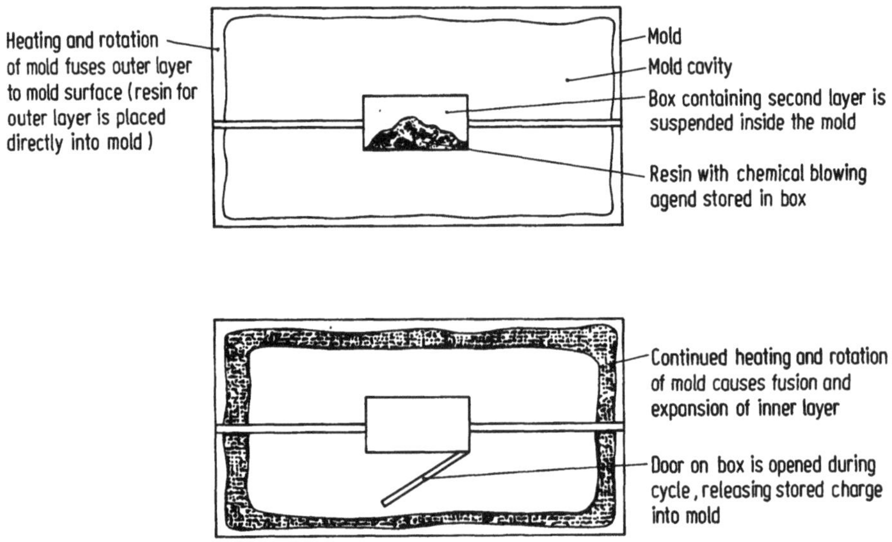

Fig. 10.3. Schematic diagram of the rotamolding Uniroyal process for producing multiple-layer IF parts (Courtesy of Uniroyal Chemicals, Uniroyal Inc., Naugatuck, USA) [6)]

In the two-drop process the virgin resin is fed to the rotational mold and fused; the second layer (resin mixed with CBA) is placed inside an insulated box, which is suspended inside the mold. After the first layer of unfoamed polymer has been distributed and partially fused, the box is opended, allowing the second charge to form a foam layer on the inner surface of the solid layer. The first layer may be crosslinked PE, the foamed layer HDPE (see Chapter 20).

The three-drop process is based on the same procedure and equipment. It should be noted that the multi-drop process can use different but compatible resins to maximize cost/performance while taking advantage of the properties offered by each resin (see Chapter 7 above).

A rotationally molded IF part can be produced in the same equipment used for unfoamed parts. When using a multiple drop process, however, it is necessary to modify the mold such that the second charge can be stored inside the mold and released during the cycle (Fig. 10.3). The mold should be vented to allow any excess gas generated by the CBA to escape, thus avoiding a pressure buildup which could possibly cause stress within the molded part. The mold must be balanced with respect to local hot or cold spots because these may cause variations in wall thickness as well as radical variations in density reduction [6)].

The *economic advantages* of this process stem from the fact that it uses conventional equipment for producing unfoamed articles, minor mold modifications (for the multiple-drop process), and rather cheap CBA's [7)].

10.2.2 The Borg-Warner Process

Rotational molding is especially convenient for fabricating large-sized (up to 50 kg) and built-up parts that cannot be produced by IM, e.g. furniture and automobile parts, boat hulls, etc. The Borg-Warner Corporation (USA) [8,9)] produces integral structures with the following process: A dried polymer is fed into the vibrating cavity of an aluminum mold heated either by steam up to 190–235 °C, a high-temperature furnace, or a circulating heat carrier. The molding cycle duration depends upon article thickness (6.3 to 12.7 mm) and the heating method and varies between 10 and 50 min. The apparent density of the articles is 400 to 800 kg/m^3.

10.2.3 Other Processes

A rotational molding process for producing IF's based on the different thermophysical properties of variously sized polymeric grains is especially interesting. Thus, if 0.15 and 2 mm PE grains, whose ratio of specific thermal capacities is 1:50, are placed in a heated mold together with a CBA, they behave differently. The coarse grains do not foam and form the surface skin, whilst the fine grains constitute the core of the article [10)].

Rotational molding can also be used to produce *multicomponent IF's* based on thermoplastics and oligomers. A process has been developed to fabricate articles with the core made of PS and the skin made of PVC in a single cycle. As the mold rotates, the PVC melt is pressed against the walls of the mold and is fused to it, whereas the granulated PS foams. By varying the temperature and volume of mold filling, very light IF's can be fabricated with an apparent density as low as 80 kg/m³. Possible skin materials include PE, latexes, and cellulose acetate butyrate [9,10].

A recent achievement using this method is the *production of tires* in a single cycle without requiring air pumping. The external surface is made of corrugated rubber, and the interior is filled with elastic PUR [10].

10.2.4 Technical and Economic Analysis

A detailed economic analysis of rotomolding IF processes has been carried out by Seifert [4]. He used single-station and six-station rotary machines as the basis of comparison and an IF furniture panel with the following parameters: length 762 mm, width 406 mm, thickness 17.5 mm, weight 2.8 kg, density 850 kg/m³, and cooling time 210 sec. It was found that the panel cools in 210 sec without any warpage after demolding. This may not be the minimum cooling time, but it was used as the basis for the analysis:

	Single Station	Six Station
Cooling time	210 sec	210 sec
Cycle time	232 sec	45 sec
Mold strip time	15 sec	35 sec
Cycles per hour	15.5	80
Parts per hour	31	180

This comparison demonstrates that a six station rotary machine will outperform a single-station machine by a margin of over five to one; the cooling time dictates the cycle time and therefore the output of a machine.

For his cost analysis Seifert [4] used a machine hour rate which was derived from the following data:

	Three Shift Yearly Cost, $	Three Shift Hourly Cost, $
Machine cost	44,200.00	8.85
Maintenance (10% of yearly cost)	4,420.00	0.88
Operation cost	5,000.00	1.00
Floor space	1,500.00	0.30
Three operators	21,000.00	3.50
Burden rate (125% direct labor)	26,250.00	4.38
Add 5% contingency	5,118.50	0.95
Total	$ 107,488.50	$ 19.85

A breakeven rate for the described machine is $ 19.85 per hour. Based on the output analysis for the furniture panel, for a cooling time of 210 sec at 80 cycles per hour and 2 cavities per mold, we find an output of 160 parts per hour, and a cost per part of $ 0.124. This cost does not include material and mold costs. Seifert did not compare the molding cost per part of $ 0.124 for the six station machine against the molding cost on a single station machine, because there are too many machines with different price ranges on the market.

On the other hand, given an estimated 232 sec button to button time for a single station machine, we come to 15.5 shots per hour, or 31 parts per hour using a two cavity mold. The operation cost alone is 11.2 cents per part, and this cost does not include machine costs, overhead, power consumption, maintenance, floor space or contingency costs.

Summing up this analysis, it can be said that multi-station rotary machines are important for the IF industry because they increase output, bring the work to the operator, pace the operator, have very low power consumption, can be easily incorporated into a production center in which molding, inspection, packing and shipping can be handled right at the press because of the high output rate [4, 9].

10.3 Free-Foaming Process

Integral foams can also be produced using conventional molding processes for pressed foams. A composition consisting of granulated polymer, CBA, and other admixtures is fed into a cold press-mold. This is then heated (generally by steam) and then again cooled (by water). Foaming proceeds at an extra pressure of 0.15–1.0 MPa; the foaming time is 3 to 15 sec, and the cooling time ranges from 20 to 60 sec [11]. The price of an IF fabricated by this method is about ten times lower than that using IM methods. The weaknesses of the method are low quality surfaces, i.e. roughness and nonuniform thickness, and low productivity.

10.4 Conventional Molding Process

Free foaming (in an open mold cavity) or foaming in a closed cavity may, in principle, be used to produce integral PO, PS, ABS and PUR foams. As in conventional molding, compositions are fed into cold molds and the foaming takes place either in externally heated molds (for thermoplastics) or using the heat evolved in the exothermic curing reaction (for oligomers). Note that compositions intended for IF production by free foaming differ somewhat from those designed for ordinary foams. Integral foam compositions contain more BA and have higher melt or solution viscosities.

Although the method can produce large-sized articles with very complicated configurations, it is not widely applied as yet because of the low quality of the articles, rather than because of the long molding cycle (up to 50 min). The point is that materials produced by free foaming and conventional molding are, strictly speaking, *pseudo-integral* polymer foams rather than true integral foams, because their skin densities are only slightly different (by a factor of 2 to 3) from those of the core. Moreover, the compact skin layer has a porous surface of uneven thickness. For these reasons, such materials are mainly utilized as thermal insulators and in the fabrication of air and liquid filters [12–14].

If the IF is not to be exposed to large mechanical stresses, it can be used as an intermediate product, for instance laminated with other materials. Such low-quality articles are produced by conventional mold pressing. Such integral foams based on PP are made in England[12]. A cold pressmold is filled with a mixture of a powdered polymer and CBA, and held at the low temperature for 2 to 15 minutes (depending upon the thickness of an article), then it is heated to 400–450 °C and cooled. The articles produced have $\varrho = 550\ kg/m^3$. In another process [13] an integral PUR is made from a liquid oligomer composition containing freons as the PBA. This is fed into a gasproof mold cavity whose pressure rapidly builds up to 0.3–0.4 MPa. The process yields large-sized articles (ϱ = 300–850 kg/m^3) with a compact skin but of irregular thickness.

10.5 References

1. Shibata, O. et al.: Japan Plast. Age *11*, No 2, 15 (1973)
2. N.N.: Plast. Eng. *34*, No 7, 56 (1978)
3. Gross, L. H., Angell, R. G.: Plast. Technol. *22*, No 5, 33 (1976)
4. Seifert, S.: Plast. Eng. *31*, No 9, 33 (1975)
5. Kravitz, H., Heck, R.: Plast. Technol. *25*, No 11, 63 (1979)
6. Celogen Rotational Molding: Technical Bulletin of Uniroyal Chemical, Division of Uniroyal, Inc., Naugatuck, USA, ASP-5152 (1982)
7. Heck, R. L.: SPE ANTEC, San Francisco USA, 1982; Plast. Eng. *38*, No 11, 37 (1982)
8. Seymour, R. B.: Polym. News *4*, No 4, 150 (1978)
9. Blue, E. B.: Plast. Technol. *28*, No 7, 175 (1982)
10. Berlin, A. A., Shutov, F. A.: Strengthened Gas-Filled Polymers. Moscow: Khimia, 1980 (in Russian)
11. Szopa, J.: Chemik, *26*, No 6, 221 (1973)
12. Harris, W. D.: Plast. Eng. *32*, No 5, 26 (1976)
13. Hossinger, M.: Plaste Kautsch. *24*, 581 (1977)
14. Shutov, F. A.: Intern. Symposium "Polyurethane-83", London England, 1983

11 Secondary Processing

Before our short discussion of the secondary processing of IF's, it should be mentioned that the techniques (finishing, machining, sanding, bonding, and fastening) are similar to those used on unfoamed and non-integral foamed plastics. However, secondary finishing for IF's is in some ways specific because of the large size, complex shape and imperfect surface of most of the parts [1,2].

Not all methods of producing IF's yield articles whose appearance, surface quality, and color, are completely satisfactory. The most rational and economical way of removing appearance and quality defects is, obviously, to improve the production methods, i.e. the primary processing. If, however, this does not ensure the desired quality, the IF can be improved either by special mold preparation (pre-molding operations) or by finishing the molding (post-molding techniques).

11.1 Pre-Molding Processes

In order to prevent surface roughness, swirls, sags, shrinkage cavities, and other surface defects, pre-molding treatments can be used. These include coating of the inner surface of the mold with fusible polymer films which smooth out (fuse) the defects arising when the material is being formed, and *in situ* lacquering and/or dyeing by pre-coating (before the mold is charged with the composition) the inner surface of the mold with lacquering of dyeing compositions (with subsequent drying, curing, etc.). These materials should adhere poorly to the metal, but well to the surface of the foam produced.

An excellent example of this technology is the Inmold-Coating process that was developed by Desma-Werke (West Germany). It was designed for producing instrumental panels made from PUR integral foam [3].

Orlando [4] reported the production of furniture from an IF based on polyurethane resin by Pozzi-Arosio (Italy). This company developed a special polyisocyanate-based composition for pre-treating the inner surface of the mold with a 0.5–6 mm layer. After spraying, the layer is fully cured in 20–30 sec, at 30–40 °C. After demolding the furniture does not require any post-molding surface operation.

11.2 Post-Molding Processes

11.2.1 Training

After demolding, a raw IF part will be suitably finished for its final application. First, the flash and the attached gate must be removed. The release agent that had been applied to the mold surface before the foaming, and which is partially transferred to the molding, must then be removed. To achieve smooth self-release, a smooth, pore-free metal surface and special "inner" release agents should be used. Fairly simple part design is also necessary.

It has been pointed out [5] that prior to painting, some skins require the following surface preparation: hand deflashing, two washing operations with a mild cleaning solvent to remove sanding and epoxy resin remains that fill voids on the surfaces, application of a base coat of urethane primer, sanding and refilling some voided areas after drying, a second coat of urethane primer, resanding after drying, and the final application of the required finish or texturing coat.

To improve the appearance and to prepare the surface for subsequent finishing operations, freshly produced IF articles are blown with air to take crumbs away, and are mechanically polished to get rid of swellings and cavities [6]. To obtain very smooth surfaces, IF's can be treated by open flame or electric (corona) discharge [7]. The most widespread commercial way of removing surface defects, however, is surface fusion (after the cooling cycle) in the mold, raising the temperature to 80–100 °C [8].

11.2.2 Painting and Coating

Painting is the most widely used method of finishing IF parts. Where high quality paint finishes are required, it is necessary for the surface to be prepared with priming, followed by high-build base coats and possibly sanding, before the final finishing coat. These are expensive operations and considerable savings can be achieved by improving the surface appearance of the moldings before finishing. Norgan [9] pointed out that recent data show that higher equipment cost, or more costly and longer molding cycles, can be more than offset by the attendant reductions in finishing costs. Much work is therefore currently being devoted to improving surface finish by modifying the polymers and additives, the molds, and the molding techniques. These efforts may be summarized as follows [9]:

Polymer modification and additives: higher flow materials, addition of fillers, fluorocarbon blowing agents, and cell control additives.

Processing techniques: increased density, increased mold temperature and injection speed, mold surface texturing, and modified or alternative processes.

LaPlaca [10] conducted two molding trials to determine whether PS integral foam parts could meet aesthetic requirements for a top cover. The first trial, using a high-pressure IM process (Farrel-USM), showed that paint finishing on the expanded-mold cover was satisfactorily completed half an hour after demolding, without any surface preparation. In the second trial the top cover was produced by a low-pressure nitrogen IF process, and the processing results were comparable to the earlier trial. However, the major difference was that the material continued to outgas after painting. Parts treated with a two-paint-coat system (color and texture) showed surface blemishes and blistering 72 hours after demolding. The same parts were painted with three coats (one a prime filler) in an effort to correct the problem.

It is recommended that the lacquering and dyeing of IF articles is done not earlier than two to three days after they have been molded; over this period most of the foaming gas is substituted by air. If done earlier, surface cracks, swellings, etc., i.e. secondary surface defects may result. Finished articles are generally lacquered and dyed using hot-cure compositions based on polyurethane, acrylics, or polyester [11]. The quality and durability of the color of a finished IF article is dictated, to a great extent, by the chemical nature of the polymer constituting the skin. Thus, articles based on PS, ABS-plastics, PC, dye better than those based on PE and polyurethane [7]. The lower the density of the skin and the higher its quality, the deeper and more uniform is the coloring. The most popular methods of dyeing are cold-dip, hot-dip, brushing, swabing, and spraying.

Because rigid PUR tends to yellow under UV-light, it should be protected with an *appropriate coating* or pigmented black or very dark brown. Especially those IF parts which will be exposed to the weather must be given an appropriate coating. There are a variety of chemically different coatings which are suitable for rigid PUR parts. However past experience has shown that PUR coatings of 100–200 μm provide the best surfaces [12].

Bayflex IF's, like all PUR parts based on aromatic isocyanates, yellow particularly strongly when exposed to UV-light. This drawback requires very dark or black pigmentation for unpainted parts. These pigments are applied as pastes, in inert carriers or polyols, and normally blended into the polyol component of the Bayflex system. Coloring can also be achieved by directly injecting the pigment paste into the mixing chamber of high or low pressure machines. Recently, however, light, stable, and non-yellowing *Bayflex parts* have been successfully produced using aliphatic isocyanates, permitting lightly colored IF parts to be produced.

On the other hand, light colors, especially pastel shades and metallic effects, can also be obtained by *painting*. For flexible PUR foams, the flexibility of the coating must match the flexibility of the substrate for proper functioning. Polyurethane coatings are especially well suited to this purpose and can be applied directly or via "transfer-coating". This latter method involves substituting the mold release agent by a paint or coat which is sprayed onto the mold surface and serves as the release barrier. Normally, these agents are primers and require subsequent application of a top coating. The conventional, direct coating technology requires washing or vapor degreasing in order to remove the release agents before painting [13].

Table 11.1. Coating problems, causes and solutions, encountered in the finishing of thermoplastic IF articles for buisiness-machine applications [18]

Problem	Probable cause	Solution
Swirls	Compound or mold surface too hot	Adhere to recommended melt temperatures for the material; control mold temperatures carefully; use high-volume solid primer filler and textured coating
Pinhole craters	Solvent attack with accompanying release of gas	Use only mild solvent or water borne primers
Blisters	Absorption of solvent with subsequent release of gas	Air dry, prebake, or store foam part at room temperature for 72 hours before coating
Bubbles in substrate	Inadequate filling of mold; inadequate foaming of part	Fill mold as fast as possible keeping careful control of material-fill levels
Adhesion with coating	Too much mold release	Use only small amounts of mold-release agents
Wicking	Mold too hot; material too viscous; solvent too active	Adhere strictly to recomended melt temperature for material; keep careful check of mold temperature through the careful placement of gates; use only small amounts of mold release

Textured IF surfaces are either formed in the molding process itself (see the foregoing discussion) or by secondary processing such as spraying, machining, thermal treatment, etc.

Some typical problems and their possible solution, encountered with secondary processing, are presented in Table 11.1.

11.2.3 Modification of Properties

Finishing can also be employed for functional reasons: to improve weatherability, to promote adhesion for metallizing, to impart electrical conductivity, to improve chemical and abrasion resistance, etc.

In order to enhance their strength properties, finished integral foam parts may be reinforced with glass fiber and metal wire, in the course of secondary processing. In some cases, a number of IF sheets may be jointed together by (threaded) metal rods. To make the skin more rugged and reduce its coefficient of thermal expansion, it can be impregnated or sized with compositions containing glass fiber. For better rigidity, IF articles may have stiffening ribs on their surface (see Chapter 22)[14]. Articles with complicated configurations are manufactured by gluing[15]; nailing and bolting[16], welding in a high-frequency electromagnetic field, fusing, etc.[17].

11.2.4 Electromagnetic Interference Shielding

The problem in today's business machine and medical instruments housing market is electromagnetic compatibility. Not a problem with the old metal housings, the nonconductive plastics being used now provide no inherent shielding from electromagnetic interference (EMI) and radio frequency interference (RFI). Electrostatic discharge and dissipation (ESD) is also a problem: static builds up on the polymeric surface and may discharge into expensive electronic components[18,19].

These problems are so important that in USA the Federal Communications Commission (FCC) requires that all electronic enclosures be properly shielded to ensure that EMI is not transmitted from the unit[20].

EMI and RFI are "electronic noise" and may be internally or externally generated. The "noise" is natural or man-made and can wipe out computer memories or cause arithmetic errors in the same way as power fluctuations and failure influence computers; shielding is required to guard against such catastrophies[18].

There have been many attempts to solve these problems, the most effective being the coating of the surface of the plastic with a metal, by vacuum metallizing, sputtering, zinc arc spraying, or with conductive nickel/acrylic or urethane coating. A detailed analysis of the EMI-RFI problem for IF parts was presented by Storms[18].

11.3 Pseudo-Integral Polymer Foams

To conclude this section, we shall give some thought to several methods for IF production by secondary processing of non-integral foams. More properly, these materials should be classified as pseudo-integral foams, because the molding process includes several stages (rather than one), and because their macrostructure is characterized by a sudden (not a gradual) change in density from the core to the skin.

A promising and very cheap version of pseudo-integral foam technology is the surface impregnation of non-integral foams with hot high-polymer melts or cold-curing oligomer solutions using equipment that is conventionally used for the impregnation of wood. It has been demonstrated [21,22] that *sandwich-like plastic* materials can be obtained, with very smooth external surfaces and good mechanical properties.

Pseudo-IF's based on thermoplastic polymers can also be fabricated from classical (nonintegral) foams by compressing them. For instance, 1.5–2.5 cm thick PE foam sheets ($\varrho = 30\ kg/m^3$) have been compressed at 170 °C and then cooled to 20 °C while still under compression. The tensile strenght of the resulting material is twelve times that of the starting foam ($4.6 \cdot 10^5$ Pa · versus $3.7 \cdot 10^4$ Pa), and the ultimate elongation value increases from 115 to 140%. Another way of producing pseudo-integral structures is to melt the surface of a thermoplastic isotropic foam by means of heated metal plates at 250–280 °C. The plates are then rapidly cooled [23]. This method provides a dense and smooth surface but can only be applied to articles with a simple geometry.

11.4 Technical and Economic Analysis

Finishing costs tend to be lower for polyurethane IF's than for thermoplastic IF's for two reasons. Firstly, paintable release agents, both internal and external, can be introduced into molds for polyurethanes, and secondly, the labor-intensive filler step can be eliminated because of the excellent surface quality of polyurethane IF's. In most instances, IF parts can be primed in the as-is molded condition without reworking the surface. Painting is still the preferred method of finishing, and pigments of dark color such as brown, brown-black, or black have been used successfully. Ultraviolet discoloration, formerly unique to polyurethanes, is also a problem for thermoplastic IF's containing high levels of flame retardants [24–26].

According to Wendle [19], a conductive coating can boost the cost of a surface treatment by 26%, and with masking, which is often required, the cost can be even higher.

A typical cost breakdown per 929 cm^2 (1 sq. ft) of a 6.35 mm (0.250 in) thick, modified PPO (Noryl) integral foam with conductive coating is as follows [19]

Production costs	Without conductive coating, %	With conductive coating, %
raw material	31	24
molding costs	7	5
general service and administration	8	7
finish (exterior)	39	17
conductive finish (interior)	—	34
margin of profit	15	13
	100	100

Note here, that EMI protection of IF parts with conductive fibers and flakes (see Chapter 2) have not caught on because of their high cost and difficulties in processing.

Table 11.2. Comparison of finishing processes for integral foams and metals [24]

Finishing process	Integral foam	Aluminum		Steel	
		Cast	Sheet	Cast	Sheet
Surface preparation	–	+	+	+	+
Vapor degrase	–	+	+	+	+
Rinse	–	–	+	–	+
Phosphate treatment	–	–	–	–	+
Drying oven	–	–	–	–	+
Anodize	–	–	+	–	–
Primer	–	+	+[a]	+	+[a]
Filler coat	+	+	–	+	–
Color coat	+	+	+	+	+
Texture	+	+	+	+	+

[a] For exterior use, a prime coat is necessary

The finishing cost as part of the overall economics of IF's will be discussed in Chapter 12. We will, however, compare here the plastic foams with such traditional materials as cast and sheet metals (Table 11.2). The metals require surface preparation (removal of flash, surface grinding and filling of porous areas), vapor degreasing (for cleaning from oil and lubricants), followed by a wash primer. Moreover, since cast metals are porous, two or three coatings of paint must be applied, similar to what is used for IF parts. Thus, the finishing of cast metals requires twice as many steps, and hence more labor, factory space and equipment than is necessary for IF finishing. The number of finishing steps for sheet metals are three times those required for IF's.

It goes without saying that the availability of so many methods for secondary IF processing is evidence of the imperfection of some of the technologies which cannot, as yet, produce articles ready for immediate use. Therefore, one of the urgent problems facing integral foam physical chemists and technologists is the reduction of secondary processing and, in the long run, its complete elimination [26].

11.5 References

1. Sneller, J.: Mod. Plast. Int. *9*, No 10, 54 (1979)
2. Duchane, D.: Engineering, Sept., 720 (1981)
3. Method Inmold-Coating: Techn. Bull. Desma-Werke Hermeskeil GmbH, Hermeskeil, FRG, 1982
4. Orlando, O.: Poliplast. plast. reinfor. *29*, No 282, 70 (1981); Riv. Color. *14*, No 154, 47 (1981)
5. LaPlaca, J. P.: J. Cell. Plast. *16*, 36 (1980)
6. N.N.: Plast. World *36*, No 2, 56 (1978)
7. Thompson, W. R.: Plast. Technol. *21*, No 1, 42 (1975)
8. N. N.: Plast. World *38*, No 7, 38 (1980); *41*, No 12, 94 (1983)
9. Norgan, M. R.: Call. Polymers *1*, No 2, 161 (1982)
10. LaPlaca, J. P.: Plast. Design Forum *5*, No 1, 73 (1980)
11. Menges, G., Schweisig, H.: Chem. Eng. (London) No 330, 194 (1978)
12. N.N.: Kunst. J. *13*, No 3, 24 (1979)
13. N.N.: Europ. Plast. News *7*, No 4, 29 (1980)

14. Gross, L. H., Angell, R. G.: Plast. Technol. *22*, No 5, 33 (1976)
15. Nicolay, A., et al.: Gummi-Asbest-Kunst. *30*, No 1, 37 (1977)
16. Zwolinski, L. M.: J. Cell. Plast. *12*, 34 (1976)
17. Coniglio, J. J.: 7th Structural Foam Conf., Norfolk USA, 1979
18. Storms, C. D.: Plast. Eng. *36*, No 8, 36 (1980); 10th Structural Foam Conf., New Orleans USA, 1982
19. Wendle, B. C.: Intern. Conf. Foamed Plastics, Düsseldorf FRG, 1983
20. Colangelo, M.: Plast. Technology *29*, No 6, 41 (1983)
21. Shutov, F. A.: Conf. Reactive Oligomers, Alma-Ata USSR, 1979
22. Shutov, F. A.: USSR Conf. Gas-Filled Plastics, Suzdal USSR, 1982
23. Throne, J. L.: Polym. Eng. Sci. *17*, 682 (1977)
24. Reinhard, D. L.: Post Molding Operations. In: Engineering Guide to Structural Foams. Wendle, B. C. (ed.). Westport: Technomic, 1976, pp. 52–65
25. Horvath, M.: J. Cell. Plast. *12*, 289 (1976)
26. Shutov, F. A.: Intern. Symposium on Plastics on Building, Liège Belgium, 1984

12 Comparison and Selection of Integral Foam Processes

12.1 Comparison of Processes

At the beginning of our discussion, several points must be mentioned. There is, on the one hand, the company that has already had experience with plastic production, and on the other hand, the producer of non-plastic parts or the consumer of plastic (or non-plastic) parts. Plastic companies are almost always interested in the technical aspects of an IF process, and, of course, in cost comparisons with similar parts made from unfoamed or foamed (but non-integral) plastics. Non-plastic producers and consumers are mainly interested in cost comparisons with wood, metal or concrete solutions to their problems [1].

Let us try to answer both types of enquirer.

In order to compare different IF molding processes, we shall consider those aspects of an IF article that are obviously affected by process variations. The most detailed and informative comparisons have been made by Eckardt[2], Semerdjiev[3], and Throne[4], and we shall follow their data.

12.1.1 Process Parameters

The basic factors influencing end product quality are different for the various IF processes. Table 12.1 gives a schematic résumé of these factors and their effect on product quality.

12.1.2 Density Reductions

Let us consider the minimum obtainable IF densities; this parameter is frequently used to illustrate the differences in foamed plastic processes. As it can be seen in Fig. 12.1, the density profiles across IF's produced by different IM processes are very different. The ICI and Hanning processes, i.e. two-component foams (TCF), yield a skin thickness that is controlled by the ratio of foamed to unfoamed plastic. There is a practical limitation to the thickness of the skin that depends on the rheological matching of the materials and the flow length within the mold [4]. Note that a gas counter pressure (GCP) process, such as the Allied, Bulgarian, or Dow-TAF processes, produces a skin nearly as thick as that from a TCF process. The skin thickness can be controlled by the gas pressure (inside the mold), or by mold movement. Figure 12.1 shows also that low pressure foaming of an amorphous polymer (5a), as produced by the UCC process, results in a gradual decrease of foam density toward the core. A very distinct but somewhat thinner skin is formed with a crystalline polymer (5b).

Table 12.1. Effect of various processing parameters on end product quality of integral foams, according to Eckardt [2)]

Injection molding process			Gas counter-pressure process			Two-component process		
Injection time	short	long	Injection time	short	long	Melt viscosity	high	low
	↓	↓		↓	↓		↓	↓
Density	low	high	Density	low	high	Proportion of core material	low	high
Injection time	short	long	Injection time	short	long	Injection time	short	long
	↓	↓		↓	↓		↓	↓
Surface finish	smooth	rough	Weld lines	slight	marked	Density	low	high
Melt temperat.	high	low	Mold temperat.	low	high	Injection time	short	long
	↓	↓		↓	↓		↓	↓
Surface finish	smooth	rough	Weld lines	marked	slight	Surface finish	glossy	matt
Mold temperat.	high	low	Tightness	good	poor	Mold temperat.	high	low
	↓	↓		↓	↓		↓	↓
Surface finish	smooth	rough	Surface streaks	none	marked	Surface finish	glossy	matt
Venting	poor	good	Gas pressure	low	high	Section thickness	thick	thin
	↓	↓		↓	↓		↓	↓
Density	high	low	Surface streaks	marked	none	Density	low	high

It is evident, however, that a thermoplastic foamed article of uniform density (i.e. non-integral polymer foam) cannot be produced by injection molding into a cold mold, as IF parts must have some surface skin.

On the other hand, the thinner the skin, the less it contributes to the bulk density of the IF (Fig. 12.2). TCF processes produce the thickest skins. According to recent data, usable TCF articles may have no more than 25% foam. The LPF processes consistently yield articles having the lowest densities, with useful articles having as much as 50% foam; this is very important for the economics of IF parts. It is apparent,

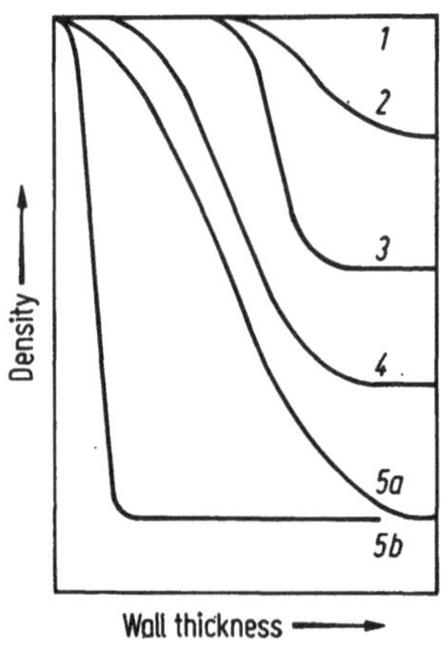

Fig. 12.1. Schematic diagram of the density profiles of IF parts produced by various IF processes; 1: conventional unfoamed injection molding, 2: gas-counter pressure injection molding without foam egression, 3: two-component injection molding, 4: gas counter pressure injection molding with foam egression (TM-process), 5a: low pressure injection molding of amorphous polymers, 5b: low pressure injection molding of crystalline polymers (Courtesy of the Society of Plastics Engineers, Inc. Brookfield Center, Conn., USA)[13]

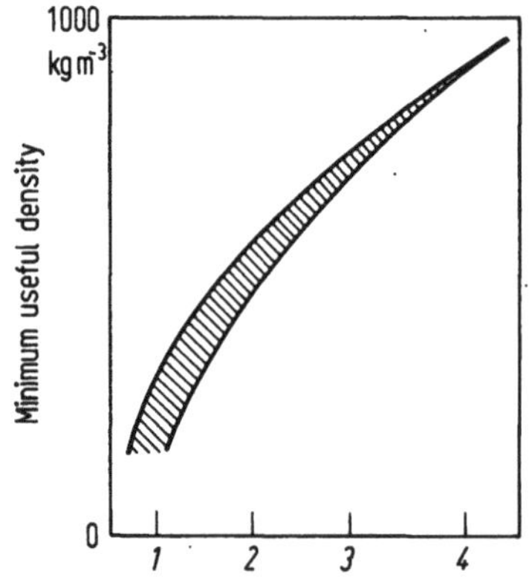

Fig. 12.2. Schematic diagram of typical minimum attainable density for the various IF processes; 1: low pressure injection molding, 2: gas counter pressure injection molding, 3: two-component injection molding, 4: conventional unfoamed injection molding (Courtesy of Technomic Publishing Company, Inc., Westport, Conn., USA)[4]

Table 12.2. Average density of integral foams produced by different molding processes, expressed as a percentage of raw (unfoamed) material density

Polymer	Unfoamed, %	Injection molding, %	Gas counter-pressure, %	Two component, %
PS	100	80	95	90
PE	100	75	85	80
PP	100	75	85	80
ABS	100	80	95	90
PPO mod.	100	85	95	90
PC	100	75	95	90

therefore, that if low density and high rigidity are important, and surface finish is less important, a LPF process offers a greater potential for cost savings than a TCF process.

Differences in raw material costs result mainly from the different densities, which can be achieved by the different techniques. In Table 12.2 the reduction of density and consequently of weight is shown as a percentage of the density of raw polymer. Apart from the reduction in weight, the multi-component process can considerably reduce the raw material costs, in some cases, by using a cheap polymer for the core (or the skin), such as plastic scrap.

12.1.3 Surface Roughness

Surface roughness can differ considerably for different IM processes, as indicated schematically in Fig. 12.3. It has been shown by Throne [4] and by Semerdjiev [3]

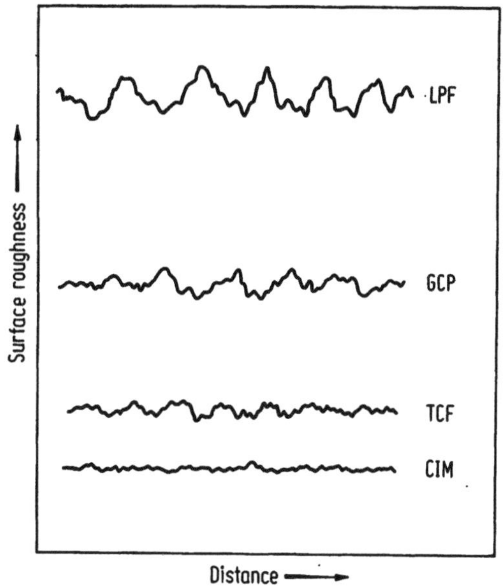

Fig. 12.3. Schematic diagram of surface roughness for various IF processes (Courtesy of Technomic Publishing Company, Inc., Westport, Conn., USA) [4]

Table 12.3. Comparative calculation[a] for a furniture carcass made from integral foam based on polystyrene, with a gloss finish (first number) or a textured finish (second number)

Production method	Raw material cost %	Production cost with raw material cost[b] %	Finishing cost %	Tooling cost %	End product cost with mold amortization for production runs of 20,000 %	50,000 %	100,000 %	400,000 %
Injection molding	100/100	100/100	100/100	100/100	100/100	100/100	100/100	100/100
Gas counter-pressure	104/104	104/104	62/94	117/117	91/106	87/104	85/103	84/102
2-component (virgin material)	97/97	115/115	62/94	100/100	91/106	90/108	89/108	89/108
2-component (virgin/regrind)	81/81	105/105	62/94	100/100	87/103	85/103	85/103	84/103
Reaction injection molding	124/–	119/–	100/–	76/–	109/–	110/–	111/–	111/–

[a] based on an integral foam injection molding process = 100%;
[b] excluding mold amortization

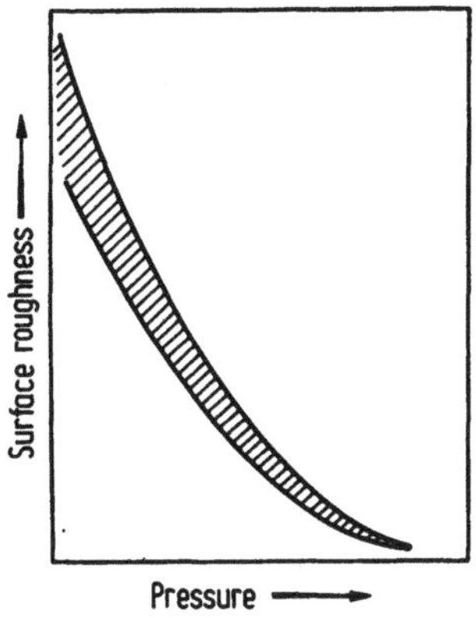

Fig. 12.4. Schematic diagram of the effect of pressure on surface roughness (Courtesy of Technomic Publishing Company, Inc., Westport, Conn., USA) [4]

that the applied pressure is a factor in surface improvement (Fig. 12.4). The crystallinity of the polymer plays also an important role in surface roughness. The cost of improving surface quality results from the labor and material required to sand, fill, and paint the LPF surface, or from the higher amortization costs for more expensive molds and equipment.

Eckardt [2], using a polystyrene furniture carcass (framework) as an example, demonstrated how the required properties influence the choice of process (Table 12.3). The production cost for a smooth versus a textured surface is also compared in Table 12.3. The table shows that a gas counter-pressure or a two-component process (the latter with regrind) is preferable for egg-shell surface, whilst a low pressure process is preferable for a textured surface.

12.1.4 Process Cost Analysis

In the literature there are very few economic analyses for different IF processes, and even less comparative cost analyses, although one such analysis was given by Throne [4]. According to his data, presented in Table 12.4, machine costs differ considerably for IF molding processes. The machine cost reflects the necessary hydraulic modification to a standard 200 oz (5.67 kg) conventional injection molding machine. Note that the special purpose machines are depreciated over a 5-year term, rather than the more typical 10–15-year span. As a summary of the analysis, the machine-hour-cost is given in Table 12.4. This analysis was based on machine costs, and direct and indirect labor costs (it was assumed that the labor and administration costs are identical for all processes). Note that LPF hourly machine costs are comparable to those for CIM.

Table 12.5 gives a quantitative comparison of mold costs for IF processes. These data show that molds for two-component processes may be slightly more expensive than those for LP processes, because of the higher quality surface finish required. In the case of gas-counter-pressure molds the price of the seals very much depends on the particular molding, and therefore these molds may be from slightly to very much more expensive (see Chapter 3) [2].

Table 12.4. Machinery costs for different IF molding processes [4)]

	CIM, 300 oz[a]	CIM, 200 oz[b]	CIM-Mod[c], 200 oz[b]	LPF	GCP	TCF-Sim[d]
Basic machine cost	325,200	248,800	274,000	185,200	210,200	306,000
Ancilliary machine cost (grinders, hopper, conveyors, mixers, chillers)	37,000	37,000	37,000	37,000	37,000	37,000
Non-production machine cost (10%)	32,000	24,900	27,000	18,000	21,000	30,600
Installation	*32,500*	*32,500*	*32,500*	*32,500*	*32,500*	*32,500*
Fixed capital	427,200	335,600	365,800	259,200	289,200	404,200
Depreciation time, years	10	10	10	5	5	5
Depreciation cost, $/year	42,720	33,560	36,580	51,840	57,840	80,840
Machine cost, $/year (3,840 hr/yr, 100% efficiency)	11.13	8.74	9.53	13.50	15.06	21.05
Machine hour, $/hr[e]	37.27	33.96	33.75	35.63	38.47	46.15

[a] 300 oz = 8.51 kg; [b] 200 oz = 5.67 kg;
[c] CIM-Mod refers to a high pressure IF process similar to the USM process;
[d] TCF-Sim refers to the two-component machines of Siemag or Battenfeld;
[e] 100% man, machine efficiency

Table 12.5. Comparison of tooling costs for various methods of producing integral foams [2)]

Molding	Method of production	Tooling cost, %
Toilet cistern	low pressure	100
	gas counter-pressure	107
	two-component	103
Hi-fi cabinet	low pressure	100
	gas counter-pressure	123
	two-component	100
Furniture	low pressure	100
	gas counter-pressure	117
	two-component	100

Eckardt gave very useful information on cost estimation for different IF molding processes [2)]. Table 12.6 lists hourly costs of machines with a maximum shot weight of 4.5 and 9.5 kg polystyrene, with injection pressures of 100 and 110 MPa. The cost values are expressed as percentages for easier comparison. Clamping units with comparable distances between tie bars were used in the machines examined. Since the shot weights of the machines compared are not identical, the hourly costs of the machines were calculated per kg shot weight, for better comparison. As expected, the lowest cost per h is achieved with LP machines, whilst that of multi-component machines is the highest. However, if related to kg shot weight, more favorable figures are obtained.

Table 12.6. Hourly machine costs for single and multi-component Battenfeld machines with different shot weights of Polystyrene and comparable distance between tie bars [2)]

Process[a]	Type of injection molding machine	Max. shot weight PS approx. 4.5 kg				Max. shot weight PS approx, 9.5 kg			
		Machine model	Max. shot weight for PS, kg	Cost per h, relative units[b]	Cost per h and kg shot weight, relative units[b]	Machine model	Max. shot weight for PS, kg	Cost per h, relative units[b]	Cost per h and kg shot weight, relative units[b]
A, B	low pressure	TSG 5000/ 500 S-C-220	4.930	88.8	76	TSG 12500/ 1050 S-C-500	10.930	88.7	79
A, B, C	high pressure	BSKM 5000/ 500 S-C	4.250	100	100	BSKM 1250/ 1050 S-C	9.750	100	100
A,B,C,D	multicomponent	BM 2 × 3200/ 500 S-C	5.100	129.0	107	BM 2 × 800/ 1050 S-C	13.280	127.3	93.5
						BM 2 × 5000/ 1050 S-C	8.500	115.9	132.9

[a] A: low pressure process; B: gas counter-pressure process; C: high pressure process; D: multi-component process.
[b] referred to high pressure injection molding = 100%

It is important to note that for the same capital invested, an IF injection molding machine (properly loaded) gives more shippable mass per hour than any other injection molding machine [5]. A more detailed cost analysis for actual IF's based on different polymers will be given below (see Sect. C).

12.2 Selection of Processes

To ascertain the best process for producing a particular IF article it is necessary to take into account the criteria which will affect the quality and profitability of the end product. The main criteria are: density and surface finish of the IF parts, equipment and tooling costs (these have been discussed in the preceding section); physical and chemical properties, and applications (these criteria will be discussed below); aesthetic criteria (this matter is outside the scope of this book). Other criteria which are important for process selection are dimension/weight ratios and number of parts.

12.2.1 Dimensions and Weight of Parts

The dimensions and weight of the future article are important factors when choosing a suitable process. Differences in the hourly machine and tooling costs will clearly affect the production costs and thus affect profitability (Tables 12.3 and 12.6). The differences between LP and HP machines increase in absolute figures as the machine size increases. The advantages of the LP machine are evident, especially for articles with greater shot weights because of the lower machine cost. As Table 12.6 shows, this does not necessarily mean that multi-component machines become less economic with increasing shot weights. The example of a machine with a 9.5 kg PS capacity demonstrates that the cost per kg shot weight can be very favorable for multi-component IM machines [2].

12.2.2 Number of Parts

Another important factor when choosing the most suitable process is the number of pieces to be produced. For short runs (small number of pieces), the mold cost is more important, whilst for long runs it becomes significantly less important. Therefore different processes may be more economic for short or long runs (Fig. 12.5). For the calculated example of a radio cabinet, the two-component process is much more favorable for runs of up to about 280,000 whilst the gas counter-pressure process is a bit more advantageous for longer runs [2]. The point where the two curves in Fig. 12.5 intersect is indicative of the fact that mold cost for the gas counter-pressure process is higher, whilst for the two-component process the machine cost is higher.

For any IF molding process, in the case of long runs one should check whether it would be more economical to use multi-cavity molds.

The utilization of machine capacity should be considered in connection with the number of articles to be produced. If the number is small, the most economic solution would be a high pressure IM machine equipped with an extra module for processing foamed thermoplastics, since this type of machine can be used for conventional

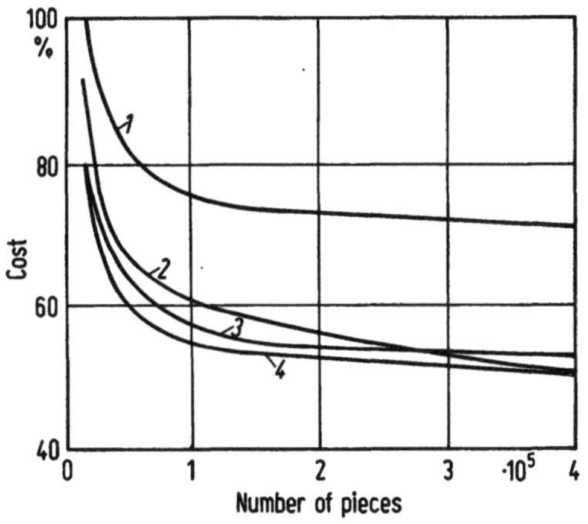

Fig. 12.5. Effect of number of IF items produced on the production cost, when using various IF processes, as illustrated using a hi-fi cabinet as an example; 1: injection molding, 2: gas counter pressure, 3: two-component process (virgin material), 4: two-component process (using regrind as core material)[2)]

injection molding the rest of the time. If, however, the number of articles to be produced justifies the use of a low pressure IM machine, this will clearly be more economical in terms of machine cost[2–4)].

12.2.3 Single/Multi Cavity Alternative

The most profound analysis of these alternatives was developed by Brooks and Colbert[6)]. According to their work, when considering single-cavity versus multi-cavity molds, the determining variables are: the machine cost per hour, the tool amortization period, and the number of parts molded per year. As an example, take a simple two-piece IF card cage, and assume that each half weights 0.45 kg (1 pound), and that it takes 3 minutes to process. The total amortization period is assumed to be 3 years. It can then be calculated that for a single cavity mold each half of the card cage could be run on a small, single nozzle machine, which is assumed to cost $ 30/hour. To obtain a complete card cage would then take 2 cycles, or six minutes, which gives a total machine cost of $ 3; the tool will be $ 15,000. Compare this to a double-cavity mold, which will need a slightly larger machine costing $ 45/hour. One three-minute cycle would produce a complete card cage and cost $ 2.25; the tooling would be $ 25,000. Thus, the question is at which volume will the $ 10,000 extra tool cost justify the 75 ¢ unit saving[6)]:

	Single cavity	Double cavity
Machine cost	$ 30/hour, or 50'/minute	$ 45/hour, or 75'/minute
Machine cost/Card cage	$ 3.00	$ 2.25
Tooling	$ 15,000	$ 25,000

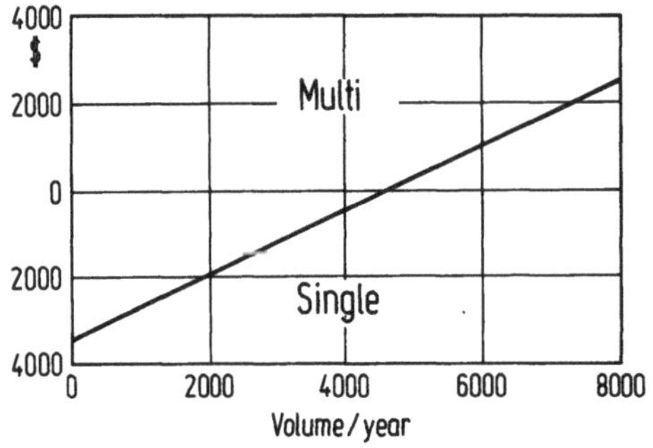

Fig. 12.6. Annual savings for single/multi mold cavity alternative; breakeven-point at 4500 card cages/year [6)]

As the graph shows (Fig. 12.6), the breakeven point for this example is 4500 card cages/year. Below this volume, a single cavity mold is more cost effective, whilst with more than 4,500 cages/year, the more expensive double-cavity mold pays off.

If multiple cavities of the same form and dimensions are required, castings become very attractive cost-wise due to the reproduction of these cavities from the same pattern.

12.2.4 Single/Multi Nozzle Alternative

In order to examine the single nozzle versus multi nozzle machine decision, we use again the Brooks and Colbert estimate [6)]. Let us assume that each individual part could be run on a single nozzle machine. Many IF applications have size and flow length constraints which require the use of equipment with some type of manifold system, but for this example, we shall assume the part size does not require the application to such system. We then need to analyze which sort of equipment is the most economical. The tooling cost is the same for either type of machine, therefore the critical variables are: machine cost per hour, number of molds per machine, set-up time, and the number of parts required per molding run (not the total volume per year as in the single or multi cavity mold comparison)[7–8)].

Brooks and Colbert took an IF cover and base of an electronic device, as an example, and assumed a 3-minute cycle time for each part. The numbers used here are not industry standards or averages, but are intended merely to illustrate how the analysis would be performed to calculate the effect on part cost. For this example, the single nozzle machine costs $ 90/hour while the multi-nozzle machine costs $ 150/hour. The machine time per set, that is cover and base, is 6 minutes for the single nozzle versus 3 minutes for the multi-nozzle. This means the single-nozzle machine cost is $ 9/set, but only $ 7.5/set for the multi-nozzle machine. Set-up times vary widely depending on the machine; if we assume four hours per tool for the single nozzle, we have a set-up cost of $ 720, which is eight hours × $ 90/hour. The multi-nozzle machine, on the other hand, because of the required balancing of the nozzles will need 24 hours to set up, or $ 3,600 [6)].

The decision in this case is between the extra set-up charge required for the multi-nozzle machine versus t $ 1.5/cycle-time saving per cover-and-base:

	Single nozzle	Multi nozzle
Machine cost	\$ 90/hour, or \$ 1.5/min	\$ 150/hour, or \$ 2.5/min
Machine cost/set	\$ 9.0	\$ 7.5
Set up time	4 hours per tool 8 hours 2 tools = \$ 720	24 hours for 2 tools together = \$ 3600

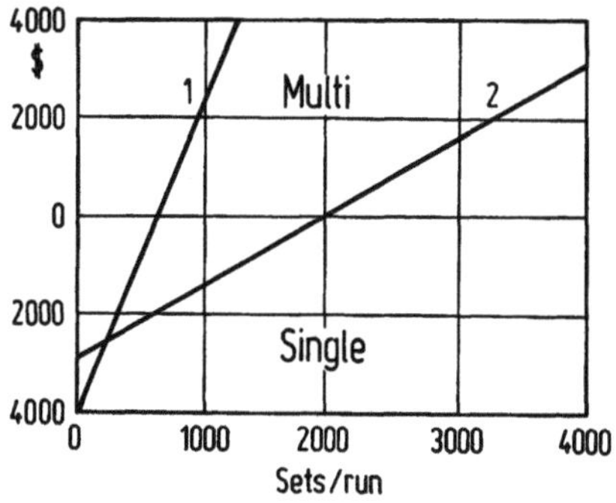

Fig. 12.7. Savings per run for single/multi nozzle machine alternative; 1: breakeven at 600 sets/run for 3 parts/set, 2: breakeven at 2,000 sets/run for 2 parts/set [6]

In this particular case, Fig. 12.7 shows that 2,000 sets per run are required before the multi-nozzle machine is cost effective.

The most critical variable in this type of analysis, however, is the number of IF parts that can be molded together per cycle on the multi-nozzle machine. If we assume there are 3 parts, instead of 2 parts, per complete unit (a cover, a base, and a bazel, for example) then, because of the even greater machine-time savings, the breakeven point drops to 600 sets per run (Figure 12.6). While the numbers used here are mere examples, tooling decisions based on this type of analysis can affect part cost considerably.

Concluding this section on principles of IF technology, let us state the main problems that technologists, chemists and designers are still faced with [7–9]:

- shortening of the processing cycle;
- fabricating large-sized (up to several meters) and heavy (100 kg or more) IF articles with complex configurations;
- reduction of the density of the foamed core;
- manufacturing thin (from 1 to 4 mm) articles;
- fabricating smooth skins with equal thickness and density.

12.3 References

1. Jentet, P.: J. Cell. Plast. *15*, 151 (1979)
2. Eckardt, H.: Ind. Prod. Eng. No 2, 170 (1980)
3. Semerdjiev, S.: Introduction to Structural Foam. Brookfield Center: Society of Plastics Engineers, 1982
4. Throne, J. L.: J. Cell. Plast. *12*, 261 (1976)

5. Hendry, J.: 6th Structural Foam Conf., Bal Harbour USA, 1978
6. Brooks, R. L., Colbert, S. J.: J. Cell. Plast. *17*, 94 (1981)
7. Berlin, A. A., Shutov, F. A.: Strengthened Gas-Filled Polymers. Moscow: Khimia, 1980 (in Russian)
8. Shutov, F. A.: USSR Conf. Gas-Filling Plastics, Suzdal USSR, 1982
9. Wehrenberg, R. H.: Mater. Eng. *96*, No 6, 34 (1982)

Section C:
Integral Polymer Foam Practice

13 Integral Foam Based on Polyurethanes

13.1 Raw Material Specifications

Integral PUR foams are manufactured using the same raw materials as those used for non-integral PUR foams, viz., hydroxyl-bearing oligomers, usually oligoethers or oligoesters, and di- or polyisocyanates, with tertiary amines, organometallic compounds and metal salts as catalysts, together with surfactants and foaming agents.

The general principles and the chemistry of manufacturing integral and common PUR's are similar to each other and will not be discussed here as they have been outlined by us in a separate monograph [1]. We shall only deal with those changes in the compositions that are needed to obtain integral PUR's.

13.1.1 Polyols and Isocyanates

Almost all polyurethanes in use today are based on polyester or polyether polyols, the polyether polyols becoming the more important in recent years because of their lower price and better phase separation in elastomeric applications. The most popular isocyanates used today are TDI (toluene diisocyanate), MDI (4,4-diphenylmethane-diisocyanate), and their derivatives. TDI is extensively used in flexible and semi-flexible foams because of its lower cost, while MDI and its derivatives are generally used in elastomers and rigid IF parts in high performance applications due to their better mechanical properties.

Elastic integral PUR's are manufactured from polyethers and polyesters, various diisocyanates, and cross-linking agents based on glycols and diamines, whereas finished items are fabricated in either one or two stages from a prepolymer. A prepolymer system based on 4,4'-diphenylmethandiisocyanate and a linear polyester (molecular mass 200) containing up to 16% NCO-groups is widely used.

Table 13.1. Properties of flexible integral PUR foams based on polyethers of various molecular masses [2]

Molecular mass of polyether	Apparent density, kg/m³	Ultimate strength, MPa		Breaking elongation, %	Hardness (relative units)
		tensile	tearing		
1,080	970–1,013	8.0	1.8	235	86
2,240	900–1,007	6.0–6.3	1.9	280–320	84
3,090	1,150	9.2	2.0	210	86
3,300	1,070	8.3	2.0	260	82

The materials based on polyesters possess higher strength properties, but have lower frost resistance as compared to the polyurethanes based on polyethers. Zaks et al. [2] have shown that the molecular mass of a polyether considerably affects the strength properties of an elastic IF (Table 13.1).

The use of different polyisocyanates makes it possible to vary the properties of an IF quite widely. In particular, it has been shown [3] that the strength, thermal and physical parameters of a rigid IF improved by substituting diurethanediethyleneglycol (DUDEG) by polyisocyanate (PIC) or by a still residue of MDI (Table 13.2). The increase in impact strength and rise in softening point obtained by using the MDI still residue can be explained, in all probability, by the residue's higher functionality and better polycyclic properties [4].

Usually a 10% stoichiometric excess of isocyanate is used for best balance of material properties. The processing parameters which are normally observed include the reaction times, which are defined as the start or cream time, gel or string time, and rise time. Deviations from the expected values are usually indicative of faults in the system composition or in its processing.

The excess of isocyanate affects the properties of the IF in the following manner (isocyanate index = isocyanate groups/OH groups) [5]

Isocyanate index	1.05	1.15	1.25
Tensile strength, MPa	0.18	0.19	0.2
Breaking elongation, %	120	67	63

A considerable variation of the strength properties of the integral PUR can be achieved by varying the concentration of the amine catalyst (Table 13.3).

Table 13.2. Properties of rigid integral PUR foams manufactured using various isocyanates [3]

Properties	DUDEG	PIC	Still residue MDI
Apparent density, kg/m³	293	327	311
Apparent density of core, kg/m³	220	218	199
Ultimate strength, MPa			
compressive	3.2	4.9	6.2
flexural	3.6	9.4	9.1
Impact strength, kJ/m²	0.214	0.270	0.410
Softening point, °C	91	112	137

Table 13.3. Effect of amine catalyst content upon the properties of an integral PUR foam

Content of catalyst, in mass % of polyol	Core density ϱ_c, kg/m³	Compressive strength σ_c in 25% deformation, Pa	Shore hardness of skin, scale A	Irreversible deformation in compression, %
14	145	$21 \cdot 10^3$	31	7.0
18	140	$26 \cdot 10^3$	40	5.5
22	140	$35 \cdot 10^3$	57	5.7

13.1.2 Blowing Agents

Integral PUR's are foamed mainly with freons, carbon dioxide (produced by the reaction of the isocyanate and water), nitrogen (introduction as such, or as a product of the thermal decomposition of the CBA's), or compressed air [4]. To lower the density, mixtures of PBA and CBA, or PBA and water, or a mixture of two types of PBA's (for example, F-11 and F-113) [5,6] are often used [6a].

An increase of the content of PBA above the optimum value causes a deterioration of the strength of rigid IF foams (Figure 13.1), which is the greater the higher the density of the product. In all probability, uneven skin thickness produced by the excess PBA is responsible for this. With optimum amounts of PBA, IF's have a more homogeneous macrostructure, with a high content of closed gas-structure-elements (GSE) (up to 97%), a denser skin and a higher strength than foams generated by the water-isocyanate system (Table 13.4) [7]. It should be rembered, however, that the thermal resistance of the latter is higher.

Zmolinski [8] has stressed that isocyanate integral PUR foaming systems impose more stringent requirements upon the concentration of water, as compared to composition for obtaining isotropic PUR foams. In order to suppress fully the evolution of CO_2, and to effect foaming at a predetermined temperature, 0.2 to 0.5 mass % of water (referred to polyol) have to be added to a stoichiometric mixture of isocyanate

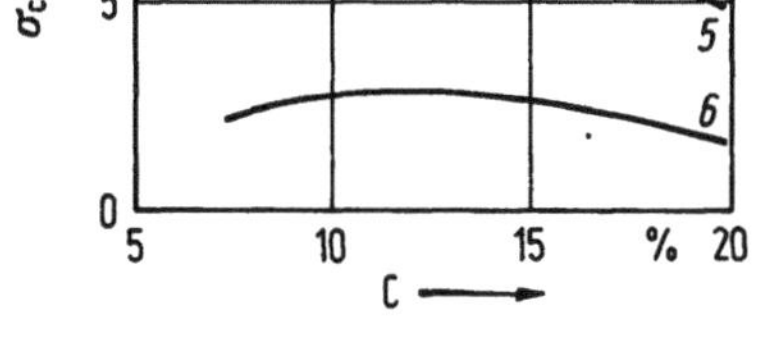

Fig. 13.1. Dependence of flexural strength σ_f (1, 3, 5) and compressive strength σ_c (2, 4, 6) upon the concentration of freon C, for rigid integral PUR foams with varying article thicknesses: (1 and 2) 400 kg/m³, (3 and 4) 300 kg/m³, (5 and 6) 200 kg/m³ [3]

Table 13.4. Properties of integral PUR foams, foamed either by freon or by water [7]

Blowing agent	Apparent density, kg/m³	Thermal resistance^a, °C	Shore hardness, scale D	Modulus of elasticity, Pa	Ultimate load (MPa) and deformation (%) in bending
Freon-11	394	69/58	62	63,000	17.5/5.8
(5 mass parts)	494	71/57	69	95,000	26.5/5.8
Water	427	74/57	54	59,700	17.0/6.0
(0.8 mass parts)	503	82/68	61	82,660	22.1/6.3

^a determined under the load of 4.5 kgf (first number) and 18.0 kgf (second number), according to ASTM-D-648-56 and polyol. An increase of the water content above the admissible amount makes it impossible, besides, to control film thickness. The effect of the PBA and water contents upon the density and core/skin thickness parameters was studied by Campbell [9] using a computer and dynamic mathematical models.

Another feature of IF manufacturing is that generally the concentration of surfactant is somewhat lower than for non-integral PUR foams [1,5].

A typical formulation for obtaining an integral PUR by injection molding includes (parts by weight): polyester, 100; aromatic diamine, 20; silicone surfactant, 0.8; catalyst (tertiary amine), 0.3; organotin catalyst, 0.02; TDI, 21.3; and freon-11, 12 to 16 [8]. It should be stressed that the manufacture of quality IF items calls for compounding the components to within 1% (by mass).

13.1.3 RIM Parts

To date, the major commercial RIM plastics are polyurethanes. This is because polyurethanes provide fast, complete reactions with no byproducts and a wide degree of modulus variability. The latter is due to the domain-forming properties of segmented polyurethanes and the introduction of internal and external reinforcing. Over the last 2–3 years RIM processes have been developed for polyester, silicone and epoxy formulations [10], as well as for interpenetrating polymer networks (IPN) based on nylon block copolymers (see above Part 8.3) [11,12,12a].

The basic chemistry of RIM polyurethanes is not new, in fact it is as old as conventional polyurethane chemistry. But in order to develop useful RIM foams, formulations taking into account the special features of RIM technology had to be developed. For example, the automobile industry required a self-supporting product with good low temperature characteristics and therefore chose elastomeric RIM parts. While cast and RIM elastomeric parts share the same basic chemistry, the sequence of reactions is different because of process technique differences. In general, the chain extender and isocyanate play a greater role in RIM than in casting with respect to dictating final product properties.

The basic chemical reactions for producing elastomeric RIM foam are summarized below [1]:

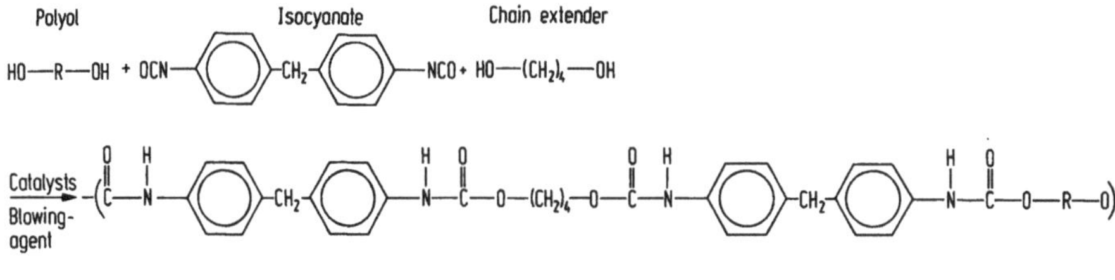

These reactions lead to the formation of "hard" and "soft" segments which determine product performance. The polyol forms the soft segments, and the isocyanate forms the hard segments. The finished product has randomly spaced hard and soft segments which have limited compatibility; the degree of compatibility controls the polymer performance as a function of temperature [13]. A detailed analysis of the chemical process on the properties of RIM parts was made Metzger [14], Mann [15] and others [52–55].

For the last few years many *experimental techniques and theoretical modeling schemes* have been used on PUR's, but most of these studies have concentrated on the physical structure and mechanical properties of previously reacted samples. Only a few workers have treated the molecular characteristics such as molecular weight and its distribution, sequence length and its distribution, and the degree of polymerization of the polyurethanes. On the other hand, RIM urethane reactions are not only complex because of the many side reactions and a high exothermicity, but also because of the many phase changes during polymerization. As Lee has indicated [16], these changes include domain formation (a change from a single phase to multiple phases because of the thermodynamic incompatibility of polymer segments or blocks), gelation (a change from a viscous fluid to network gel with chemical or physical crosslinking) and other phase changes (from rubbery material to glossy polymer or from amorphous material to semi-crystalline polymer, with increasing degree of polymerization and packing of polymer segments). These changes will certainly be influenced by any ongoing chemical reaction. Therefore, they depend strongly on the chemical composition of the reactants and on the temperature. However, until recently there has been little understanding of how the polyurethane structure develops during molding and how the process variables influence gel time and green strength at demolding time.

The raw materials for producing RIM integral PUR foams are supplied by Bayer, BASF-Elastogran, Imperial Chemical Industries, Dow Chemical, Rhône-Poulenc, Shell, Union Carbide, Upjohn, and others.

One of the most popular systems — *Baydur* — illustrates the progress in chemistry and technology of RIM formulation over the past years. The first generation Baydur system (1968) had 10–15 min demolding time, and 80 °C temperature resistance; the second generation (1971) had 5–10 min and only negligibly by the density of the item. For example, increasing the thickness of 2–4 min and 180 °C, respectively [17]. The latest generation of Baydur system (Baydur 726, developed by Mobay Chemical, USA) has 1–1.5 min demolding time, and excellent strength properties (see Table 13.10, below). Baydur 726 is a two-component polyurethane consisting of a formulated polymeric isocyanate based on MDI and a formulated polyol resin; the system is priced at $ 1.30/lb [17a, 53, 55].

13.1.4 RRIM Parts

The principal reasons for using fiber reinforcements in rigid RRIM parts are to meet the design stiffness requirements at part thickness below 3 mm, to reduce the TEC (Thermal Expansion Coefficient), and to improve high temperature properties for best paintability [5].

When selecting reinforcing materials for RRIM, a number of factors have to be considered, according to McBrayer [13]:

1) *Fibrous materials* are preferred to particulate materials; longer fibers give the greatest reinforcement, but there are limits on the aspect ratio (length/diameter) so that viscosity build-up with loading is minimized. Lower viscosity increases the possible loading and reduces the tendency to plug impingement nozzles.

2) To increase the level of loading, *the filler* must be compatible with both, the resin and isocyanate components.

3) *Good adhesion* must be obtained between the matrix and the filler to maximize reinforcement and prevent in-use failure because of separation of filler and matrix. Several studies have shown a silane treatment to be particularly effective in improving adhesion, as well as improving compatibility with the resin and isocyanate.

4) *Filler density* should be low, to keep part weight low. Since RIM products have a cellular core, adjustments in density may be made varying the blowing agent level.

5) *Filler cost* should be as low as possible consistent with performance to maximize benefits.

The most popular fillers for RRIM parts are the following [4, 5, 13]:

Milled glass fibers are continuous glass filaments which have been hammer-milled through screens into a range of short lengths (usually 0.8–6 mm). Actual fiber lengths represent a distribution with more than 50% of the fibers being less than a tenth the size of the screen.

Chopped glass strands are bundles of glass filaments which have been mechanically cut into discrete lengths, typically 3–6 mm. Because the chopped strands are bundles of fibers, they are more difficult to disperse in a slurry than milled fibers.

Wollastonite is a naturally occurring calcium metasilicate with a needle-shaped form. Depending on the grade used, the properties of a RRIM IF are similar to those obtained with milled glass fibers.

Mica (surface treated) can provide reinforcement in a plane instead of along a single axis. For this reason the modulus at a given loading is higher than that obtained with milled glass fiber.

Silica (amorphous) is a particulate filler and, as such, appears to be used more as an extender than as a reinforcing agent.

Microspheres are inert, hollow mineral or polymer spheres of varying particle sizes. They have the potential of giving larger modulus increases than milled glass fibers. This is due primarily to the fact that they are high volume, low modulus fillers, rather than low volume, high modulus as glass. In addition, they reduce the density of a RRIM part. On the negative side, microspheres cannot be loaded to the same weight level as milled glass because of the high viscosity build-up. The recommended loadings are to a volume less than or equal to that of glass. Microsphere loaded RRIM articles are as cold resistant as syntactic-integral RRIM polyurethane foams[4,5a,64].

Fillers may also be made from renewable resources. It has been shown [18] that starch granules extracted from ordinary crop plants, such as rice and maize, can be used as fillers for most common-used plastics, including polyurethanes for RRIM products.

13.2 Processing Parameters Specifications

Integral PUR foams are manufactured by all known techniques for producing IF's — injection molding, extrusion, and rotational molding — using either standard or special equipment and tooling. Currently, the IF industry in many countries is producing rigid, semi-rigid and flexible integral PUR foams, with a wide range of density (150–1,200 kg/m^3), weights (0.1–100.0 kg), shapes, surface quantities, and technical and aesthetical properties.

Whereas the principles and the commercial processes for manufacturing these materials have been discussed in Chapter 3–11, we shall now deal with those processing particularities of integral PUR foam manufacture which are important in practice.

13.2.1 Molding Time

All other process parameters held constant, the molding time depends on the composition (component reactivity), as well as on size and shape of the item. Usually, the dwelling time of an IF composition in the mold is no longer than 8 or 10 min; the items harden completely in 48 hours.

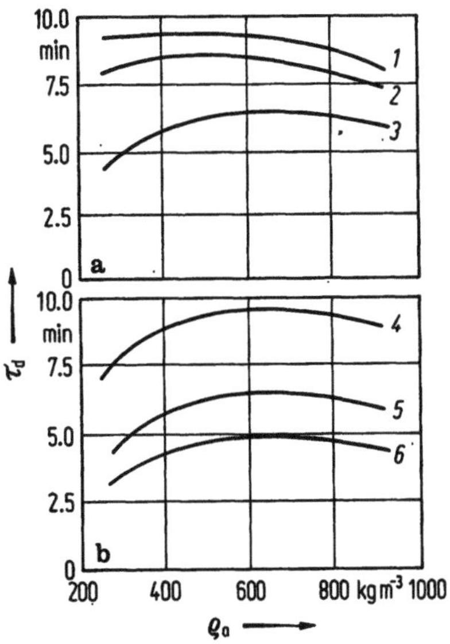

Fig. 13.2. Relationship between the density ϱ_a of integral PUR foam and the dwelling time τ_d, (a) at various polyol concentrations and (b) at various mold temperature; 1, 2, 3: respectively 9, 6, 3% (mass) of polyol at a mold temperature of 50 °C; 4, 5, 6: mold temperatures of, respectively 70, 50, 30 °C, with polyol concentration of 3% (mass) [19]

The dwelling time (or molding time) (τ_d), starting from the moment the composition is injected, represents the largest portion of the forming time of any IF. It was shown [19] that τ_d is greatly affected by the thickness of an item, to a lesser degree by mold temperature and blowing agent concentration, an item from 10 to 30 mm almost triplicates τ_d from 3.5 to 10 min, while increasing the thickness concentration, and only negligibly by the density of the item. For example, increasing the thickness from 5 to 25 mm lengthens τ_d from 1 to 5 min (Fig. 13.2). This is explained by the more difficult heat dissipation from a thicker mold.

In order to reduce τ_d, it is necessary to cool the mold rapidly. Thus, an advisable practice when manufacturing a PUR foam is to heat the cold mold initially with water at 80 °C, then (after the mold has attained the temperature of 50 °C) to reduce the temperature of the water to 50 °C, and finally to cool the mold with water at 20 °C. This combined cooling not only accelerates the hardening of the item involved, but also shortens τ_d from 10 to 3.5 min.

13.2.2 Molding Temperature

Depending on the manufacturing method the temperature of an integral PUR composition ranges from "cold" (20 °C) to "hot" (60 to 80 °C). However, it should be emphasized that, whatever the procedure and the starting temperature of the mold, the temperature of the composition must always go high enough to evaporate the PBA as the article hardens.

The absolute temperature attained in an exothermic PUR hardening reaction is, of course, independent of the mass being foamed. However, the amount of heat evolved is directly related to the mass and the density of the item involved. This explains an interesting fact observed by Abramov et al. [20], who found a "scale effect" when manufacturing IF items, meaning that the skin of a large IF item is always thinner than that of a small item, because of the greater amount of heat accumulated in the mold which increases the time the composition remains at a given temperature. These data illustrate once more the effect of temperature upon the skin of an integral structure.

Using dynamic and mechanical methods and IR-spectroscopy, Fritsch et al. [21] have shown that despite a temperature profile across the mold, the degree of hardening of an integral PUR part is fairly uniform throughout the volume of the item, although the surface layers always harden less than the core. Subsequent thermal treatment at 70 to 110 °C results in an appreciable increase of the rigidity of the material (by up to 40%) because of this uniform hardening throughout the volume of the item.

13.2.3 Skin Density

One of the most effective ways of increasing the strength and rigidity of an IF is to increase the thickness of its skin. The easiest way to achieve this goal is to lower the molding temperature (Fig. 13.3). It should be remembered, however, that decreasing the molding temperature increases the molding time. For example, at 49 °C the molding time is 2 to 3 min, while at 27 °C it is 10 min. Another way of increasing the thickness of the skin — by creating denser and heavier materials — has practical limitations, as the skin fails to gain thickness above a certain value of IF article density ϱ_a (400 kg/m³) (Fig. 13.4).

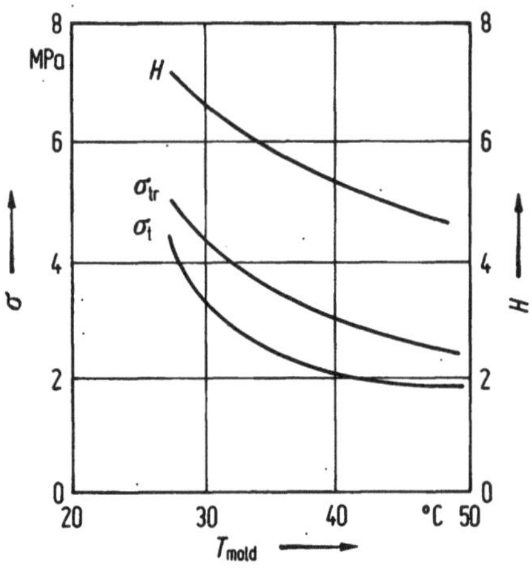

Fig. 13.3. Dependence of the strength properties upon the mold temperature for flexible integral PUR foams: Shore hardness (H ×10); tearing strength (σ_{tr} ×1.78); tensile strength (σ_t ×7, MPa) [22]

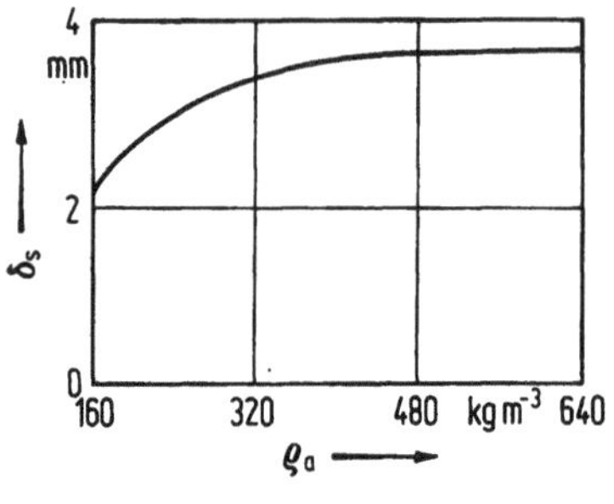

Fig. 13.4. Dependence of the skin thickness δ_s upon the density ϱ_a of integral PUR foams produced by injection molding [23]

Essipow et al. [24] investigated in detail the distribution of density, strength properties and fluidity of polyester-based PUR formulations and IF's as functions of the chemical composition and process factors. They indicated that an increase of the number of hydroxyl groups in the polyol, a rise in the activity of the catalyst and a decrease in the freon boiling point brings about a larger non-uniformity of the density throughout a cross section. This is mainly caused by an increase of the density of the external layers. By contrast, an increase in mold temperature produces the opposite effect.

Depending on whether the compressive load (σ) is applied parallel or perpendicular to the skin, the absolute values of the corresponding *collapse load* differ. Moreover the absolute values of $\sigma^{\parallel}$ and $\sigma^{\perp}$ change appreciably with the mold wall temperature. Wernicke [25] reported that the value of $\sigma^{\perp}$ for integral PUR increases linearly with the increase in T_{mold} from 30 to 60 °C, whereas $\sigma^{\parallel}$ has a maximum at 30 to 40 °C (Fig. 13.5). These results bear out once more the established principle that low mold temperatures raise the density of the skin and lower the density of the core of the IF.

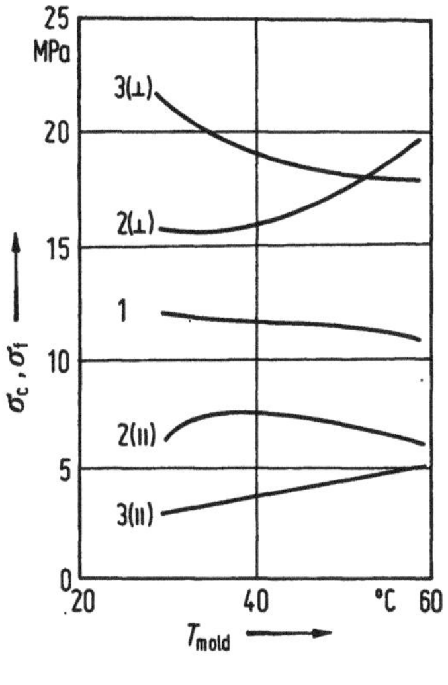

Fig. 13.5. Effect of mold temperature upon the flexural strength σ_f and compressive strength σ_c parallel $\sigma_c^{\parallel}$ and perpendicular $\sigma_c^{\perp}$ to the skin of an integral PUR foam [25]

13.2.4 Molding Pressure

For an IF item of given composition, dimensions, and shape, the molding time T_d is determined mainly by the temperature and pressure in the mold, during heating and cooling.

Hossinger [26] investigated these effects with the universal composition Syspur-SD (GDR). The studies were conducted with an aluminum mold having a volume of 13.8 l and with IF items 18 and 42 mm thick. The temperature of the composition was varied from 20 to 30 °C, and the mold temperature varied from 40 to 55 °C; the mixing pressure ranged from 15 to 23 MPa, and τ_d amounted to 9 min. A rise in ϱ_a from 400 to 650 kg/m^3 (this corresponds to an increase in the degree of compaction from 2.25 to 3.65) entails: growth of the maximum pressure in the mold from 0.17 to 0.35 MPa, increase of the time it takes for the pressure to fall to atmospheric, from 7.5 to 9 min, rise of the temperature at which the item is produced, from 25 to 45 °C. It was found that the pressure developed by this system during foaming lies between 0.75 and 0.95 times the degree of compaction [4].

As for other types of IF, the maximum temperature in the mold depends on ϱ_a, whereas a mixing pressure between 15 and 23 MPa has little effect upon the temperature and pressure inside the mold. Another empirical relationship is that increasing the thickness of the item (from 18 to 42 mm) causes a negligible increase of the temperature of the IF in the mold [26]. In turn, the temperature of the item decreases at a lower rate at high (20 MPa) than at low (15 MPa) mixing pressure. Accordingly, higher mixing pressures correlate with greater probabilities of cracks appearing in the core of an item, lower static strengths of the material and more wear in the equipment (nozzle and by-pass valve).

As we have mentioned, the kinetics of the variations of pressure and temperature, as the integral structures are forming, are as important as the maximum absolute values of pressure and temperature of the PUR in the mold. Thus, it is possible to determine the optimum molding time by using pressure variation curves, and to find the best start times and heating and cooling durations by using temperature variation curves, for any given composition (Fig. 13.6).

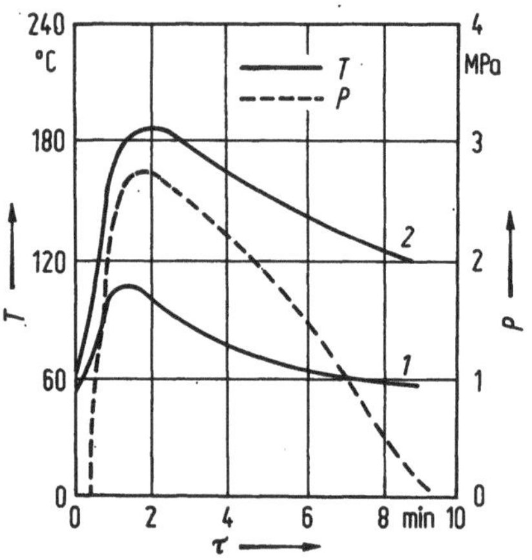

Fig. 13.6. Temperature and pressure variation during the formation of an integral PUR foam based on a polyester system; temperatures of the composition at the wall (1) and in the centre of the mold (2) [26]

13.2.5 Selection of Optimal Parameters

The main factors influencing the molding of an IF part — temperature and time — were studied in detail by Essipow et al. [24] for PUR materials based on polyester (Syspur-SD, made in the GDR). The authors assumed that the molding time can be determined using two criteria: one physical in nature, namely the capacity of the material to take up internal stresses at elevated temperature, the other *chemical*, namely the time necessary for terminating the polycondensation reaction in the boundary zone (near the walls) of the material. In fact, the structure of an IF (thickness of skin, distribution of density throughout the cross section, and other parameters) is dictated by the temperature and pressure fields in the mold. The pressure developed in the mold is taken up by the walls (membranes) of the foam cells and produces stresses in them which may deform the item if prematurely demolded. On the other hand, the temperature, governed by the exothermic hardening reaction, also induces internal stresses, which are particularly high in the central zone of the IF part, where the molding temperature is at its maximum. A temperature gradient across the mold during foaming causes different reaction rates in the different cross section zones. Near the walls, the reaction rate is considerably lower than in the center, and is mainly a function of the temperature of the walls.

In an ideal case, *an item should be demolded* after the hardening reaction has gone to completion in the zone near the wall (chemical criterion) and when the pressure in the central zone of the item drops to that of the surrounding medium (physical criterion). Actually, these conditions are never fulfilled completely, as this would require too long a molding cycle. The actual molding time has to be as short as is compatible with good quality.

Let us consider now, how, the IF molding times can be decreased without sacrificing process conditions [24]. The molding conditions have been studied for a standard two-component Syspur-SD-4502 system, in heated steel molds measuring 200 × 200 mm in area and 10 to 40 mm in depth. The degree of conversion in the boundary zone was determined from the residual content of isocyanate groups. The molding time was determined on the precondition that the thickness of the item should not change by more than 1% after demolding.

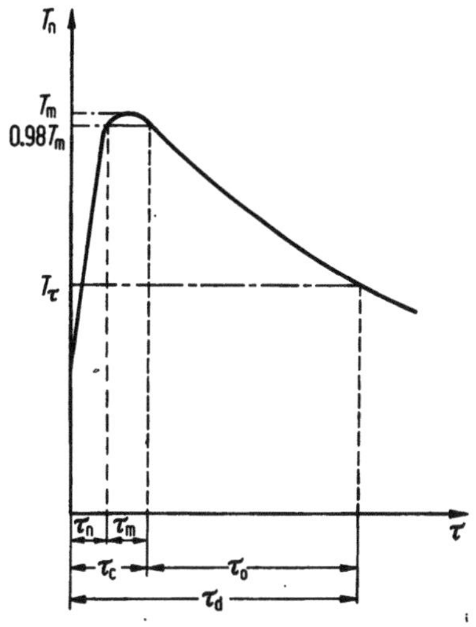

Fig. 13.7. Temperature variations in the composition during the molding of an integral PUR foam (for explanation, see text) [24]

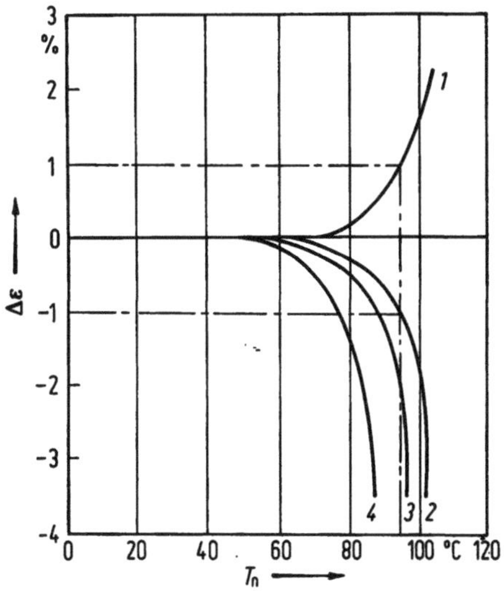

Fig. 13.8. Thermomechanical behavior of an integral PUR foam (ϱ_a = 580 kg/m³, heating rate 45 °C/hour) under various compressive loads; 1, 2, 3, 4: respectively 0.01, 0.44, 0.88, and 2.24 MPa [24]

A typical curve of the variation of the *temperature in the center of the mold* is presented in Figure 13.7. The curve may be divided into three sections corresponding to the different stages of the manufacturing process: τ_n is the time for the temperature in the center of the mold to attain 98% of the maximum temperature (T_m) on the temperature rise side of the curve; τ_c is the time for the temperature in the mold center to drop to 98% of T_m on the temperature fall side of the curve, and τ_0 is the time required to cool the material to the temperature at which the change of the thickness of the item (after demolding) is no more than 1%.

Thus, the dwelling time (τ_d) can be found from the equation

$$\tau_d = \tau_c + \tau_0 = \tau_m + \tau_n + \tau_0 \tag{13.1}$$

The values of τ_n and τ_m were determined from the temperature curve (Figure 13.7) and τ_0 from the thermomechanical curve (Figure 13.8). The thermomechanical data show that the deformation depends on the compressive load. For example, a load of 10^4 Pa corresponds to a positive deformation, i.e. the item expands because the gas pressure in the cells increases as the temperature rises. At greater loads (Figure 13.8, curves 2, 3, 4), the items are compressed at higher temperature. A deformation of $\Delta\varepsilon = 1\%$, as assumed in the definition of τ_0, may be attained at various combinations of loads and temperatures. For each chemical composition, the values of the load σ_c and the mold temperature T_n should be chosen such as to meet the most stringent requirements of molding this particular type of IF. For the sample under consideration this was $\sigma_c = 0.88$ MPa and $T_n = 89$ °C.

Table 13.5. Effect of the temperature and the thickness of an IF part upon the dwelling time ($\varrho_a = 580$ kg/m³) [24)]

Mold temperature, °C	Mold thickness, mm	τ_d, s	τ_n s	τ_n % of τ_n/τ_d	τ_m s	τ_m % of τ_n/τ_d	τ_0 s	τ_0 % of τ_n/τ_d
20	19.5	215	114	53.0	42	19.5	59	27.5
50	19.5	435	102	23.4	48	11.0	285	65.6
80	19.5	1,100	100	8.7	68	5.9	982	85.4
50	10	–	60	–	33	–	–	–
50	20	480	93	19.3	51	10.6	336	70.1
50	30	1,320	103	7.8	110	8.3	11,307	83.9
50	40	2,240	112	5.0	203	9.4	1,925	85.9

Table 13.5 lists data for evaluating the contribution of each of the time parameters upon τ_d. As expected, an increase of the temperature, or of the thickness of the item leads to an increase in τ_d. The main contribution to τ_d is that of the cooling time (65 to 85%), while the parameter τ_n is important only at low mold temperature; the contribution of τ_m is the smallest. An empirical relationship between τ_0 and the temperature parameters of the process has been found:

$$\tau_0 = A \lg \frac{0.98\,T_m - T_{mold}}{T_\tau - T_{mold}} \tag{13.2}$$

where T_τ is the temperature in the center of the item for the shortest molding time, T_{mold} is the temperature of the mold, and A is an empirical coefficient that depends on the density and thickness of the item, and on the heat removal from the mold walls (for the case at hand, A = 960).

When foaming under adiabatic conditions, the maximum foaming temperature (T_m^{ad}) is directly proportional to the initial concentration of the hydroxyl groups in the system, viz.

$$T_m^{ad} = T_0 - \frac{(n_{OH})_0}{C \cdot \varrho_0} (\Delta H_1 \cdot \varphi_{OH} + \Delta H_2 \cdot \varkappa \cdot \varphi_{H_2O})(1 - Q) \tag{13.3}$$

where T_0 is the temperature of the item under adiabatic conditions, in K; $(n_{OH})_0$ is the concentration of the initial hydroxyl groups, in mole/cm^3, C is the specific heat capacity of the composition, in cal/g degree, ϱ_0 is the average density of the components of a mixture, in g/cm^3, ΔH_1 and ΔH_2 are the heat of the exothermic reactions of isocyanate groups with hydroxyl groups and with water, respectively, φ_{OH} and φ_{H_2O} are the degree of conversion of the hydroxyl groups and water respectively, $\varkappa$ is the ratio of the initial concentration of water to the initial concentration of hydroxyl groups, and Q is the fraction of the total heat expended for evaporating the blowing agent.

Equation (13.2) shows that τ_0 can be reduced by decreasing the maximum foaming temperature. For example, with $T_m^{ad} = 160$ °C, $T_{mold} = 50$ °C, and $T_\tau = 110$ °C, we have $\tau_0 = 70$ s, whereas with $T_m^{ad} = 120$ °C, we get $\tau_0 = 6$ s. However, under real conditions such decrease of τ_0 is unattainable: it would require a considerable reduction of the concentration of hydroxyl groups (see equation 13.3) and this would greatly affect the properties of the foams.

In practice, the two major factors affecting τ_0 are the mold temperature (T_{mold}), and that temperature at the item's center at which the deformation of the item after demolding is not greater than 1% (T_τ). For example, a drop of T_{mold} from 50 to 20 °C reduces τ_0 by a factor of three. However, a low T_{mold} slows down the hardening reaction near the walls. Therefore, one expedient way of lowering τ_0 is to increase the heat resistance of the foam. For example, if T_τ can be raised from 90 to 120 °C, and $T_{mold} = 50$ °C, the value of τ_0 falls from 251 to 6 s.

Obtaining a good quality skin surface requires that T_{mold} be much lower than the temperature at the item's center. On the other hand, too low a T_{mold} may result in the foam failing to mold at all, as the hardening reaction near the walls is inhibited, and the material will stick to the mold walls. Therefore, the hardening time near the walls is not only a major process parameter, but one which dictates the quality of the integral structure. Jessipow et al. [24] have calculated this parameter for the Syspur-SD-4502 system by assuming that the boundary zone is the only site of polyurethane formation, that the material is in this zone from the beginning of the process, and that the temperature of the foam in the boundary zone is constant and equal to that of the mold.

It is common knowledge that the formation of a polyurethane can be described by the equation:

$$\frac{d\varphi_{OH}}{d\tau} = k_T(n_{OH})_0 (1 - \varphi_{OH})^2 \tag{13.4}$$

which, after transformation and integration, can be converted to

$$\tau_\varphi = \frac{\varphi_\tau}{(1 - \varphi_\tau)\, k_T(n_{OH})_0} \tag{13.5}$$

$$k_T = k_0\, e^{-E/RT} \tag{13.6}$$

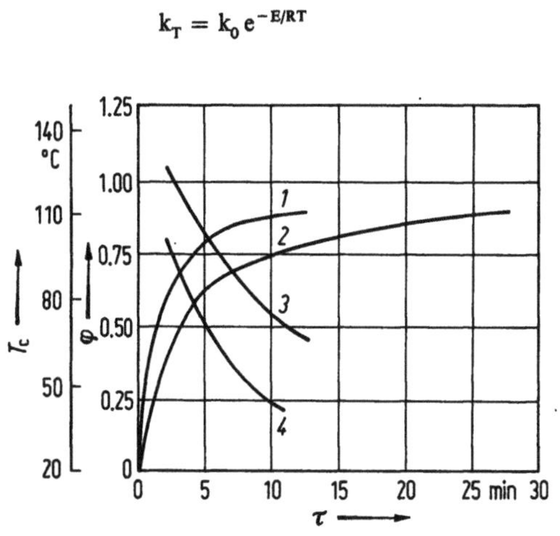

Fig. 13.9. Dependence of the degree of conversion of isocyanates groups (φ) in the zone near the mold wall and the variation of the temperature in the mold centre (T) upon the mold wall temperature. (1 and 3) wall temperatures of 50 °C and (2 and 4) 20 °C; $\varrho_a = 600$ kg/m^3, $\delta_a = 20$ mm [24]

where τ_φ is time of reaction from $\varphi = 0$ to $\varphi = \varphi_\tau$, and is measured in s; k_0 is a constant; $(n_{OH})_0$ is the concentration of the hydroxyl groups in the initial composition, in g — eq./cm³, and φ_τ is the degree of conversion at the time τ.

Therefore, knowledge of the parameters k_T, $(n_{OH})_0$, and φ_τ is essential for finding the optimum value of τ_φ. Spectroscopic methods have indicated that the degree of conversion (in terms of isocyanate groups) in the zone near the mold walls is always lower than that in the center of an IF item (integrally or freely foamed), amounting to 91% (IF skin), 94% (IF center), and 99% (non-IF center).

It follows from Figure 13.9 that at $T_{mold} = 20$ °C, the molding time is 3.6 min., since at that point $T_n = 89$ °C is reached which corresponds to the physical criterion for manufacturing quality IF's. However, under these conditions, the degree of conversion of the isocyanate groups is merely 0.55, which is insufficient to fulfil the chemical criterion. If T_{mold} is raised to 50 °C, the time necessary to attain $T_n = 89$ °C increases to 7.4 min., but then $\varphi_\tau = 0.84$, which almost meets the chemical criterion.

13.2.6 Practical Recommendations

These data support the general principles underlying the formation of integral structures. For integral PUR foams, account should also be taken of their chemical and processing particularities. These can be formulated as follows [1,4,5]:

1) *Variations in the temperature* of the components between 20 and 30 °C materially affect temperature and pressure in foaming.

2) *The molding time* τ_d is determined mainly, by the thickness of the IF item and can be calculated from its maximum thickness.

3) The maximum values of the *molding temperature* and *molding pressure* are determined by the density ϱ_a of the item, which in turn depends on the degree of compaction of the material in the mold. The mold temperature greatly affects item density, degree of compaction, and pressure inside the mold. Doubling the degree of compaction (from 2.5 to 5) raises the pressure by a factor of 4 (from 0.1 to 0.4 MPa).

4) To *minimize molding time*, the mixing pressure should be maintained between 15 and 18 MPa. The higher the mixing pressure, the slower is the pressure drop inside the mold, and the higher the temperature of the item on demolding. Premature demolding results in cracking.

5) The *maximum filling rate* must not exceed 2 m/s otherwise the air trapped inside the mold adversely affects the structure of the skin and the core of the IF.

13.3 Properties

The possibility of adjusting the morphological parameters and the rigidity of the polymer matrix over a wide range permits the production of PUR foams with a wide

Table 13.6. Strength properties of vaious types of integral polymer foams, wood and aluminum

Material	Apparent Density ϱ, kg/m³	Tensile Strength σ_t, MPa	Flexural Strength σ_f, MPa	Tensile Modulus E_t, MPa
Integral PUR foam	100–800	10.0–25.0	3.0–50.0	0.3–1.5
Integral HIPS foam	500–950	5.0–12.0	17.5–37.0	0.65–2.7
Integral ABS foam	700–900	15.0–30.0	28.0–39.0	0.10–1.8
Wood (beech)	500–600	1,335.0	123.0	15.0
Aluminum	2,800	80.0–600.0	10.0–180.0	60.0–80.0

variety of properties and hence a wide range of applications, both as new materials and as substitute of traditional materials (Table 13.6) [4].

13.3.1 Strength Properties

Rigid Materials

Rigid PUR integral foams have outstanding mechanical and thermal properties which make them uniquely qualified for various structural applications in which they effectively compete with wood, solid or foamed thermoplastics, glass-reinforced plastics, and sheet or cast metals [4,5].

Table 13.7. Properties of polyurethane integral foams based on Baydur systems [17]

Properties[a]	Group 1	Group 2	Group 3
Mechanical Properties			
Flexural strength, MPa	40–42	40	50–52
Deflection, mm	16	12–12.5	14–16
Elasticity modulus, MPa	1,000–2,000	850–950	1,000–2,000
Tensile strength, MPa	18–20	18	16–22
Elongation at break, %	10–12	6.3–8	6.5
Push strength, MPa	18	17–20	12
Impact strength, kJ/m^2	19–21	14–15	20–22
Thermal Properties			
Heat distortion resistance, °C	70–80	100–120	115–120
Thermal conductivity, W/K · m	0.07–0.08	0.07–0.08	0.07–0.08
Electric Properties			
Electric strength, kV/cm	100	90–105	115–125
Electric surface resistivity, ohm	$1–3.10^{13}$	$1–3.10^{13}$	$1–3.10^{13}$
Electric volume resistivity, ohm · cm	3.10^{14}	$6–40.10^{13}$	$1–10.10^{14}$
Other Properties			
Water vapor transmission, g/m^3 · day	0.15	0.15	0.15
Water absorption, % vol.	<0.4	<0.7	<0.4

[a] for samples having a density of 600 kg/m^3 and a thickness of 10 mm

A series of special formulations have been developed for different applications. For example, there are currently eight commercial Baydur IF's whose properties may be divided into three groups, depending on the type of polyol (Baydur) and isocyanate (Desmodur) used for the formulation (Table 13.7) [17,53–55].

The first group is characterized by good dynamic qualities and toughness coupled with high rigidity and average heat distortion resistance. This group of materials is best applied for furniture, sports equipment, and non-thermal high load carrying components.

The second group of materials has greatly improved heat distortion resistance. These systems were developed for making thermally stable housings and structural parts for the electrical engineering industry, and for moldings which are glued under heat and pressure. Because of the excellent flow characteristics of these compounds, low density parts can be produced (300–400 kg/m^3).

The third group of materials has excellent mechanical and thermal properties and is used for statically-loaded IF articles. The main applications are in the building and electronic engineering industries. Development products are available for a variety of special interest areas, such as combustion modified systems, systems for very lightweight decorative moldings and wood carving imitations, etc.

Data on the technology and properties of rigid IF parts produced in the USSR are presented in Table 13.8 [27].

Table 13.8. Manufacturing parameters and the properties of rigid integral PUR foams produced in the USSR [27]

Parameters, Properties	Commercial grade				
	5	13/1	324	15	15/2
PUR composition					
Times					
start, sec	22	24	24	20	18
gelling, sec	49	54	54	42	38
finish, sec	77	86	86	72	60
Density at free-rise foaming, kg/m^3	198	178	178	144	150
Fluidity, mm/g	5.3	5.9	5.9	4.8	11.8
PUR integral foam					
Density, kg/m^3	609	532	367	526	457
Strength					
compressive, MPa	25.2	18.2	9.4	–	–
flexible, MPa	45.8	38.6	21.7	32.4	24.0
tensile, MPa	20.3	18.3	11.7	14.7	10.2
Elasticity modulus	126.4	120.8	88.9	–	–
flexural, MPa	1,383.3	1,200.0	683.3	979.3	714.3
Impact strength, kJ/m^2	10.2	10.1	6.0	7.5	5.8
Shore hardness, scale A	215	248	141	229	206
Time					
molding, min	5	5	8	7	9
burning, sec	–	–	–	10	12.2

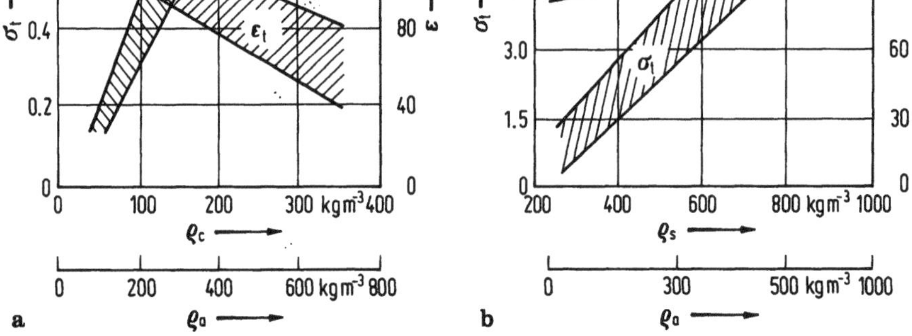

Fig. 13.10a and b. Dependence of tensile strength σ_t and breaking elongation ε_t of (a) the core and (b) the skin, upon the density of the core ϱ_c, the density of the skin ϱ_s, and the overall density ϱ_a, for an article made from an integral PUR foam [28]

A special feature of the Duromer rigid PUR integral foam (Bayer AG) is the possibility to make very thin items. Some mechanical properties are presented in Fig. 13.10; they refer to the IF body as a whole, but core and skin may differ in strength and deformation properties because the density of the core is higher.

Rigid PUR foams have a breaking elongation of 5 to 20%. Compressive deformations partially disappear once the compressive load is no longer applied. The hardness of these materials (in Shore units) varies from 40 to 55 (scale A) and from 70 to 90 (scale D)[4,56].

Flexible Materials

The flexible and elastomeric integral PUR foams can be formulated with densities as low as 200 kg/m^3, for soft products such as seats, with intermediate densities of 400–600 kg/m^3 for applications such as shoe soles, and with very high densities of 900–1,100 kg/m^3 and with microcellular structure to be used for automotive body parts.

For example, ®Bayflex IF's are available in a range of densities and hardness, by combining the different Bayflex (polyols based on branched or linear polyethers) and ®Desmodur (isocyanate compounds, usually MDI) systems [17].

Bayflex 20 and Bayflex 30 IF's are mostly open-cell foams in which the tough elastomeric outer skin protects the core from damage. Higher density Bayflex IF's correspond, in terms of their properties, to microcellular elastomers. They are made from low molecular weight glycols and liquid isocyanates based on pure MDI which give the material very good deflection and tear strengths (Table 13.9). For the stiff Bayflex systems, special emphasis is on the high modulus of elasticity at low temperatures.

The elasticity of these materials can be improved not only by changing the functionality of the starting components (see above), but also by modifying the matrix with rubbers. Dow Chemical (USA) has developed [29] an original material called 'Inner-

Table 13.9. Mechanical properties of flexible "Bayflex" high density PUR integral foams

Property	Bayflex-70[a]	Bayflex-90[b]	Bayflex-92[c]	Bayflex-101[d]	Bayflex-110[e]
Density, kg/m^3	600	950	1,050	1,000	1,100
Tensile strength, MPa	5.7	14.0	17.0	20	35
Elongation at break, %	250	300	380	180	300
Tear resistance with incision, kN/m	8.3	38	38	50	80
Shore hardness, Scale D	–	36	39	56	64
Modulus of elasticity, MPa					
−30 °C	225	475	300	950	1,400
+20 °C	27	100	90	300	600
+65 °C	13	23	60	120	360

[a] from Bayflex-V.P. PU 1935 AT and Desmodur PF; [b] from Bayflex 5331M and Desmodur PC; [c] from Bayflex-V.P. PU 0515 A and Desmodur PF; [d] from Bayflex-V.P. PU 1900 AT and Desmodur PC; [e] from Bayflex-V.P. PU 0505 A and Desmodur PF. The properties were determined for IF panels whose dimentions were: 200 × 200 × 10 mm (Bayflex 70), and 300 × 200 × 4 mm (all higher density types)

foam' which possesses a high "comfort factor" (high ratio of the strength at 65% to that at 25% compressive deformations) and a narrow hysteresis loop. This material is obtained by introducing thermoplastic pellets (3 to 15%) based on a modified PS into the starting composition. These pellets are foamed after the IF part has been manufactured, by heating the part in a high-frequency current field.

RIM and RRIM Materials

The major use of elastomeric RIM foams has been thus far for materials requiring a modulus of 140–350 MPa. On the other hand a high growth rate has been predicted for the use of high modulus RIM foams, having a modulus of at least 700 MPa and elongation of at least 40–50%. These latter foams may be classified as either semi-flexible or semi-rigid elastomers.

While high-modulus foams are only in limited use today, specific application areas that are now being considered are automobile body panels, shipping containers, storage bins, body parts for riding mowers, golf carts, tractors, and agricultural machinery, as well as appliance cabinets (see Part 13.5).

The latests achievement in RIM technology is thick-wall (6.35 mm) urethane IF, based on Baydur 726 (Table 13.10, see also Part 13.1.3). At this thickness and a specific gravity of 0.85, PUR parts have equal or better mechanical and impact properties than ABS, PS, PC, or modified-PPO integral foams of the same thickness

Table 13.10. Comparison between the thin-wall integral foams based on RIM polyurethane and various thermoplastic resins [17a)]

Properties	PUR, Baydur 726 (Mobay Chemical)	PC, Lexan FL-410 (General Electric)	Mod. PPO, Noryl FN-150 (General Electric)	PS, Styron 6087-SF (Dow Chemical)
Density, kg/m^3	850	1,000	900	850
Thickness, mm	6.35	4.00	4.00	6.35
Flexural modulus, MPa (10^3 psi)	1,680 (240)	2,840 (406)	1,910 (273)	1,925 (275)
Flexural Strength, MPa (10^3 psi)	66.5 (9.5)	77.0 (11.0)	44.8 (6.4)	39.9 (5.7)
Tensile strength, MPa (10^3 psi)	40.6 (5.8)	51.1 (7.3)	30.8 (4.4)	16.1 (2.3)
Elongation, %	10.0	3.6	4.1	20.0
Mold Shrinkage, in./in.	0.008–0.009	0.004–0.006	0.006–0.008	0.005–0.007
Gardner impact, in.lb	74	—	53	28
Heat deflection temperature at 1.8 MPa, °C (264 psi, °F)	82.7 (181)	132.2 (270)	82.2 (180)	75.5 (168)
Flammability rating, UL 94	V-0, 5 V[a]	V-0, 5 V[b]	V-0, 5 V[b]	V-0[c], 5 V[d]

[a] expected rating at 6.35 mm thickness based on supplier testing; [b] expected rating at 4.0 mm thickness based on supplier testing; [c] at 3.18 mm thickness; [d] at 6.35 mm thickness

and density reduction. This combination of properties makes Baydur 726 competitive with thermoplastic IF's for applications such as computer housing and medical equipment [17a]. In some of these applications, the RIM parts may replace other high modulus plastic products. The key is to determine the in-use performance. At the same time some problems in making high modulus RIM require solutions. These include brittleness on demolding, and the generally very high reaction rates which make molding difficult [30].

Glass fiber reinforcement strongly affects the mechanical and thermal properties of RIM polyurethane parts. This is shown in Table 13.11 [13].

In order to evaluate the differences between high and low modulus matrices in RRIM, McBrayer [13] used two PUR systems — System A (nominal modulus 640 MPa) and System B (nominal modulus 1,300 MPa). The effect of the polymer matrix can be seen from Table 13.12. Obviously a certain wanted modulus can be achieved with less glass fiber in System B compared to System A. The results in Table 13.1 were

Table 13.11. Effect of glass fiber reinforcement on low modulus and high modulus polyurethane RIM parts [13]

Property	Low Modulus		High Modulus	
	unreinforced	15% fiber	unreinforced	15% fiber
Flexural modulus, MPa	203	364	847	1490
Tensile strength, MPa	19,7	18,2	31,6	34,3
Elongation, %	123	63	83	27
CTE, 10^{-6} K^{-1} [a]	88	48	71	36

[a] CTE: coefficient of thermal expansion

Table 13.12. Effect of polyurethane matrix on RRIM properties [13]

	Glass fiber, %[a]				
	0	0	15	30	30
System[b]	A	B	A	A	B
Density, kg/m³	1.01	1.10	1.11	1.17	1.22
Tensile strength, MPa	22.3	28.5	21.9	33.2	44.3
Elongation, %	249	140	85	17	24
Tear strength, kN/m	26.9	34.1	29.9	29.2	40.5
Coefficient of Thermal Expansion, 10^{-5} K^{-1}	15.2	12.6	—	3.2	3.3
Flexural Modulus, MPa					
−30 °C	1,780	2,410	2,930	4,660	5,950
23 °C	640	1,300	1,250	2,370	3,910
65 °C	343	874	706	1,250	2,870
120 °C	129	202	258	642	681
Modulus ratio (30/65)	5.19	2.76	4.15	3.73	2.07
Shrinkage, %	1.1	0.8	—	0.2	0.2

[a] based on foam weight; [b] see text

Table 13.13. Effect of fiber orientation on properties of RRIM parts

	Glass fibers[a], %			
	0	15	30	40
Tensile strength, MPa				
∥	22.3	21.9	33.2	33.7
⊥	22.8	20.0	21.5	24.7
Elongation, %				
∥	240	85	17	14
⊥	240	123	19	22
Tear strength, kN/m				
∥	26.9	29.9	29.2	33.4
⊥	29.9	30.3	26.7	29.4
Coefficient of thermal expansion, 10^{-5} K^{-1}				
∥	15.2		3.2	
⊥	15.3		8.9	
Flexural modulus, MPa				
∥	640	1,250	2,370	3,950
⊥	630	860	1,270	1,540

[a] based on the foam weight

obtained from measurements parallel to the direction of flow of the material in the mold. The data in Table 13.13 show the differences in properties measured parallel (∥) and perpendicular (⊥) to the flow in glass fiber reinforced System A. The major differences are seen in the flexural modulus and coefficient of thermal expansion.

McBrayer[13)] summarized the effects of glass fiber reinforcing on the properties of RRIM articles in the following manner (for the density held constant, and 25% glass fiber):
tensile strength – unaffected,
elongation – decreased significantly,
Shore hardness – unaffected or slightly increased,
tear strength – slightly increased,
heat sag – significantly improved,
flex modulus – nearly tripled,
temperature sensitivity – significantly reduced,
coefficient of thermal expansion – significantly reduced,
shrinkage on demolding – significantly reduced.

13.3.2 Thermal Shrinkage

The parameter we consider now is very important for integral structures and in particular for those of polyurethane. The shrinkage of foams is not governed by their softening points only, in contrast to unfoamed plastics. The high shrinkage of foams is mainly caused by their macrostructure. There is *a constant gas exchange between the surrounding medium and the gas inside the cells*. This exchange is particularly intense in the first minutes and hours after the foam has been made, because the temperature of the foaming gas inside the cells is slightly above that of the medium (air). Because the gas diffuses out of the cells faster than the air diffuses into the cells, the pressure inside the cells falls, and as a consequence the material shrinks.

These phenomena, which are characteristic of any gas-filled plastic, are more complicated for an IF because of its more compact or monolithic surface layer which prevents "natural" gas exchange. As a consequence additional *internal stresses* (along with the usual process stresses) appear in the structure of the material and are "frozen in". In subsequent use, particularly at higher temperatures, the relaxation of these stresses leads not only to considerable shrinkage, but also to cracks in the skin. We note here that the slow gas exchange may sometimes be a favorable factor, for example, when these materials are used as heat insulators. In particular, blowing gases with low thermal conductivity coefficients, such as the fluorocarbon freons, may remain inside the material for several months [5].

For most integral PUR foams, the shrinkage is small, no more than 0.4 to 0.6% (6 hours, 70 °C); nonetheless these values are several times the dimensional tolerances (0.1 to 0.2%) demanded of these materials. The shrinkage phenomena in PUR foams also depend on the ratios $\delta_s/\delta_{iz}/\delta_c$ and ϱ_s/ϱ_c (See Fig. 1.3) and on the composition of the blowing gas [4,5].

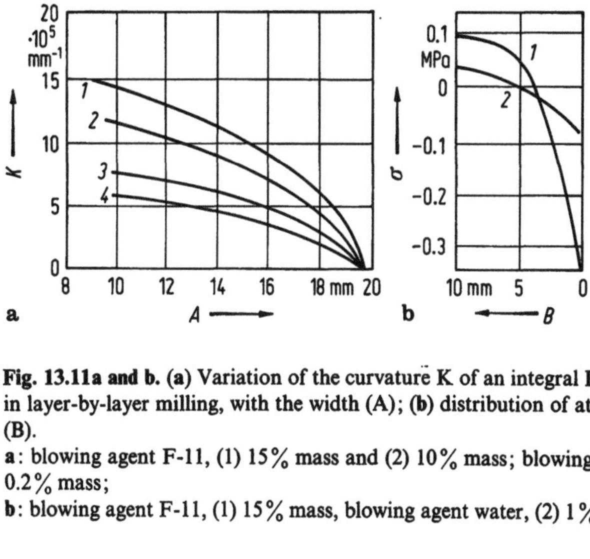

Fig. 13.11a and b. (a) Variation of the curvature K of an integral PUR foam part (200 × 10 × 19 mm) in layer-by-layer milling, with the width (A); (b) distribution of attendant stress σ with the thickness (B).
a: blowing agent F-11, (1) 15% mass and (2) 10% mass; blowing agent water, (3) 1% mass and (4) 0.2% mass;
b: blowing agent F-11, (1) 15% mass, blowing agent water, (2) 1% mass [31]

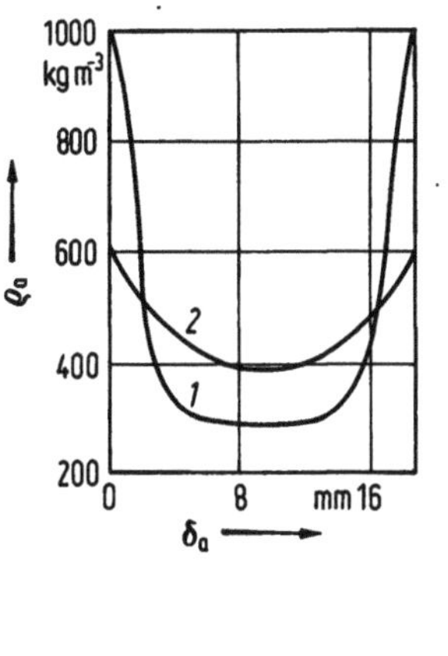

Fig. 13.12. Distribution of apparent density ϱ_a throughout the width δ_a of integral PUR foamed with F-11, 15% mass (1) and water, 1% mass (2) [31]

Unfortunately, very few investigations have been devoted to finding the mechanisms behind shrinkage in IF's, and to methods for preventing them. However, the solution of these problems is very pertinent, particularly if it is realized that today's technology makes it possible to manufacture IF's with very high dimensional accuracy.

Some very important results concerning the causes of shrinkage of IF were recently obtained by Vasiliev et al. [31]. They investigated the absolute value and the distribution of residual stresses in the structure of a polyether-based integral PUR foam, using the "mechanical method". This method comprises essentially the determination of the sag (by means of a comparator) of a specimen after it has been cut layer by layer. It was found that the resistance of a specimen to shape-change, evaluated in terms of the variation of the specimen curvature, is reduced as its thickness decreases (Fig. 13.11). The curvature of the specimens from materials foamed by freon is greater than that for materials foamed by CO_2 (i.e. by the addition of water). Moreover, compressive stresses along the width of a specimen in its skin change to tensile stresses in its core. The stress in a material foamed by Freon-11 is less uniform than that in a material foamed by CO_2. This difference is caused by the different integral macrostructures created by the two gases. For example, foaming by freon gives rise to a structure having a sharp change from the bulk weight of the skin to that of the core, leading to greater residual stresses (Fig. 13.12). By contrast, a structure foamed by CO_2 is characterized by a smooth density change from skin to core. The explanation for smaller residual stresses in CO_2-foamed structures is probably that the linear expansion coefficients of the skin and core differ negligibly. Moreover, it was found that the absolute value of the residual stress increases with the apparent density of the material. Thus, the residual stress at ϱ_a = 320 kg/m^3 amounts to 1.1 MPa, whereas for ϱ_a = 720 kg/m^3, it may be as high as 2.2 MPa. The specimens obtained by cutting an IF part layer by layer increase their curvature even more on heating, i.e. heating enhances the development of the stresses (the "de-freezing" effect). This is why the thermal treatment of an IF part – keeping it in the mold at 90 °C and subsequently cooling it slowly – considerably lowers the residual stresses [31]:

Time in mold at 90°C, hour	1	2	3
Specimen curvature, 10^5 mm^{-1}	10.2	4.2	2.7
Modulus of elasticity, MPa	1400	1400	1400
Residual stress, MPa	1.4	1.5	0.34

However, the character of the residual stress depends not only on an IF's macrostructure, i.e. on physical causes, but also on *chemical factors*, i.e. its chemical composition. In particular, it was reported [31] that the type of *catalyst* used greatly influences the residual stress, much more than does its concentration. Stresses are higher when using triethylamine (TEA) as catalyst, than when dimethylethanolamine (DMEA) or dimethylcyclohexylamine (DMCHA) are used. The polyester used has a lesser bearing upon the value of the residual stresses.

It was shown by McBrayer and Carver [32] that the shrinkage of a glass fiber reinforced RRIM part is significantly reduced as compared to an unreinforced part (Table 13.14). The effect is, however, highly anisotropic. Part design and gating must be carefully considered to take advantage of this effect.

Table 13.14. Glass fiber and shrinkage of PUR systems

Urethane system	Milled glass fiber, %	Mold shrinkage, %
Flexible, high modulus	0	1.2
	15	0.6
Rigid	0	2.1
	15	0.5

13.3.3 Thermal Expansion

While increased stiffness is an important reason for the wide application of RRIM integral foams, the major advantage of these materials resides in the lowered coefficient of thermal expansion (CTE). This is important because a major design problem is the fact that the CTE's of various components may vary widely, as can readily be seen from the following typical values [13]:

Material	CTE, $10^{-6}/°K$
steel	11
aluminum	24
RRIM polyurethane	150

Due to the nature of the RRIM process there is a natural tendency for the *rod-like fibers to orient themselves* in the direction of the flow and in the direction of the foaming. A final RRIM part thus has an anisotropic structure. Figure 13.13 shows the behavior of the CTE and flexural modulus as the concentration of 1.6 mm screen size milled glass fiber increases. The CTE parallel to the flow direction is seen to decrease with increasing glass content. It should be noted, however, that in the perpendicular direction the CTE, going through a maximum, decreases to a value slightly less than that of unfilled foam, at 20% by weight glass. MacGregor and Parker [33] supposed that the increase is the result of a stress built up in the polymer by the reduced expansion in the parallel direction. The increase in flexural modulus in Fig. 13.13 is a further illustration of the directional nature of glass fiber reinforcement. A microscopic examination of RRIM parts has shown that fiber orientation is random in the skin, but oriented in the flow direction in the core zone, though fibers were found to be uniformly distributed throughout the sample. These trends have been found in samples of different thickness (2.54–6.35 mm). It seems that orientation, while being a limitation, is not a critical problem. The proper choice of the location for liquid injection, as well as part design, can be used to provide reinforcement in the desired direction [30–33].

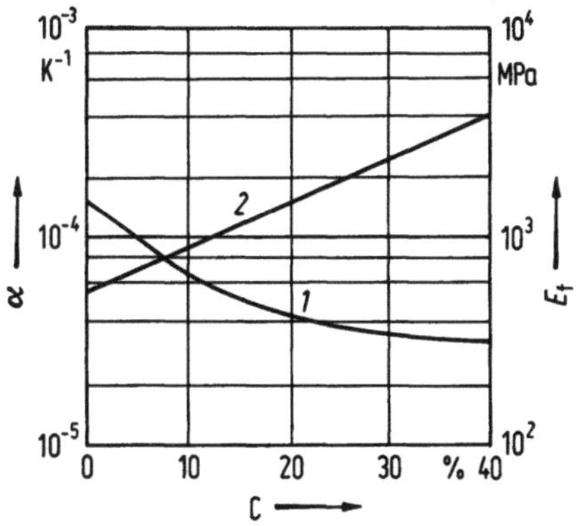

Fig. 13.13. Effect of glass fiber concentration C on (1) the coefficient of thermal expansion α parallel to the flow direction and (2) the flexural modulus for integral RIM part [38]

It should be mentioned that these results are in contradiction with data[5] on the morphology of reinforced integral foams, but these particular RRIM parts had a very low gas volume, and their specific density was almost equal to that of unfoamed parts. This means that in this system the effect of fiber orientation in the foaming direction is practically absent.

13.3.4 Thermal Stability

The thermal stability of integral foams is not better than that of the usual foams manufactured from the same compositions. The solid skin slows down the diffusion of oxygen to an IF, reducing the rate of thermal-oxidative destruction of the polymer, and therefore enhancing the fire resistance of the material. The substitution of the blowing gas (CO_2, N_2, freons, etc.) by air may require several months [35,36]. The thermal stability of an IF can be increased by changing the chemical structure of the starting polymers, using, for example, polyurethane compositions based on polyisocyanates and diphenylmethanediisocyanate.

Isoderm (Upjohn, USA [4]) presents good thermal properties, better than those of, for example, the material Duromer: a service temperature of 150 °C (versus 100–125 °C), a heat conductivity coefficient of 2.5×10^{-5} (3.5×10^{-5}) $°C^{-1}$, a specific heat capacity of 1380 J/kg, and weight losses at 200, 400, and 700 °C amounting to, respectively, 2, 8, and 70% [17].

Most integral PUR are combustible, but incombustible grades such as Isoderm and Syspur-SD-4505, are also available. The fire resistance of different PUR foams is discussed later on in more detail (see Table 21.2).

13.3.5 Light and Color Stability

Integral PUR foams, as polyurethanes in general, are characterized by poor light stability. The search for improvement has led to the development of new types of resins that are used as protective coats. The following three systems were developed by Michaels and Kane [37]. The first system is the product of the pre-polymerization of an isocyanate with a relatively large oligomer, which retains some reactive isocyanate groups. When applied to a surface, this composition hardens through the action of air moisture. The main component of the second system is a polyurethane syrup of low molecular mass which is mixed with polyols before being applied. The

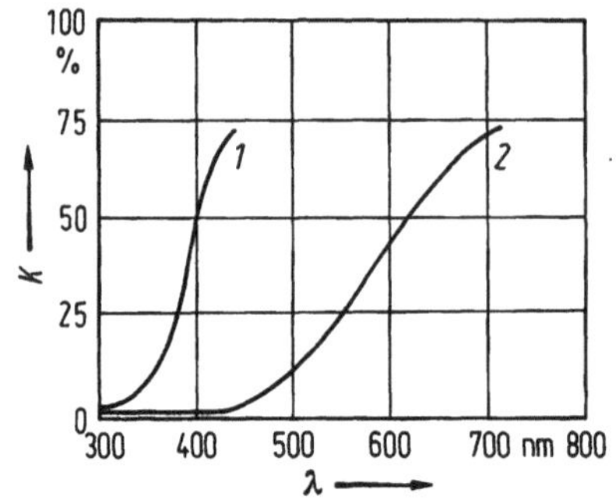

Fig. 13.14. Ultra-violet reflection K of integral PUR foam surfaces, (1) before, and (2) after, 1,000 hours weathering at different wavelengths λ [38]

third system is based on the use of a 10 to 25% solution of unreactive PUR elastomer. The resultant polyurethane varnish hardens completely after the evaporation of the solvent. What is important here is that the tensile strength (36 MPa) and the breaking elongation (450%) of these coats are higher than those of the IF, thus enhancing the IF's structural reliability. To obtain stronger coats IF's based on two-component compositions should be used [4].

If the light and color stability of an PUR-based IF is achieved through a change in the ratio of the amine to isocyanate groups in the composition proper, it is good practice to maintain the isocyanate index between 1.05 and 1.10, i.e. to ensure a small excess of isocyanate in the starting composition [4].

Rigid PUR integral foams, based on aromatic isocyanates, absorb strongly in the short UV region (Fig. 13.14). Continued exposure to light causes the absorption maximum to shift towards longer wavelengths. *Intramolecular reactions displace the absorption spectrum* into the visible range, and the material yellows. This phenomenon is also observed when PUR foams are weathered in the open air. The color changes and the surface becomes dull. Although the mechanical properties remain largely unchanged, the appearance is spoiled [17].

Recently [57] Cannon USA Inc. and Recticel Foam Corp., USA demonstrated a marriage of new technologies in PUR processing and chemistry for producing an array of colored IF parts that would not fade under sun and other sources of UV light (Table 13.15). The marriage was based on the new Cannon Color System (CCS) used on Recticel's new Colo-Fast® foams [58]. Light stability is achieved by modifying the typical chemical formulation of the polyurethane. Non-aromatic isocyanates, such as aliphatic alicyclic types, are substituted for the aromatic isocyanates (MDI or TDI) commonly used. These materials cost 30–35% more than comparable non light-stable products [59].

Table 13.15. Physical properties and light stability of RIM integral polyurethane foams [59]

Properties	Colo-Fast IS (Semi-flexible)	Colo-Fast HM (elastomeric)	Colo-Fast HR (reinforced elastomeric)	Durofast (rigid)
Density, kg/m³	304–608	891–1,104	1,104	496–704
(lb/cu ft)	(19–38)	(56–69)	(69)	(31–44)
Shore hardness	20–70 A	50–70 D	60–75 D	70–80 D
Flexural modulus,	—	203–707	504–1,526	609–1,015
MPa (10³ psi)		(29–101)	(72–218)	(87–145)
Light stability:				
Xenon test, hr	1,000	1,500	1,500	1,500
Fadeometer test, hr	600	1,000–1,500	1,000–1,500	1,000–1,500
Weatherometer, hr	500	1,000–500	1,000–1,500	1,000–1,500

13.3.6 Dielectric Properties

The dielectric properties of integral PUR foams are as a whole comparable to those of non-integral foams, with the exception of the dielectric strength (E_{ds}), which is considerably higher for IF's, and is close to the strength of unfoamed plastics

Table 13.16. Electro-physical properties if integral PUR foams [4, 17]

Property	Apparent density ϱ_a, kg/m³	
	650ᵃ	850ᵇ
ε, 50 Hz	1.9	—
100 Hz	—	2.22
800 Hz	1.8	—
10^6 Hz	1.6	2.18
tg δ, 50 Hz	0.024	—
100 Hz	—	0.012–0.047
800 Hz	0.007	—
10^6 Hz	0.016	0.039–0.061
E_{ds}, V/mm	50	216–220
$\varkappa_s$, ohm	10^{11}	—
$\varkappa_v$, ohm · cm	$2 \cdot 10^{14}$	10^{15}–10^6

ᵃ Duromer, testing temperature 70 °C

(Table 13.16; see also Table 13.7). It is interesting that E_{ds} is higher for an IF foamed by a freon than for one foamed by carbon dioxide [5].

As ϱ_a rises, the dielectric properties of IF's deteriorate. For example, the values of ε and tg δ for elastic integral PUR foams of densities of 230, 650, and 850 kg/m³ are 1.32, 2.18, 2.45 and 7×10^{-3}, 1.8×10^{-2}, and 2×10^{-2}, respectively.

13.3.7 Other Properties

Because of their cellular core, PUR integral foams have better noise insulating properties than their unfoamed counterparts. High void content decreases the sound propagation, and PUR integral foams have noise insulating properties which are superior to wood in the audible frequency range [4].

The water absorption of integral PUR foams is not higher than 1% by mass. The high stability of polyurethanes with respect to water and wet media (see Table 13.7) makes PUR IF's suitable for filtering elements and for sorbing petroleum and oil from water [17, 39]. The vapor permeability of an IF is very low, for example less than 1 g/min for Duromer (ϱ_a = 650 kg/m³) [4].

High abrasive resistance is characteristic of all IF's, and of PUR foams in particular.

13.4 Applications of Rigid Parts

Rigid, semi-rigid and flexible PUR integral foams can be applied in every field of human activity. The last few years have seen the importance of these materials grow because of the development of RIM and RRIM integral foams[62, 63, 62–64].

Rigid PUR integral foams offer significant advantages in application areas where high mechanical strength is required. These advantages may be summarized as follows:

- almost unlimited design possibilities,
- high stiffness in combination with relatively low weight,
- density variation between 200 and 800 kg/m^3,
- part weights as high as 100 kg,
- no sink marks with changing cross-section,
- no dependence of flow path upon wall thickness,
- good aging and chemical resistance.

13.4.1 Construction Industry

IF materials satisfy many of the demands of the construction industry, e.g. for flat roof drains, skylights, window frames, and window profiles [63]. Rigid PUR integral foams have become increasingly popular for window construction (see Chapter 23) due to the outstanding suitability of these materials for the purpose (Table 13.17) [38].

The following properties of polyurethanes are also important for construction applications: economical production, heat and sound insulation, corrosion resistance, low weight, and high mechanical stiffness[39–42,62–65].

Table 13.17. Physical properties of a window based on a rigid PUR integral foam

Properties	Method of measurement	Value
Main coefficient of thermal expansion	VDE 0304/1	$18 \cdot 10^6\ K^{-1}$
Resistance to driven rain	DIN 18055, sheet 2	$\geqq C^a$
Thermoconductivity coefficient with double glazing having a 12 mm air gap		$\leqq 1.8$ kcal/m^2 h °C
Sound insulation with double glazing having a 12 mm air gap	VDE 2719	$\geqq 28$ dB
a-value[b]	DIN 18055, sheet	$\leqq 0.2$

[a] The C-value is a measure of the water tightness of the joints between the inner and outer frame. It means the frame will withstand a wind of force 11 on the Beaufort scale, and may be used in buildings up to 100 m high.

[b] The a-value is the rate of air permeation between the window joints.

13.4.2 Furniture

Rigid foams are used for plastic furniture and as components in furniture which utilizes both plastic and traditional materials. Integral PUR articles for furniture are always painted and pleasant to use. In contrast to other plastics, these foams are thin-walled and are not cold to the touch.

13.4.3 Office Equipment

Rigid PUR integral foams are well introduced in the electronic industry, and are used for making housings and components for computers, copiers, telex machines, etc. Stiffness coupled with low weight, low heat distortion, combustion modifications (in order to achieve UL-Subject 94 Classification V-O, see below Table 22.1), sound deadening, and design freedom are some of the characteristics which allow these materials to be applied in office equipment.

13.4.4 Audio and TV Equipment

The basic reasons for using rigid IF's in this application are the favorable sound and tone qualities, good accoustic damping qualities, and the possibility of controlling frequency reverberation [44]. These applications also require low heat distortion [5].

13.4.5 Electrical Industry

Combustion modifications and electrical properties make PUR IF's suitable for cable distribution boxes, house connection boxes, cable joints and other parts [4, 5, 60].

13.4.6 Sports Equipment

Sporting applications are especially demanding with regard to mechanical properties. IF's are used in ski cores, tennis rackets, and table-tennis bats.

13.4.7 Sanitary Fixtures

Criteria such as design freedom, good thermal insulation, avoidance of condensation, and the possibility to foam-in accessories, account for the many applications of PUR IF's for bathroom fixtures, such as wash basins, bidets, shower bases and bath enclosures.

13.4.8 Other Applications

Different applications in industry exert widely varying demands upon IF's. The possible variations in both the physical properties and construction possibilities are the reason why IF's can be used for such varied applications as filter plates, machine covers, conveyor rollers, windmill blades, pinball machines, automatic sales machines, special pallets, and fish net floats [13, 17, 35, 38–43, 62–63].

Kitchen tooling, musical and electro-musical instruments, wood- and metal-like pictures, bas-reliefs, flower bowls, statuettes, candlesticks are some further applications of IF's.

Rigid integral PUR foams have been applied in the insert-molding of a large (864 × 2,337 mm) glass cover in a RIM frame, for solar heating systems [45]. Other new applications of rigid RIM parts have been in the shipbuilding [46] and automotive industry [52–56, 61].

13.5 Applications of Flexible Parts

According to Wirtz [46], flexible PUR foams can be produced in the following density ranges for different automobile applications: 200–300 kg/m^3 for safety and comfort padding, steering wheel coverings; 300–500 kg/m^3 for bicycle and motorcycle seats, commercial vehicle seating; 700 kg/m^3 for exterior vehicle protection, i.e. bump strips and rib strips; 900 kg/m^3 for bumpers, fenders, structural parts; 1,000–1,100 kg per m^3 for flexible body sections (fenders), front and rear ends.

13.5.1 Passenger Cars

Since the mid-1960's, flexible integral PUR foams have been widely applied in the automobile industry, especially for interior comfort and safety items which utilize both soft and semi-rigid low density IF's. Currently, soft PUR parts are used, e.g., for head supports and chest protectors, while stiffer IF's are used for steering wheels, arm rests, knee rails, door and post covers, as well as complete crash-padded instrument panels [15].

As to exterior body parts, RIM materials have been used since the late 1960's. The first applications for passive car body protection were bumper overriders and bumper end caps from Bayflex 70, in the density range 400–800 kg/m^3. In 1973, the US regulations for bumpers, MVSS-215, were quickly met by the use of flexible PUR foams. A bumper design called the "Hi-Flex" bumper satisfied these regulations when produced from Bayflex 90; this impact resistant bumper survived the 8 km/h (5 Mph) barrier, as well as pendulum impacts, with no resultant permanent damage, and it did so throughout the operational temperature range of −30 °C to 65 °C.

Today many automobile manufacturers protect the front and rear ends of the vehicles from minor impact damage through the use of flexible coverings called facias. The flexible facia absorbs minor impacts with no visual damage to the car. In order to satisfy the impact requirements of MVSS-215, energy absorbing PUR foam cushions and hydraulic shock absorbers are mounted behind the flexible facia.

Aside from bumper applications, stiff PUR foams are employed for window cranks, wheel-arch moldings, side trim, and others (Figure 13.29). These articles require good dimensional stability for shape retention and must provide some impact resistance.

All over the world, automobile drivers today are protected from injury by a variety of passive safety restraints [47–49].

13.5.2 Vans and Trucks

Van and truck cabs are upholstered with low density flexible PUR integral foams in a similar fashion as passenger cars. Crash pads, arm rests, head restraints, steering wheels and post covers are all conventional areas of application. The use of these materials as exterior trim combines passive safety with good protection from normal abuse, such as that encountered on construction sites. Systems such as Bayflex 20 are readily applied as bumpers, flexible panelling, grill inserts, wind deflections and fenders on both busses and trucks.

13.5.3 Two-Wheel Industry

The two-wheel industry has used the abrasion resistance properties of flexible IF parts for seats for bicycles, mopeds, and motorcycles. The durability of these IF articles, coupled with favorable production economics have made this a very successful application. The harder foams can also be applied on motorcycles for instrument panels and aerodynamic enclosures. The impact resistance and flexibility offer advantages over the traditionally used steel and glass-reinforced plastics.

13.5.4 Prospects of the Automotive Industry

In USA many cars contain examples of RIM technology. Starting in the 1975 model year, RIM parts had an average thickness above 4 mm; by the 1978 model year, thickness was reduced to 3.8 mm. Today, many parts are 3 mm and below (Chrysler Omni, Chevrolet Camaro, Pontiac Le Mans). For instance, the typical part produced by Davidson Rubber Div., even for large full fascias, is 2.5 mm thick (Ford Mustang).

European RIM parts have generally been 5 mm thick or more, and this may be one factor in the slower growth of RIM in European markets. Today a few European and Japanese manufacturers have begun to take advantage of more recent RIM technology (Toyota Celica, Peugeot 505, Mercedes S-Sedan).

Some years ago, both Ford and General Motors announced their intentions to commercialize rigid reinforced RRIM fenders, and the term "Friendly" fenders has been used to describe these products. Production trials were expected to take place in 1980–81 and production programs were expected for 1982–83. Industry sources have predicted growth rates for RRIM foams that are comparable to those for elastomeric RIM foams [50].

Almost 23 million kg of RIM foam were produced in the USA in 1978; projections are for 110–140 million kg of polyurethane RIM foam in automobile flexible fascia applications alone, by 1985 [16]. In general, the use of plastics in USA automobiles is growing rapidly, and both General Motors and Ford have projected the use of plastic components to almost 130 kg per car by 1985. This is almost twice the 1979 use of plastics. If the reduction in average car size is taken into account, the use of plastics as a percentage of vehicle weight is even more dramatic: it will almost triple. It is very important to emphasize that RIM polyurethane parts, and especially integral RIM products have been leading this growth and are expected to continue to provide low weight, low cost answers to the automobile industry's need for improved materials. Elastomeric RIM fascias for front and rear ends were expected to be used on 32% of all cars produced in the USA by 1981; this represents over 30 million kg of polyurethane chemicals [50].

13.5.5 Shoe Soles Industry

Flexible and semi-rigid non-RIM and RIM integral PUR foams based on both polyether and polyester, are used for shoe soles, often as completely blended mixtures.

For the various requirements of the shoe industry, a variety of soles with a wide range of properties is available.

Polyether based systems are used for the production of flexible, semi-rigid, and rigid soles and inserts. A semi-rigid PUR system permits property variations according to special application needs such as wedge heels, wedge heels with directly molded soles, and orthopedic shoes. Wedge heels, for example, require good edge stiffness which can be obtained with a 90 Shore A product. In the other hand, orthopedic shoes require a softer product to encompass a wider stiffness range, and these materials can be made available upon request. Completely rigid, wood-like products can also be used for wedge heels, particularly for imitation-wood soles. Good surface finish without air entrapment and excellent nail holding qualities are desirable. A few examples of other applications of polyether-based systems are: twin colored soles for leisure shoes; boots; soles of higher density, e.g. for rural boots; solid, transparent soles and imitation crepe soles; ski and skating boots; liners for ski boots.

A wide range of products based on polyester systems is also available for the shoe industry. Significant increases in the sales of soles for leisure and sport shoes have been recently obtained in many countries. The most popular sport shoes have two-color soles, representing two different materials with the outer sole surface being higher in quality and density in order to provide better abrasion resistance, especially for hiking and mountain shoes. Polyester systems are used for inner liners for ski boots, soles for safety shoes with some electric conductivity, and high abrasion resistant soles for tennis shoes [17].

13.5.6 Other Applications

Aside from the vehicle and shoe industries, there is a variety of other applications for flexible PUR integral foams, especially in the furniture industry, e.g. for office chairs and domestic arm-chairs, private and public transportation seating, mattresses, flexible kitchen furniture fronts, etc.

13.6 Technical and Economic Analysis

13.6.1 PUR versus PS Integral Foams

It is very useful for engineers and consumers to compare the economic data for commercial grades of IF's based on the most popular thermosetting polymer, PUR, with those based on the most popular thermoplastic polymer, PS.

A detailed economic analysis between PUR and PS integral foams produced by both a RIM and a low-pressure IM process was carried out by BASF (FRG) for two IF parts: a chair bottom (600 × 750 × 720 mm) and a writing-table post (760 × 540 × 360 mm) [51]. It was found (Tables 13.18 and 13.19) that making the chair bottom from PUR is more profitable than making it from PS. By contrast, the writing-table post is more profitably made from PS. These two examples are illustrations of the necessity of a detailed analysis of the equipment, raw material and labor costs for a correct choice of the polymer and IF process before an IF part is produced.

The commercial and marketing situation for IF parts based on PUR and PS will be discussed further in Chapter 23.

13.6.2 RIM Foams versus Other Materials

With the advent of high performance elastomeric (HPE) RIM parts with high flexural moduli, IF part costs are close to steel, not even taking into account the assembly advantages. Moreover, a RIM part has at least a 2 to 1 weight advantage compared to steel, whilst compared to aluminum, a RIM foam is significantly lower in cost.

Table 13.18. Comparison of economic and technical parameters for producing two IF parts (chair bottom and table post) from PUR or PS, by RIM or IM technology, respectively [51)]

Parameter	Chair bottom		Writing-table post	
	PS part IM process	PUR part RIM process	PS part IM process	PUR part RIM process
Equipment cost, DM				
total, including	586.0	350.0	420.0	220
molding machine	436.0	—	380.0	—
metering unit	—	75.0	—	75.0
mold clump unit	—	100.0	—	80.0
mold	150.0	150.0	40.0	40.0
thermocontrol unit	—	12.0	—	12.0
mixing head	—	13.0	—	13.0
Productivity, unit/hour	4	4	15	9.2
Molding cycle, min	15	15	4	6.5
Weight of part, kg	9.6	7.2	7.2	5.4
Average density of part, kg/m³	800	600	800	600
Raw material cost[a] per one part, DM	21.97	22.48		
Labor cost[a] per one part, DM	27.50	17.20		

[a] data for writing-table post see Table 13.19

Table 13.19. Raw material and labor costs for an IF part (writing-table post) based on PS or PUR, produced by IM and RIM processes, respectively

Parameter	Productivity, parts/year				
	60,000	120,000	240,000	500,000	750,000
IM process, PS part					
material cost total, DM	21.50	21.50	21.50	21.50	21.50
polymer	14.40	14.40	14.40	14.40	14.40
lacquer	7.10	7.10	7.10	7.10	7.10
labor cost, DM	0.75	0.75	0.75	0.75	0.75
RIM process, PUR part					
material cost total, DM	27.40	27.40	27.40	27.40	27.40
polymer	21.30	21.30	21.30	21.30	21.30
lacquer	6.10	6.10	6.10	6.10	6.10
labor cost, DM	0.81	0.54	0.54	0.54	0.54

Compared to either steel or aluminum, RIM foams are damage and corrosion resistant, and are more energy efficient (Table 13.20) [50)].

By now RIM productivity, in terms of cycle time, is about 1 to 2 minutes per part, and paint costs are reduced by the use of high solid paints. At the same time, RIM parts are lower in cost than steel parts even without taking into account the assembly advantages of RIM [50)].

Table 13.20. Comparison of RIM integral foam parts with metal alternatives

Parameter	RIM	HPE-RIM	Steel	Aluminum
Weight ratio	1.2–1.6	1	2.4	1
Material cost ratio	1.2–1.6	1.1	1	1.6
Design	1-piece	1-piece	multiple piece	multiple piece
Resistance to damage	excellent	excellent	poor	poor
Features	industry standard	improved assembly, cost	corrodes	high energy usage to manufacture

RRIM parts are not as stiff as steel. Rigid RRIM fascias can be made with a strength in the range of 2,000 to 3,000 MPa, but fenders are considered the most important RRIM application. The main reasons for this are [50]:

1) The potential weight savings.
2) Fenders have few structural requirements.
3) RRIM parts can be made as single integrated components.

In addition to low weight, many of the factors which made elastomeric RIM foams useful also favor RRIM foams, i.e. damage and corrosion resistance, low tooling and retooling costs, competitive economics, process advantages including ease of molding and distortion resistance, as well as energy efficient materials and process.

The RRIM process and IF parts appear to be ready to compete on a commercial basis with SMC (compression molded sheet molding compound) and BMC (injection-molded bulk molding compound) for exterior automobile body applications such as fenders, hoods, and doors.

Perhaps the most important factor favoring RRIM technology is the *excellent paintability* of the product. Compared to other plastics, PUR RRIM parts have an excellent and durable surface finish, the surface is glass-free, and color as well as gloss can be made to match metal surface finishes [50].

While RIM polyurethane can be used as a replacement for metal and wood, RRIM foams may also find application as a replacement of SMC. This is particularly true in cases where the rigidity advantage of SMC is not critical. RRIM parts are lighter and more impact-resistant than SMC and the RRIM material is not susceptible to the surface defects which occur with SMC's [13].

Table 13.21. Comparative fender data for polyurethane IF's, manufactured by RRIM and SMC processes, and for steel and aluminum fenders

Parameter	Fender material			
	PUR (RRIM)	Steel	PUR (SMC)	Aluminum
Thickness, mm	2.5	0.8	2.5	1.2
Specific gravity, kg/m³	1,200	7,800	1,500	2,700
Weight, kg	2.0	4.0	3.2	2.2

Based on prices in USA, the price of a PUR RRIM part is close to that of a corresponding steel part. But the former weighs only one half as much as the latter (Table 13.21). The RRIM part is lower in cost than the other competitive materials, SMC and aluminum [50)].

13.6.3 Non-RIM versus RIM Foams

A comparison between "classical" (non-RIM) IF polyurethane and RIM polyurethane IF parts was carried out by Ferrari [50)], for rigid parts. There are many similarities in the chemistry and technology of high-density (heavy) non-RIM parts and high modulus RIM parts, though there are some significant differences as well (Table 13.22). "Classical" IF polyurethane technology has made a total commitment to rigidity, at the sacrifice of elongation and non-brittleness. In addition, it is limited with regard to part thickness, because sections much below 6.4 mm develop friable skins and require longer molding cycles. High-modulus RIM parts, on the other hand, have been developed from resilient polyurethane elastomer technology, and they retain a high level of elongation and non-brittleness. RIM PUR is almost two-thirds as rigid as the "classical" IF polyurethane of the same density, and it can be molded in sections of 2.5 mm or less.

Table 13.23 shows a comparison of three types of PUR integral foam (assuming parts with similar size and design), which are produced by three different polyurethane processes. Thermoplastic PUR is not very competitive because of much higher raw material costs, even though it is made by a very efficient and clean production process. Cast PUR has the lowest material costs, but compared to RIM it has approximately three times the cycle time and twice the labor cost. The combination of longer cycle time and higher labor cost results in a significant cost disadvantage when compared

Table 13.22. Comparison of rigid polyurethane materials

Material	High Modulus Rigid RIM	Integral Rigid Polyurethane Foam
Nominal specific gravity, kg/m^3	1,000	650
Typical molded thickness, mm	3.2	6.4
Typical flexural elastic modulus, MPa	1,050	1,050–1,400
Typical elongation, %	60–70	5–10

Table 13.23. Comparison of polyurethane processes

Process	RIM	Cast	Thermoplastic
Specific density, kg/m^3	950	700–1,000	1,100
Cost/kg (raw materials only)	$ 1.32	$ 1.10–1.32	$ 2.64
Cycle time, min	3.0	9.0	1.0–2.0
Labor content, relative units	1.0	2.0	0.5

to RIM, not only in terms of IF unit costs, but also in terms of RIM facility and tooling costs (see Chapter 8, Tables 8.1 and 8.2).

According to Harrick [61], in USA an average annual gain of 10% is forecast for microcellular PUR parts during the 1983–1987 period. Rigid foam will rise by about 6% per year, while flexible foam is expected to increase 5% annually. These estimates include a leveling in 1986, when a readjustment expected in the economy will affect the rigid and flexible sectors of the industry. Flexible rigid, and microcellular PUR foams increased by 8.9% in 1983 (over 1982), and a combined 32.6% gain by 1987 (over 1982) is expected.

It was estimated that by the 1985 model year, 75% of all domestic (US) cars would have PUR front and rear ends, and that RRIM auto components would reach the 9-million kg level by 1987[61].

Additional growth for non-automotive RIM applications is expected in farm equipment, housings for electronic equipment, and beverage-containner applications. Based on expected increases in residential housing starts (sheathing) and appliance sales, rigid PUR foam was predicted [61] to hit the 263-million kg level by 1983.

13.7 References

1. Berlin, A. A., Shutov, F. A., Zhintinkina, A. K.: Foam Based on Reactive Oligomers. Westport: Technomic, 1982
2. Zaks, Y. B., et al.: Plastmassy No 4, 59 (1976) (in Russian)
3. Kragis, I. G., et al.: Mekhanika Polymerov No 2, 346 (1977) (in Russian)
4. Berlin, A. A., Shutov, F. A.: Strengthened Gas-Filled Polymers. Moscow: Khimia, 1980 (in Russian)
5. Shutov, F. A.: USSR Conf. Chemistry and Technology of Polyurethanes, Vladimir USSR, 1979

5a. N.N.: Modern Plast. Intern. *13*, No 1, 16 (1983)

6. Kiesewetter, M., et al.: IFL-Mitt. *19*, No 4, 133 (1980)

6a. N.N.: Plast. Technology *29*, No 4, 16 (1983)

7. Matonic, D. E., Kaminski, A.: Plast. Technol. *22*, No 10, 57 (1976)
8. Zwolinski, L. M.: Rubber Age *107*, No 7, 50 (1975)
9. Campbell, G. A.: J. Appl. Polym. Sci. *16*, 1387 (1972)
10. Wehrenberg, R. H.: Mater. Eng. No 3, 35 (1982); No 3, 47 (1982)
11. Kubiak, R. S.: Plast. Eng. *36*, No 3, 55 (1980)
12. Pernice, R., Frisch, K. C.: J. Cell. Plast. *18*, 121 (1982)

12a. Frisch, K. C.: Plast. Rubber Intern. *8*, No 1, 17 (1983)

13. McBrayer, R. L.: J. Cell. Plast. *16*, 332 (1980); SAE Techn. Paper, No 80015 (1980)
14. Metzger, S. H.: Chemical Technology – Microcellular Elastomers. In: Reaction Injection Molding. Becker, W. E. (ed.). New York: Van Nostrand Reinhold, 1979, pp. 17–55
15. Mann, M. F.: J. Elatom. Plast. *14*, No 1, 11 (1982)
16. Lee, L. J.: Rubber Chem. Technol. *153*, 342 (1980)
17. Bayer-Polyurethanes: Handbook. Bayer AG, FRG, 1979

17a. N.N.: J. Cell. Plast. *19*, 135 (1983); Plast. World *41*, No 6, 68 (1983)

18. Webb, S.: Proc. Chem. Eng. *33*, No 12, 33 (1980)
19. Dunkley, C. D.: Ingenieursblad *42*, No 7, 173 (1973)
20. Abramov, S. A., et al.: Plastmassy No 11, 29 (1975)
21. Fritsch, P., et al.: Plaste Kautschuk *22*, 489 (1975)
22. Seefried, C. G., Withman, R. D.: J. Cell. Plast. *8*, 256 (1972)
23. Grive, R. L., et al.: J. Cell. Plast. *5*, 358 (1969)
24. Jessipow, J. L., et al.: Plaste Kautschuk *22*, 489 (1974); IFL-Mitt. *14*, No 3, 85 (1975)
25. Wernicke, H.: IFL-Mitt. *14*, No 3, 77 (1975)
26. Hossinger, M.: Plaste Kautschuk *24*, 581 (1977)
27. Deberdeev, L. I., et al.: Plastmassy No 9, 59 (1982) (in Russian)
28. Wirtz, H.: Kunststoffe, *60*, 303 (1970)

29. Geijen, P., Moore, T. L.: Kunststoff-Berater *20*, 345 (1975)
30. Methven, J. M., Dawson, J. R.: Reinforced Foams. In: Mechanics of Cellular Plastics. Hilyard, N. C. (ed.). London: Applied Science, 1982, pp. 323–358
31. Vasil'ev, A. S., et al.: Plastmassy No 1, 48 (1977) (in Russian)
32. McBrayer, R. L., Carver, T. G.: Elastomerics *113*, No 6, 30 (1981)
33. MacGregor, C. J., Parker, P. A.: SAE Techn. Papers No 790166 (1976)
34. N.N.: Poliplasti *27*, No 264, 78 (1979)
35. Börger, H.: 5th Internat. Conf. Cellular and Non-Cellular Polyurethanes. Strasbourg France, 1980
36. Bayer, L., Börger, H.: Kunststoffe *72*, 359 (1982)
37. Michaels, C. R., Kane, R. P.: SPE J. *15*, No 9, 28 (1969)
38. Kleimann, H.: Cell. Polym. *1*, No 2, 105 (1982)
39. Oder, H.-J.: Plaste Kautsch. *27*, No 2, 88 (1980)
40. Klepek, G.: Konstruieren mit PUR-Integralhartschaumstoffen. München: Carl Hanser Verlag, 1980
41. N.N.: Mod. Plast. Int. *11*, N o3, 55 (1981)
42. Obrziet, J. J.: Iron Age, *1980*, June 9, 29
43. Harter Polyurethan-Integralschaumstoff: Technical Pamphlet, Bayer AG-Chemiewerkstoffe. Bestell-Nr.: PU 53045, 1976
44. N.N.: J. Cell. Plast. *17*, 172 (1981)
45. Martine, R.: Mod. Plast. Int. *10*, No 2, 44 (1980)
46. Wirtz, H.: The Application of the RIM Process in Europe. In: Reaction Injection Molding. Becker, W. E. (ed.). New York: Van Nostrand Reinhold, 1979, pp. 180–198
47. Kleiner, G.: Chem. Ind. *22*, 793 (1981)
48. N.N.: Mater. Plast. Elastomeri No 3, 17 (1982)
49. Weller, P. A.: J. Cell. Plast. *16*, 273 (1980)
50. Ferrari, R. J.: J. Cell. Plast. *16*, 338 (1980)
51. N.N.: Plastverarbeiter *12*, 337 (1976)
52. Gill, W. A.: J. Cell. Plast. *19*, 168 (1983)
53. Daniels, W. G., Harasin, S. J.: J. Cell. Plast. *19*, 179 (1983)
54. Masi, P., et al.: J. Appl. Polym. Sci. *28*, 1517 (1983)
55. Ostfield, H. G.: J. Cell. Plast. *19*, 141 (1983)
56. Lifshitz, J. M.: Polym. Eng. Sci. *23*, No 3, 144 (1983)
57. N.N.: J. Cell. Plast. *19*, 134 (1983)
58. N.N.: Cannon News, Polyurethane Technology, pp. 2–4 (1983–1984)
59. N.N.: Plast. World *41*, No 1, 78 (1983)
60. Reisser, A.: Swiss Plastics *2*, No 10, 23 (1980)
61. Harrick, G. T.: Mobay Chemical Corp., Polyurethane Division, Pittsbourgh, USA (1984)
62. Shutov, F. A.: Intern. Symposium "Polyurethane-83", London England, 1983
63. Shutov, F. A.: Intern. Symposium on Plastics in Building, Liege Belgium, 1984
64. Shutov, F. A.: Adv. Polym. Sci. *73/74* (1985)
65. Shutov, F. A.: 4th Conf. Mechanics and Technology of Composites, Varna Bulgaria, 1985

14 Integral Foam Based on Polystyrene

Integral foams based on polystyrene (PS) account for a sizeable quantity of commercial thermoplastic IF's. Up to 1980 the output of these materials grew very quickly. For example, in the USA alone, the output increased 40-fold over the period 1975 to 1980 and amounted to 136,000 tons per year [1]. The situation is now changing: integral PS foam production is decreasing, whereas the development of other IF materials, especially PUR integral foams, is increasing rapidly (see Chapters 13 and 23).

14.1 Technology Specifications

Integral PS foams are manufactured in many countries using general-purpose and high-impact polystyrene (HIPS) resins, as well as copolymers and mixtures with PE, ABS, and other thermoplastics.

The BA's used are mainly organic CBA's such as azodicarbonamide (0.1–2%), 4,4′-oxybis(benzosulfonylhydrazide) (0.1–2%), and n-toluenesulfonylsemicarbazide (0.1–5%), while inorganic CBA's are less often used, for instance the one-to-one mixture of sodium bicarbonate and citric acid [2].

PBA's have been introduced industrially only recently, and currently many PS integral parts are being produced with freons and gases (hydrogen, nitrogen, pentane) [3,4].

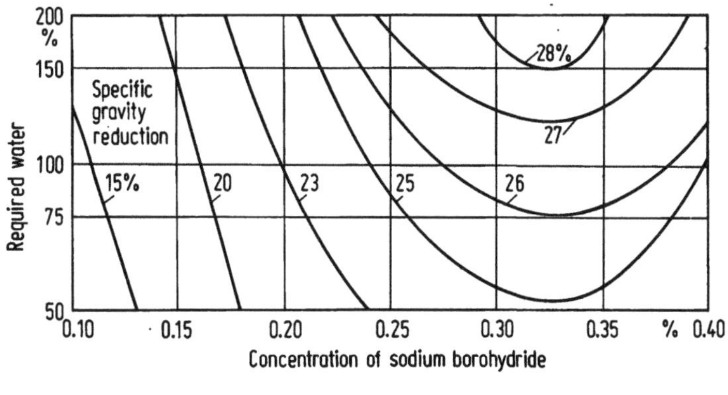

Fig. 14.1. Specific gravity reduction of PS integral foam parts, as a function of sodium borohydride and water activator concentrations is represented by a family of curves [5]

Recently, Gribens and Rei [5] demonstrated the possibilities of a new CBA – sodium borohydride ($NaBH_4$) – for producing IF's based on HIPS. This CBA has the following advantages over conventional CBA's: it produces 10 to 20 times more gas (H_2), leaves no toxic or staining residue, and is chemically activated. Excellent IF's with density reductions of around 25% can be obtained using $NaBH_4$ concentrate and an activator system containing water or polyacrylic acid and fused silica (Fig. 14.1). By using this CBA, most conventional injection molding machines can be easily converted to IF production, without any additional equipment, such as a nitrogen sparge tube on the plasticating barrel or a compressed gas metering system to control the gas flow.

14.2 Commercial Processes

Most integral PS items are manufactured by injection molding and extrusion, though some are produced by rotary molding and press molding.

14.2.1 Injection Molding

The two-component injection molding method (the ICI technique) is used to manufacture IF items having cores of PS and skins of a thermosetting plastic or polyurethane [2].

The SSF-process of Japan's Sumitomo manufactures thin items (not thicker than 3 mm). The foaming is done with nitrogen (2 to 12%) using an injection pressure of 90 MPa and an injection rate of 3 kg/s, to yield items measuring 1,500 × 100 mm [3]. Another method is known under the trade mark Vestyron-TSG (GWH, West Germany) which uses pre-expanded PS pellets [4].

Injection molding is employed to produce three main types of items: thick-walled (thicker than 4 mm), lightweight (ϱ_a no greater than 400 kg/m^3), and items imitating wood or metal. Either special, or modified standard, injection molding machines are used. The modifications are necessary to increase the foaming coefficient and provide smooth surfaces on the fabricated items. This is achieved by raising the rate of injection, changing the plasticizing temperature (higher heating rate), and providing devices for accurately volume-metering the melt. The metering is usually carried out under a pressure that is half the maximum possible, with injection times of 0.5 to 1 s, and the injected volume less than the volume of the mold. During low pressure injections, a lighter locking block may be used, notwithstanding the large surface and volume of the molded item. The mold differs in design from those employed for the more usual foams in that it is thicker than 4 mm, and made of a tool steel, aluminum and polyester, or an epoxy resin with an aluminum filler, the latter being 50% cheaper than usual molds. Injection molding machines can be improved by the addition of a hydraulic accumulator to increase the injection rate.

Special-purpose medium-pressure and high-pressure machines can foam PS with foaming coefficients of 4 to 5 and can produce items with very smooth surfaces. A typical installation for manufacturing a PS-based IF has the following characteristics [2]: a single injection assembly combined with two die mold clamping assemblies (clamping pressure of 295 tonnes), maximum injected batch of 3.7 kg, maximum plasticizer capacity of 500 kg/h, maximum molding pressure of 130 MPa, and injection rate of 1420 cm^3/s (see Fig. 5.6).

The relationship between the melt temperature and the density of a finished item is very complex (Fig. 14.2), and the optimum temperature lies between 128° and 135 °C. The cause of the increase of ϱ_a at low temperatures is evidently the high

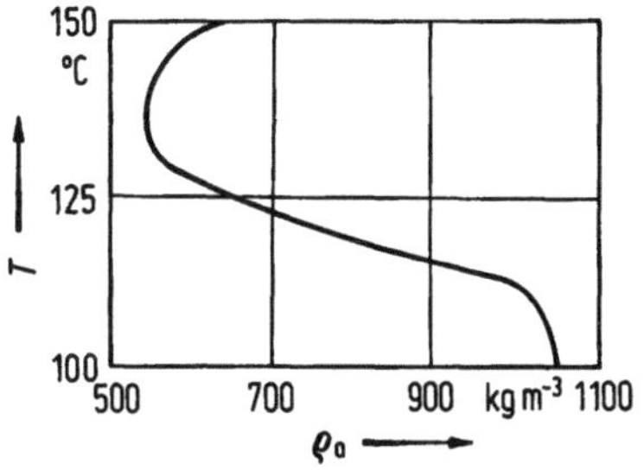

Fig. 14.2. Effect of melt temperature upon the density ϱ_a of an integral PS foam produced by extrusion [6]

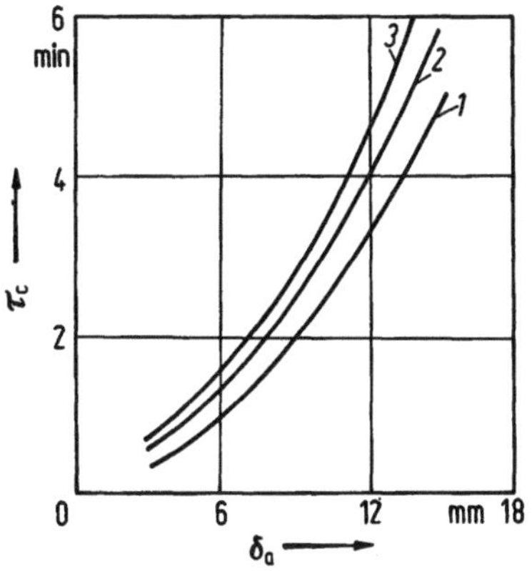

Fig. 14.3. Cooling time τ_c required for an integral PS article with different thicknesses δ_a and densities; 1, 2, 3: 700, 800, and 900 kg/m³ respectively [7]

viscosity of the composition, which prevents a rise in the blowup ratio. By contrast, the increase of ϱ_a at temperatures above the optimum is due to the rupture of the boundary areas of the foamed mass (because of its low viscosity) and the escape of some of the foaming gas from the system.

The molding cycle time in the high-pressure injection molding of PS may be as high as 8 or 10 min., so that this process proves to be excessively costly, despite high possible blowup ratios, i.e. low consumption of the raw material per unit volume of fabricated item (polymer material efficiency). The longest process stage, the cooling of the material, depends (all other conditions being equal) on two parameters, i.e. the density and the thickness of the IF items, and particularly on the latter (Fig. 14.3) [7]. A decrease of the mold temperature evidently lowers the cooling time, but it also increases the shrinkage of the material (Fig. 14.4), the transverse shrinkage being greater than the longitudinal shrinkage (see Chapters 2 and 12).

As the distance from the gate increases, the deviation of the local apparent density (ϱ) from the average value ϱ_a increases. For example, 400 mm away from gate ϱ_a is between 10 and 15% higher.

The injection gate should be situated in the center of the item being molded, and at its thinnest point, so as to allow the injection pressure to overcome the high flow resistance and to fill the zones with the largest cross sections while retaining a low foaming pressure. To improve the surface quality and to yield a uniform macrostructure, several gates (nested molds) are preferable.

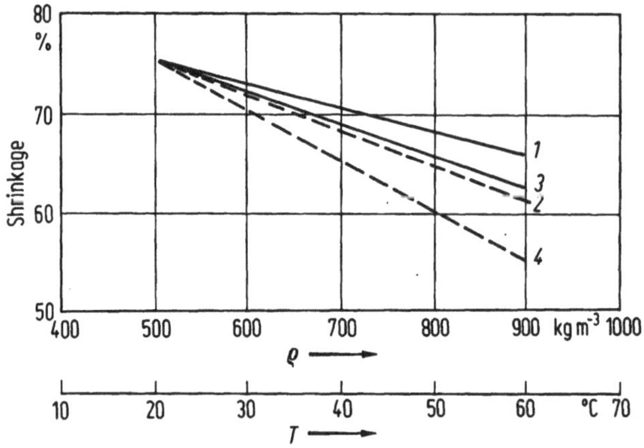

Fig. 14.4. Shrinkage versus density (1 and 2) and versus mold temperature (3 and 4), for an integral PS foam item; 1 and 3: longitudinal shrinkage, 2 and 4: transverse shrinkage [8)]

Researchers in the USSR have developed another process for producing integral foams based on polystyrene by injection molding [2)]. The method consists essentially of melting the material in the heating cylinder of a thermo-softening automatic unit and saturating it with isopentane vapor (5.5 to 6%). A fine cell structure and a high quality of the surface layer are achieved by dry mixing the starting composition, consisting of the polymer, a CBA (azodicarbonamide or a 1:1 mixture of sodium bicarbonate and citric acid, 0.2%), and plasticizers (vaseline oil and butyl stearate in a 1:1 ratio). The optimum parameters of the process are: injection pressure 48 MPa; injection molding temperature 130 to 140 °C; mold temperature 40 °C; cycle time 1 min. The resultant material has the following characteristics: ϱ_a = 180 to 500 kg per m^3; ϱ_s = (1.03 to 1.04) × 10^3 kg/m^3; and ϱ_c = 80 to 130 kg/m^3. The strength properties of the material correlate with the value of ϱ_a as follows:

ϱ_a, kg/m^3	180	450
Ultimate strength		
tensile, MPa	1.5	5.0
flexural, MPa	5.0	15.0
Impact strength, kJ/m^2	0.6	2.5

14.2.2 Extrusion

The main extrusion method for obtaining integral PS foams is the *Celuka technique*, which was developed by Ugine Kuhlmann (France), along with its modifications. It is capable of producing any profile from 4 to 50 mm thick, as well as hollow sections for the construction and furniture-making industries (see Chapter 9). Integral PS foam can be extruded in the usual types of extruders after a slight modification to provide for thoroughly stirring and dispersing of the blowing agent. The extruder head should remain at a temperature sufficient for decomposing the blowing agent, while the compression ratio in the extruder cylinder should not be lower than 2.5:1.

Process parameters for manufacturing integral PS and ABS foams are given below:

	PS	ABS
Extruder temperature, °C		
Ist zone	160	165
2nd zone	205	195
3rd zone	210	205
Head temperature, °C	210	205
Nozzle temperature, °C	195	185
Screw rotation speed, rpm	40	30
Density of article ϱ_a, kg/m^3	600	620
Content of blowing agent, mass %	0.20	0.30

When making PS-based extruded IF's, the temperature of the extruder head is usually several degrees below the decomposition point T_{dec} of the PS, and as a result the item is fairly stable even at the exit of the nozzle.

14.2.3 Other Processes

There is *a "melt" method* consisting in foaming a blank containing a CBA between heated sheets of monolithic PS; another method is for manufacturing *"double" IF's* containing a mixture of PS and PVA, PVC or synthetic rubber as the matrix. The composition includes a CBA, PS, water dispersions of the other polymers, and solid plasticizer particles (polyacrylate). This mixture is heated under pressure in an air-tight mold and foamed to form an integral elastic material [2]. It may also be noted that there is a method of obtaining *elastic IF items* based on mixtures (1:1) of PS and butadiene/styrene elastomers, as well as other "chemical" methods for creating integral structures, by dissolving the external layer of an isotropic polystyrene foam in strong solvents (ketones and ethers) and subsequently heating and compacting the material [9].

The external surface of an IF item can be made very smooth or textured by *etching with solvents*, by *coating* with varnishes based on acrylate, alkyd, or polyurethane resins, by welding on relief films of PS or PVC, by printing patterns, and by *applying foils* of aluminum, nickel, chromium, and gold [10].

14.3 Properties

14.3.1 Mechanical Properties

The properties of PS-based IF foams change widely as a function of the method of manufacture, secondary treatment techniques, composition, density, dimensions and shape (see Table 14.1 and, below, Table 15.1).

Schleith [11] has studied in detail the behavior of integral PS foams (Hostyren-SVP) under various short-time and long-time *loadings*. The investigations concerned the following grades of this material: SVP-4205, easy-flowing with a medium-impact strength; SVP-5205, medium-impact strength and high deformation resistance on heating; SVP-2405, very easy-flowing and high-impact strength; SVP-3405, easy-flowing and high-impact strength. All tests were carried out according to the procedures adopted for monolithic plastics.

Short-time tests for flexure on impact have shown that the impact resistance of these materials rises considerably in the presence of a skin 1 mm thick, the impact energy being taken up mainly by the core of the material. It turned out that the type of PS in core and skin has a substantial bearing on the impact resistance. For example,

Table 14.1. Mechanical properties of unfoamed and integral foam PS

Properties	General-purpose PS			High-impact PS		
	unfoamed	IF		unfoamed	IF	
Apparent density, kg/m³	1,050	730	560	1,050	730	560
Ultimate strength, MPa:						
tension	26.6	12.8	9.2	18.5	9.8	6.0
compression	75.0	33.6	21.0	68.5	24.5	12.8
flexure	58.5	36.4	27.3	40.0	37.3	31.0
Elasticity modulus, MPa:						
tension	2,790	1,800	940	1,975	990	645
compression	2,450	1,310	730	1,655	886	382
flexure	2,670	1,910	1,450	2,100	1,480	1,000
shear	880	500	290	640	310	183
Breaking elongation, %	12.1	9.0	6.7	22	12.6	9.84
Hardness of skin (Shore)	82	78	72	80	80	70
Hardness of core (Rockwell)	90	74	50	61	68–75	60

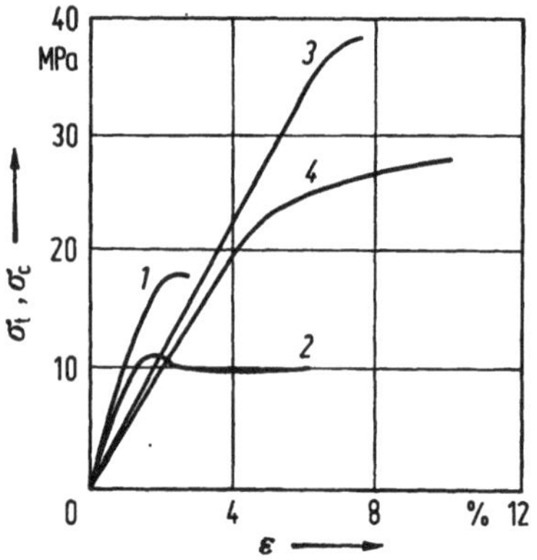

Fig. 14.5. Stress-deformation curves in tension (1 and2) and in compression (3 and 4) of integral PS parts; (1 and 3): Hostyren-SVP-5205, (2 and 4): Hostyren-SVP-3405 (ϱ_a = 800 kg/m³, relative humidity 50%, testing temperature 20 °C [11])

the impact strength of an IF manufactured entirely of a high-impact PS is twice that of a material whose core is made from a medium-impact PS, and whose skin from a high-impact PS. The flexural strength (σ_f) of integral PS foams decreases with apparent density, but to a lesser degree than the impact strength. Specimens based on the medium impact PS are stronger and less sensitive to variations of the deformation rate than the other materials. Tests in tension indicated that the material based on the high-impact PS possesses high strength and resilience, small elongation, a clear-cut yield limit, and a large margin of elasticity (Figure 14.5, curves 1 and 2). The observation that compressive strength σ_c exceeds σ_f by a factor of almost two can be explained by the fact that the stresses were calculated in terms of the initial cross sectional area of the specimens, which decreases under tension and increases under compression.

Long-term tests for flexure (up to 1000 h) have indicated that the greatest deformation (prior to rupture) is in specimens based on medium-impact PS. As the test temperature rose from 25 to 100 °C, the deformation increased substantially from 0.15

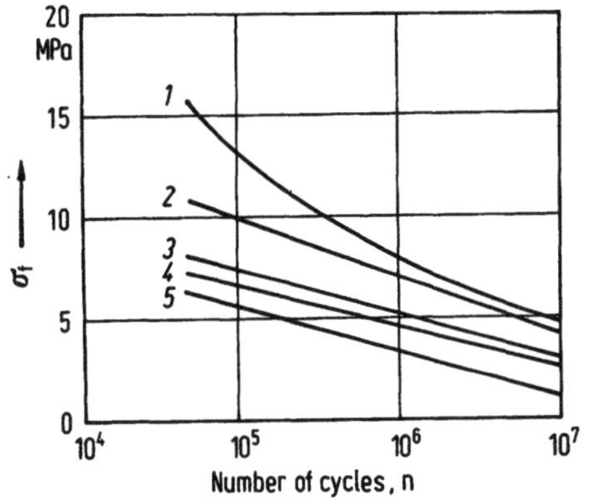

Fig. 14.6. Relationship between the flexural strength of an integral PS and the number of test cycle n; 1: unfoamed PS; 2 and 3: Hostyren-SVP-2405; 4 and 5: Hostyren-SVP-4205; 2 and 4: $\varrho_a = 900\ kg/m^3$; 3 and 5: $\varrho_a = 700\ kg/m^3$ [11]

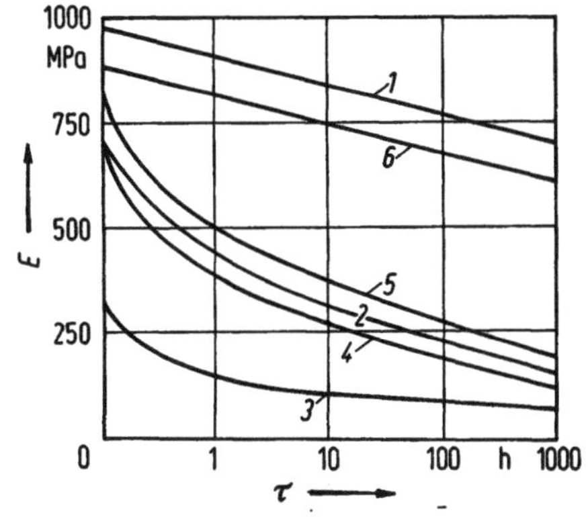

Fig. 14.7. Variation of the flexural modulus with time, for different integral foams; 1 and 2: foam based on ABS at 23 ° and 56 °C; 3, 4 and 5: foams based on PE at 23 °C, and containing, respectively, 0, 10, and 20% mass of fiber glass; 6: foam based on HIPS at 23 °C [13]

to 1.0%. Generally, the deformation under bending loads depends not only on the modulus of elasticity in flexure (a function of the duration of the test, the value of the load and the temperature), but also on the moment of inertia (a function of the cross-sectional area of the specimen). The fatigue (endurance) strength of an integral PS foam under cyclic bending loads increases with the apparent density of the material and depends on the type of PS, with the high-impact PS having the highest fatigue strenght (Fig. 14.6). For IF items undergoing cyclic loads, uneven thickness of the skin leads to earlier cracking. Schleith suggests the introduction of a "reliability factor" (equal to 3) in strength calculations for the above items [11].

The ultimate strength and modulus of *elasticity in flexure* are the main strength parameters for an IF (Fig. 14.7). For the same value of ϱ_a, σ_f is critically affected by the thickness of the skin (δ_s); in the general case, the relationship $\sigma_f = f(\sigma_s)$ is quadratic. By contrast, for the same thickness of the skin, the relationship $\sigma_f = f(\sigma_a)$ is linear (Fig. 14.8) [12, 13].

The *moment of inertia* of an IF increases with decreasing density (Fig. 14.9). However, σ_f decreases strongly below $\varrho_a = 700\ kg/m^3$ (see Fig. 14.8). This decrease cannot be compensated by the growth of the moment of inertia, and the stiffness of the material diminishes.

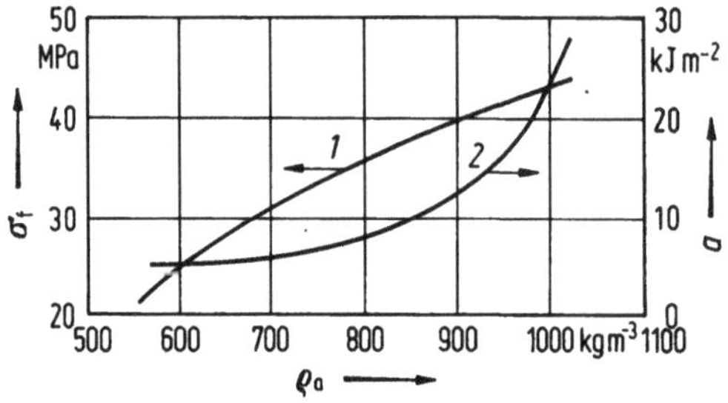

Fig. 14.8. Dependence of flexural strength (1) and impact strength (2) upon the density of an integral PS [8].

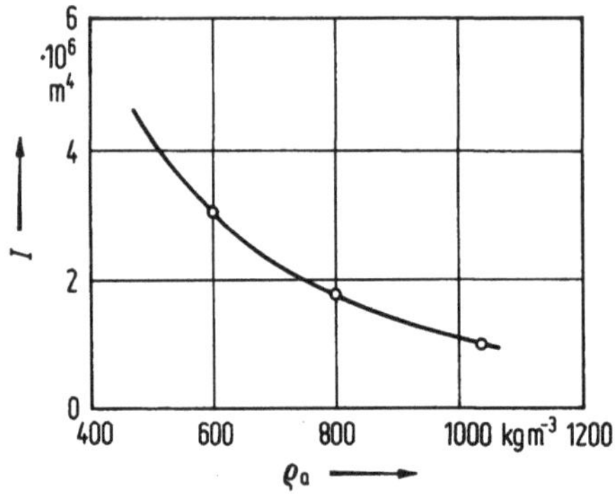

Fig. 14.9. Relationship between the moment of inertia I and the density of an integral PS (Hostyren-SVP-4205); 1, 2, and 3: parts with the same mass, but different cross sections, respectively 60 ×17, 60 ×13, 60 ×10 cm [13]

It has been shown that the *reinforcement* of an integral PS foam with glass fiber improves all mechanical properties [14]. For example, for a 0.635 cm thick IF part molded with 30% foam, a 20% glass fiber filling leads to the following results [15]:

	Reinforced IF	Unreinforced IF
Density, kg/m^3	840	730
Tensile strength, kPa	34.5	19.3
Flexural modulus, MPa	5,170	2,070
Flexural strength, kPa	58.6	34.5
Unnotched impact strength, J/m	44.0	35.0
Heat distortion temperature at 1.8 MPa, °C	190	160

With a view to evaluating the behavior of integral foams as structural components with a reasonable measure of reliability, Kramer [6] has suggested a "reliability factor" criterion. This is described as the ratio of the volume of an IF to its cross sectional area; for integral PS foams, this ratio should be four.

The strength properties of thin-wall PS integral foam (grade Styron 6087 SF, produced by Dow Chemical, USA) are presented in Table 13.10.

14.3.2 Other Properties

Integral PS foams are *combustible materials* (see Table 22.1 below) [16]. The upper limit of their service temperature is a function of the composition, the type of PS and

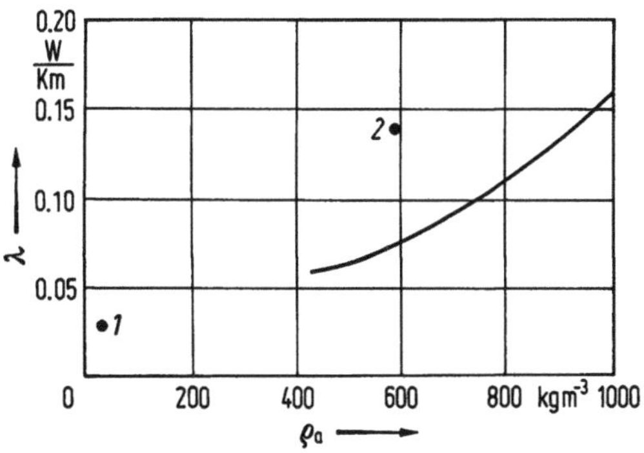

Fig. 14.10. Dependence of thermal conductivity λ upon the density of an integral PS; points 1 and 2 are, respectively, common foamed (non-integral) PS and plywood [8]

the additives, but it is never higher than 70 or 80 °C. Even at 40 °C, it is good practice to lower the admissible loads by a factor of two (see Table 13.13 above).

We have repeatedly emphasized that integral foams are *close to natural wood* in macrostructure and properties. The industrially manufactured IF's close to wood are integral PS foams, and their density, hardness and accoustic properties are very similar. In terms of heat insulation, compressive strength, water resistance, resistance to deformation in moist media, and resistance to attack by fungi, integral PS foams surpass wood (Fig. 14.10). Wood, however, has higher values of σ_f and σ_t, impact strength, and better resistance to attack by atmospheric agents. Even the best types of "artificial" wood, for example, extrusion IF's such as Mira-Wood, Fuju-Wood, and Sturo-Wood (Japan), poorly resist atmospheric exposure and UV [2,3]. Moreover, some plasticizers (for example polybutylene) added to enhance elasticity and reduce brittleness, deteriorate the resistance of integral PS foams to UV light and to photodegradation.

14.4 Applications

The major opportunities for integral PS foams will not come from replacing conventionally molded or extruded plastics, but from *competing with non-plastic materials.*

The substitution of PS integral foams for *wood* has aroused much interest in view of the continuously rising cost of timber and labor, the industry's trend toward rationalization, and the automation of production [17,18].

The main wood-replacement applications are, at present, as decorative or bulk materials, where the PS integral foams have not to compete with the good mechanical properties of wood. Because the modulus of wood is an order of magnitude higher than that of integral PS foam, an IF design must be radically different to compensate for the difference [19].

The penetration of PS foams into the furniture market has come chiefly in applications requiring complex shapes, though they are also used in decorative areas, such as drawer fronts, kitchen cabinetry, complex doors, etc. The ability to reproduce elaborate surface patterns in an IF has become highly desirable, expecially since the high production costs of woodworking is leading to the disappearance of highly carved styling in wood. On the other hand, where surface detail is unimportant, such as in interior structural parts for sofas and chairs, no surface finishing is required for PS parts.

Where wood requires machining, or the rejection rate is high (for certain woods), IF parts in general, and also those made of PS foam, become effective competitors, e.g. for paint-brush handles, broom sticks, handles for kitchen utensils, as well as for wood-like or metal-like frames, basreliefs, incrustations and complex profiles.

Integral foams based on PS are relatively free of *vibrations and resonance*, thus offering *accoustic properties* superior to those of wood. These materials are therefore widely used in audio equipment.

Plastic pallets cannot compete with wood pallets on a straight cost basis, but the inherent disadvantages of wood (susceptibility to insect attack and rotting, moisture absorption leading to swelling and subsequent shrinkage, fungi infection, etc.) compared to IF's, may be changing the picture. Plastic pallets are uniform in weight, hygienic, resistant to a wide range of chemicals; they can be sterilized, and their service life is considerably longer, particularly if used outdoors, etc. Pallets made from a PS integral foam are widely used in food handling and distribution, in the pharmaceutical industry, and in slaughterhouses [18].

A potential area of development is fiber-reinforced integral PS foam, mainly in automobile and container applications [14,15].

14.5 Technical and Economic Analysis

Towards the mid-1980's the prices for different grades of PS are expected to be as follows (¢ per kg): general purpose PS — 110, expandable regular grade — 141,

Table 14.2. Custom IM processes for IF's based on general purpose PS for a flat plaque, 30.5 x 30.5 mm (100% man and machine efficiency; 2 shift operation) [21]

	CIM, 300 oz[a]	CIM, 200 oz[b]	CIM-Mod[c], 200 oz[b]	LPF	TAF on LPF	TCF-Sim[d]
Thickness, mm	9.5	6.3	9.5	9.5	9.5	9.5
Density, $g \cdot cm^{-3}$	1.05	1.05	0.683	0.683	0.788	0.840
Part weight, g	8,350	5,570	5,440	5,440	6,250	6,660
Cycle time, min	4.25	2.0	4.25	4.25	3.25	4.75
Parts/hour	14.12	30.0	14.12	14.12	18.46	12.63
Mold material	Steel	Steel	Steel	Alum.	Alum.	Steel
Mold cost, $	95,800	76,000	95,000	51,900	70,300	105,40
Mold amort., $/hour	8.32	6.60	8.25	4.51	6.10	9.15
Material cost, $/hour	95.86	135.76	62.41	62.41	93.98	68.60
Blowing agent cost, $/hour	—	—	1.69	1.69	2.55	1.86
Manufacturing cost, $/hour	37.27	33.96	33.75	35.65	38.47	46.15
Interest, $/hour	2.58	3.63	1.72	1.72	2.54	1.88
Packaging/shipping, $/hour	7.81	11.06	5.08	5.08	7.65	5.59
Selling price, $/hour	159.29	197.80	123.03	121.57	165.38	153.16
Selling price, $/piece	11.28	6.59	8.71	8.61	8.96	12.13
Selling price, $/g	1.35	1.18	1.60	1.58	1.43	1.81
Total selling price, finished $/g	1.35	1.18	2.09	2.82	2.08	1.81

[a] 300 oz = 85.5 kg;
[b] 200 oz = 56.7 kg;
[c] CIM-Mod refers to a high pressure IF process similar to the USM one;
[d] TCF-Sim refers to the two-component machine of Siemag or Battenfeld.

expandable modified grade — 154, ignition resistant grade — 176, and integral foam grade — 178 [20].

In order to estimate the actual part cost of an IF article, consider — following Throne [21] — the custom molding cost of a flat plaque based on general purpose PS (Table 14.2). Note that the HP and LP processes yield 35% foam, whereas the Hanning TCF process gives only 20% foam. Comparing the selling price per piece, it is apparent that the 9.5 mm LP parts are the least expensive to make, with the exception of the 6.3 mm unfoamed part. It is surprising that the margin between the LP and HP IF's is no more than 10%. The cost of the Hanning process reflects the higher equipment costs.

However, the foam surface quality differs, and some parts may not be acceptable for all applications. This means that *finishing costs* should be added in. The number of steps and the cost of each step depend upon the complexity of the part and the specific application. The "total selling price" given in Table 14.2 is a comparative estimate of the finished cost for a high gloss, "wet look" lacquer furniture application. Note that parts made by the Hanning TCF process are assumed to have a suitable as-molded surface. However, since the mold pressure in the TCF process is much lower than in CIM, the as-molded surface may not be sufficiently glossy for every application. Note too, that articles prepared by the LP process are much more expensive than those made by higher pressure processes. Surprisingly, a solid 9.5 mm CIM part results less expensive than any finished foamed part. Obviously the incorporation of the labor-intensive finishing steps can dramatically influence the decision whether or not to use foamed materials [23–25].

We have already mentioned that because of the large shot capacity and multiple-nozzle arrangement inherent in IF molding, a number of smaller parts can be produced simultaneously by using either one multiple-cavity mold or several smaller molds (Chapter 12). The production of several IF parts per cycle can significantly reduce final part cost, although molding too many parts per cycle may increase the ultimate part cost. An analysis of molding costs shows (Table 14.3) that the best overall set-up is one

Table 14.3. Cost analysis of a typical business-machine housing based on PS and PPO integral foams

Cost factors, $/part	Number of molds in press		
	1	3	6
PS integral foam[a]			
Resin (0.9 kg at 1.38 $/kg)	1.26	1.26	1.26
Mold (7,500 $/mold over 50,000 parts)	0.15	0.45	0.90
Processing (60 $/hour, 24 shots/hour)	2.50	0.83	0.42
Total cost/part	3.91	2.54	2.58
PPO integral foam[b]			
Resin (0.9 kg at 2.1 $/kg)	1.90	1.90	1.90
Mold (7,500 $/mold over 100,000 parts)	0.075	0.225	0.45
Processing (60 $/part, 24 shots/hour)	2.50	0.83	0.42
Total cost/part	4.475	2.955	2.77

[a] 50,000 parts/run, halogen-modified high impact PS; [b] 100,000 parts/run

that employs three molds[22]. While the part cost saving of using a multiple-item mold compared to using a one-item mold is substantial, the difference in cost between three- and six-item mold operations is not enough to justify the extra tooling cost. This is valid for any IF process, i.e. for those based on PS and PPO (Table 14.3).

Polystyrene IF's have the following advantages over PC and modified PPO integral foams: lower cost, multi-source manufacture, excellent material flow. Their limitations are: reduced heat deflection temperature and lower physical and mechanical properties [23].

Moreover, PS parts do not have to be outgased as long as PPO parts prior to painting, i.e., gases used to create the cellular structure do not take as long to come to equilibrium with air at atmospheric pressure. With PPO parts, additional time (up to two weeks) is often required to avoid bubbles that may form during painting. The net result is less handling and warehousing cost in the case of PS, which leads to reduced overall processing costs [25].

Once considered strictly a commodity thermoplastic, PS is now finding new uses in structural or load-bearing applications, while flame-retardant grades are making further inroads into such consumer products as TV cabinets, without compromising impact strength and other important mechanical properties. Because of its cost/performance advantage, PS is expected to recapture applications once lost to ABS, modified PPO, PC and PP in selected applications.

Another benefit of PS is easy processing. Today's commercial PS resins also provide improved flame-retardant formulations. These improvements along with a brandnew styrenic polymer, poly-*para*-methylstyrene (PPMS), promise a bright future for PS products [25].

Additional cost and marketing information on integral PS foams is presented in Chapter 12, 13, and 22.

14.6 References

1. N.N.: Plast. Technol. *21*, No 5, 9 (1975); *22*, No 1, 26 (1976)
2. Berlin, A. A., Shutov, F. A.: Strengthened Gas-Filled Polymers. Moscow: Khimia, 1980 (in Russian)
3. Shibata, O., et al.: Japan Plast. Age *11*, No 2, 15 (1973)
4. Menges, G., et al.: *62*, 151 (1972)
5. Gribens, J. A., Rei, N. M.: Plast. Eng. *38*, No 3, 26 (1982)
6. Kramer, A.: Kunststoffe *64*, 350 (1974)
7. Throne, J. L.: Techn. Papers 31st Ann. Techn. Cong. SPE, Montreal Canada, 1973, 205–210
8. Posetti, J. A., Selian, M.: Mater. Plast. Elast. *15*, No 1, 39 (1978)
9. Hruby, R. F.: SPE Techn. Papers *19*, 303 (1973)
10. Voros, F., Maly, J.: Plast. Hmoty Kaučuk *9*, 297 (1972)
11. Schleith, O.: Kunststoffe *65*, 421 (1975)
12. Ponesicky, Y., Nezbedova, E.: Plast. Hmoty Kaučuk *18*, 16, (1981)
13. Domininghaus, H.: Maschinenmarkt *81*, No 47, 850 (1975)
14. Eckardt, H.: Ind. Prod. Eng. No 2, 170 (1980)
15. Wendle, B. C.: Additives and Fillers for Structural Foams. In: Engineering Guide to Structural Foam. Wendle, B. C. (ed.). Westport: Technomic, 1976, pp. 115–118
16. N.N.: Plast. Eng. *38*, No 1, 24; No 4, 46 (1982)
17. N.N.: Mod. Plast. Int. *10*, No 10, 38 (1980); *11*, No 3, 58 (1981)
18. Semerdjiev, S.: Introduction to Structural Foam. Westport: Brookfield Center: Soc. Plastics Engineers, 1982

19. N.N.: Kunststoffe *72*, 376 (1982)
20. N.N.: Plast. World *37*, No 10, 16 (1979); *38*, No 5, 11 (1980)
21. Throne, J. L.: J. Cell. Plast. *12*, 261 (1976)
22. Sauers, M. E.: Plast. Technol. *21*, No 13, 48 (1975)
23. LaPlaca, J. P.: J. Cell. Plast. *16*, 36 (1980)
24. N.N.: Plast. Technology, *26*, No 1, 15 (1980)
25. Wehrenberg, R. H.: Mater. Eng. *97*, No 2, 26 (1983)

15 Integral Foam Based on Poly(vinyl chloride)

Currently flexible and rigid IF's based on PVC and its copolymers are produced in many countries. The ready availability and the low cost of the raw materials, the absence of warping under the action of moisture, high weather-resistance, good colorability and compatibility with other polymers are the properties which underlie the remarkably rapid growth in the output of these materials, once the technological problems related to the narrow temperature range of processing of this polymer were overcome.

Most commercial integral PVC parts are extruded as prefilled items and pipes, although injection molding is also employed on a modest scale.

15.1 Technology Specifications

The main principles governing the manufacture of an IF by extrusion have been discussed (Chapter 9), and therefore we shall only deal here with the particular aspects of the process as applied to PVC. The extrusion of an integral PVC foam is only possible with very fluid compositions. Hence PVC's having Fikentscher constants K_F in the range of 57—65, as well as copolymers of vinyl chloride with vinyl acetate and high-impact PVC are used [1]. The flow modifiers mostly used are polyacrylates, in particular poly(methyl methacrylate), styrene copolymers, in particular ABS (up to 20 mass parts), and methyl methacrylate-butadiene-styrene. To stabilize the PVC in processing and in service, organotin, barium-cadmium, and lead-tin containing *stabilizers* are used, together with epoxy-soya oil and UV absorbents. The PVC compositions are mixed with considerable amounts of lubricants and a filler (upto 10 mass parts). The *pigments and dyes* used are determined by their compatibility with the other components and their resistance to hydrochloric acid, which is a product of PVC decomposition. The pigment concentration is usually 10 to 30% less than that needed when making monolithic items. If the material is to imitate wood, the dye (3 to 7 mass parts) is introduced as concentrate into the polymer, the K_F constant of which is 10 to 12 units higher than that of the processed PVC.

The CBA's used are azodicarbonamide (0.3 to 1 mass parts), sodium bicarbonate (2 to 8 mass parts), sulfohydrazide, and benzosulfohydrazide (0.5 to 2 mass parts); the last two can also be employed together [2]. It should be stressed that CBA's are almost always used in combination with decomposition activators, as practically all the stabilizers added to PVC to facilitate the processing appreciably lower the gas numbers and increase the decomposition temperatures of the CBA's. Where PBA's are employed, it is necessary to introduce nucleating agents.

The service properties of integral PVC foams (strength, heat and frost resistances, fluidity, adhesiveness, etc.) can be improved by modifying the initial polymer, such as copolymerizing vinyl chloride with other monomers, e.g. vinyl acetate, vinylidene chloride, and diallyl maleate[1,2]. These IF's have some very unusual properties. At temperatures between −60 and +75 °C they behave like elastic materials (E > 700 MPa), between 75 and 95 °C like semi-rigid materials (E = 7–70 MPa), and at 140 °C they retain their shape for 7 days. Over the same period of time their water absorption increases by 20 to 40% at room temperature [3]. To render these IF's chemically resistant, compositions based on mixtures of PVC and copolymers acrylonitrile-butadiene-styrene (ABS) are used.

A typical formulation for a rigid part is (in mass parts): PVC (K_F = 57) 100, stabilizer (tin) 2.5–3.5, lubricant 2.0, surfactant 2.0–3.0, modifier 3.0–10.0, CBA 1.0–1.5, activator-inhibitor 0.2, and filler 2.0 [4].

15.2 Commercial Processes

According to data presented by Kamiyama and Sakai [5] the best properties of stiff integral PVC foams manufactured by the *Aron-HW process* can be achieved with the following process parameters: pressure in the extruder 20 MPa (versus 16 MPa for ABS and 7 MPa for PS); length-to-diameter ratio 25:1; nozzle diameter 200 mm; initial temperature of the three zones of the cylinder 135, 140, 145 °C, final temperature in these zones 150, 155, 160 °C, and nozzle temperature, 150 °C.

Domininghaus [1] reports similar parameters, but specifies that the thread depth-to-worm pitch ratio of the screw must be 2:1 for stiff PVC's and 3:1 for elastic PVC's. The properties of the two integral PVC foams (E-501-N and E-502-N) manufactured by the Aron-HW process are [5]:

	E-501-N	E-502-N
Apparent density, kg/m³	890	700
Tensile strenght, MPa	21–23	15–18
Flexural strength, MPa	38–43	30–35
Compressive strength, MPa	25–26	16–17
Rockwell hardness, scale R	60–70	40–50

If high strength is not required, an integral structure, consisting of a thin skin and an unevenly foamed core, can be obtained using casting techniques in open or closed molds. To this end, PVC plastisol is foamed by compressed air or nitrogen[2], the air pressure should be 0.3 MPa, and the foaming rate 1 l/min. To make finer cells and stabilize the macrostructure, the composition has to contain such as surfactants the sodium and potassium salts of stearic or oleic acid. The foamed composition is poured into a mold which was preheated to 160–200 °C, the mold temperature is rapidly raised to 290 °C (4–5 min) and then slowly decreased to 160 °C (in 100–120 min). The resultant items have ϱ = 340–350 kg/m³ and $\hat{E}_s$ = 0.1–0.3 mm[1,3].

A major trend in the technology of profiled items (ϱ_a = 400–700 kg/m³) is the *coextrusion of foamed and unfoamed PVC*. It has been shown [4] that the manufacture of smooth surface items requires a minimum compression ratio in the combined flow

Table 15.1. Properties of IF parts produced by the Celuka process [6]

Properties	PS		PVC			HDPE
	general purpose	high impact	general purpose	high impact	super high impact	
Density, kg/m^3	350–550	500–600	330–700	600–700	400–500	540–500
Modulus of Elasticy, MPa	1,000–1,600	1,100–1,400	1,400–1,800	1,300–1,400	450–550	450–550
Impact strength, kJ/m^2, at 22 °C	110–150	300–320	390–410	—	530–580	—
Brinell hardness	20–35	35–45	25–40	35–45	15–20	—
Max. operating temperature with no stress, °C	80–85	75–80	70–80	70–75	70–75	85–90
Coefficient of linear thermal expansion from −60° to +60 °C, × 10^5	5–7	6–7	5–7	6–7	7–8	11–12
Coefficient of thermal conductivity, W/m · °K	0.044–0.052	0.058	0.030–0.058	0.058	0.050	0.100
Oxygen index	0.185	0.185	0.305	—	0.275	0.175

channel; improving the "gloss" of the surface calls for higher melt temperatures at the nozzle outlet. However, at temperatures above 190 °C the PVC melt flow is erratic due to sticking to the mold (the "slip-stick" effect). The extrudate surface then becomes uneven, and the density of the item increases sharply because of loss of some of the foaming gas in the extruder spinneret [4].

IF items with *very smooth ("mirror"-like) surfaces* are manufactured by a method developed by Vinatex Ltd. This is a "mobile density" molding process using plasticized PVC [3,4]. The high quality of the surface is achieved by adjusting the foaming kinetics. The mold in which the foaming takes place is provided with a movable core which is removed during the cooling period of the material. The blowing agents generate gas in a narrow temperature interval. The polymer is plasticized in the usual screw kneading machine but at temperatures lower than those usual for plastication. The extrusion nozzle has a number of small (0.38 to 0.63 mm diameter) holes through which the material is injected very quickly under high pressure just when the mold temperature rises sharply, this resulting in the rapid decomposition of the CBA. The injection pressure is 140 MPa, and the plunger travens at 10.2 cm/s. The mobile core is made from a heat-insulating material, such as reinforced phenolic resins. In the course of the foaming, the core is gradually withdrawn from the mold, and this controls both the degree and the rate of cell formation. The surface layer of the material is formed through contact with the cold surface of the mold. To ensure quick filling of the mold, the cross sectional area of the gates must be at least twice that usually employed. This process lowers the density of the plasticized PVC from 1200–1350 kg/m^3 to 850 kg/m^3 (with Shore hardness of 45 to 90). The manufactured items have a maximum mass of 227 g, a thickness of 6.35 mm, and a skin of 1 mm.

15.3 Properties

The main properties of PVC extruded IF parts, compared with other thermoplastic profiles, are presented in Table 15.1, while their fire resistent characteristics are given in Table 15.2 (see also Table 22.1) [6]

As mentioned in Chapter 9, the main *market applications* of extruded IF's are profiles, having maximum rigidity with minimum cost. For a fixed overall density, the deflection stiffness (rigidity) increases with the skin thickness; and if the stiffness is fixed, the maximum skin thickness is required, if the profile is to have minimum

Table 15.2. Fire resistance of PVC Celuka profiles according to British standards [7]

Test Method	Test for	Result
BS.476:1968 Pt.5	Ignitability	Performance P, self-extinguishing
BS.476:1968 Pt.6	Fire propagation	Performance under 11.6, sub-index 4.10, slow propagation
BS.476:1971 Pt.7	Surface spread of flame	Class 1, slow spread
BS.2782:1970, method 508-D	Flammability	Very low flammability

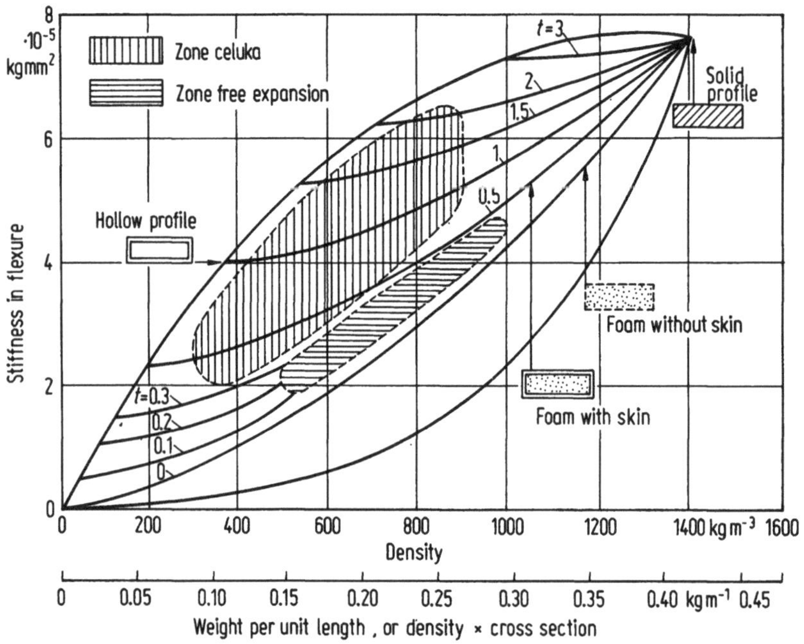

Fig. 15.1. PVC Celuka profiles (30 × 10 mm) with different skin thicknesses (t, mm). Stiffness versus density and weight [8]

weight and, therefore, minimum cost. Figure 15.1 shows schematically the relationship between rigidity and the cellular/skin structure of an IF profile according to Jentet [8] (see also Figure 1.2).

To take an example, let us consider a 30 × 10 mm rectangular PVC profile, whose rigidity should be within the range 4×10^5–5×10^5 kg · mm^2. A free expansion profile would have an apparent density of 800–900 kg/m^3, whereas a Celuka skinned profile needs only a density between 450 and 750 kg per m^3. In practice, the Celuka density would, in fact, be between 450 and 600 kg/m^3, because a skin of about 1 mm thickness can easily be obtained.

It should, however, be mentioned that the same rigidity for the same polymer can be acquired from a rigid, hollow profile with a 1 mm wall thickness, but in many applications it is simply not practical to work with a hollow profile, and the slightly higher cost of a non-hollow Celuka profile is easily outweighed by its other advantages. For example, it is much easier to join a Celuka profile to other items, because of its large surface area. It also has better thermal and accoustic properties for some applications than does a hollow profile.

Jentet[8] calculated the *mechanical properties* of Celuka IF profiles with equations derived for sandwich constructions (see Chapter 21), and he showed that the theoretical calculations are confirmed by practical measurements (Table 15.3).

Steingerwaldt [2] reported that the drop in the apparent density of an integral PVC foam to 700 kg/m^3 (from 1,380 kg/m^3 for the solid plastic) is accompanied by a decrease of the heat resistance of the item by 10 to 15 °C; the flexural and the tensile strengths, as well as the breaking elongation, fall by a factor of 2:

Apparent density, kg/m³	600	700	800	900	1,430 (solid)
σ_f, MPa	18	24	31	38	86
E_f, MPa	920	1,200	1,400	1,700	3,300
T_m, °C	72	73	74	75	80

Considerable efforts are now going into the development of extruded IF *pipe with an inner cellular structure* and good unfoamed outer surface. Such pipes would have lower material costs with only a slight reduction in strength properties. The improved thermal insulation of cellular structures, as well as their lower level of residual stress, would be favorable for drain pipes. It is possible with the extrusion process to produce pipe with a skin either on the inside, or on the outside, or on both surfaces. Research is underway on low pressure pipe for drain, waste and vent-pipes, and for telephone conducts [9]. For example, pipes obtained by the Armocell process (see Chapter 9), and containing 20% glass fiber as filler, have ϱ_a = 830–920 kg/m³, λ = 0.72 W per (m · K), a shrinkage of 2% at 150 °C (15 min.), E = 1,250–1,450 MPa, and σ_t > 21 MPa [3].

Table 15.3. Comparison of experimental and calculated mechanical properties for extruded IF profiles (30 × 10 mm), PVC Celuka [8]

Overall density,	Skin thick-ness,	Experimental		Calculated	
		Stiffness in flexure (E × I),	Flexural modulus,	Stiffness in flexure (E × I),	Flexural modulus,
kg/m³	mm	kg × mm²	kg × mm²	kg × mm²	kg × mm²
500	0.8	365,000	146	362,600	145
670	1.2	480,000	192	478,400	194
800	1.8	500,000	236	594,000	237
1,020	3.2	725,000	290	722,600	289
680	0.5	370,000	148	367,200	146

15.4 Applications

The present markets for extruded IF's are predominantly the building and *construction industries*. The applications range from moldings, architraves, skirting, baseboard, to outdoor sidings and cladding for interior partitions, window and door frame-systems, building panels, walling and roofing [10–13].

Furniture applications include drawer sides, cabinet components, hollow panels, decorative strips and moldings, picture frames, park benches, etc. For these applications PVC extruded foams offer the following advantages: a broad range of color and surface appearance; more intricate design and shapes including embossed surface imprints; ease of assembly by the common fabrication methods used for wooden articles (i.e., sawing, drilling, routing, nailing, stapling, etc.) and also those used for plastics (i.e., welding, adhesive bonding, heat shaping, etc.).

Industrial uses include material handling items such as pallets and fabricated boxes. The good insulation properties of PVC articles are used to advantage as thermal insulation in cold storage rooms and as supports for air conditioners. The good resistance to moisture and chemicals is utilized in elements in the chemical industry [11].

In comparison with unfoamed PVC pipe with the same crush resistance or stiffness, extruded IF pipe based on PVC currently offers polymer savings of 15%, and in the near future this value can reach 30%. Such pipes have a diameter range of 160 to 350 mm, and offer the following advantages over asbestos-cement, tile and unfoamed PVC: low weight reduces shipping and handling problems and possibly allows longer lengths; low density reduces material costs, and foam structure has better thermal insulation which offers energy conservation possibilities and reduces condensation on cold water pipes. But these pipes are limited in application to underground non-pressure drain or sewer systems for medium temperature fields, because of their low mechanical strength and existing industrial standards [8,9,13].

15.5 Technical and Economic Analysis

A discussion about possible investment in a particular material is always a discussion about the future, and scrutinizing the evolution of the prices of competitive materials is interesting. With regard to Celuka IF articles, a number of Celuka licensees were asked by Ugine Kuhlmann (France) to quote the respective costs of a hypothetical profile realized in wood, aluminum, or PVC, in their particular countries (Fig. 15.2) [8]. The results are given in Table 15.4. Moreover, the energy required for producing the same profiles was considered (Table 15.5). The fabrication cost is influenced both by the raw material and manufacturing costs, especially when considering industrial operations like extrusion or automatic machining in which labor is only a small proportion of the costs. In the future PVC Celuka profiles will become more and more competitive with wood and metal because of their low manufacturing energy demands.

Any attempt to draw definite and complete conclusions about the future competition between different solutions from the two tables (Tables 15.4 and 15.5) would be ambitious but, country by country, and according to each particular situation, constructive discussions could start from the above basis, which was developed by Jentet [8].

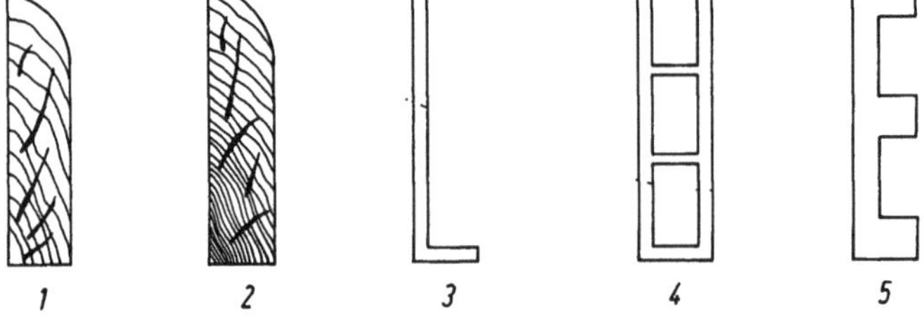

Fig. 15.2. Various types of profiles for cost estimates (Table 19.4); 1: soft wood, 2: hard wood, 3: extruded aluminium 1.5 mm thick, 4: unfoamed PVC 1.5 mm thick, 5: integral PVC Celuka foam 6.0 to 8.0 mm thick [8]

Table 15.4. Comparative price estimates for wood, metal and PVC profiles[a] (150 × 25 mm) in different countries [8)]

Material	UK	USA	FRG	Australia
Soft wood	£ 113/m³	Southern Pipe Grog 2, $ 188/M board food	East Europe Redwood, DM 158/m³	Western Red Cedar, $ 102/m³
Hard wood	Iroko, £ 161/m³	Poplar 2B Grade, $ 175/M board foot	Shipo Mahagoni, DM 175/m³	Merranti, Iroko, $ 142/m³
Extruded aluminium	Ingot, £ 630/t	$ 950/t	DM 4350/t	$ 1770/t
PVC unfoamed, or PVC Celuka	Polymer, £ 340/t	$ 628/t	DM 1600/t	$705/t

[a] see Fig. 15.4

Table 15.5. Total Energy Consumption in the production of Profiles (100 × 25 mm) from Different Materials [8)]

Material	Density, kg/m³	Bulk Material				Energy Consumption	
		TDE/T[a] for feed stock	TDE/T for conversion	Total TDE/T	Total Kcal/cm³	Kcal/cm of profile	Finishes[b], Kcal/cm
Aluminum	2,700	–	5.6	5.6	2,600	409.5	11.7
PVC (unfoamed)	1,400	0.55	1.4	1.95	465	140.6	–
PVC (Celuka)	500	0.55	1.4	1.95	166	130.6	–

[a] TDE/T = Ton of Oil Equivalent per ton of material produced;
[b] the finish is 25 micron anoding

In general, the most frequently discussed point is the comparison of Celuka profiles as wood and metal substitutes. The Celuka product is not so much a competitor for other kinds of plastic, but rather opens up a new family of materials and can be introduced into product areas where unfoamed plastics or conventionally foamed (non-integral) plastics are not able to penetrate for technical and/or economical reasons.

15.6 References

1. Domininghaus, H.: Plast. Mod. Elastom. *26*, No 8, 110; No 9, 91 (1974); Maschinenmarkt *81*, No 47, 850 (1975)
2. Steingerwaldt, F.: Plastverarbeiter *26*, 588 (1975)
3. Berlin, A. A., Shutov, F. A.: Strengthened Gas-Filled Polymers. Moscow: Khimia, 1980 (in Russian)

4. Barth, H.-J.: Europ. J. Cell. Plast.: *2*, 111 (1979)
5. Kamiyama, T., Sakai, K.: Japan Plast. Age *9*, No 2, 39 (1971)
6. Celuka: Technical Pamphlet of Ugine Kuhlmann, Paris France, 1976
7. Saundseal Window: Technical pamphlet of the Crittal Window Ltd., Catalogue No 14, Manor Works, Braintree, Essex, England, 1977
8. Jentet, P.: J. Cell. Plast. *15*, 151 (1979)
9. Kiessling, G. C.: J. Cell. Plast. *12*, 337 (1976)
10. N.N.: Plastverarbeiter *30*, No 7, 404 (1980); *31*, No 4, 237 (1981)
11. N.N.: Plastverarbeiter *31*, No 8, 940; No 9, 1094 (1981)
12. N.N.: Mod. Plast. Intern. *13*, No 6, 34 (1983)
13. Shutov, F. A.: Intern. Symposium on Plastics In Building, Liège Belgium, 1984

16 Integral Foam Based on Polyolefins

16.1 Technology Specifications

16.1.1 Injection Molding

Integral foams based on PE, PP, and their copolymers can be manufactured by all processes designed for obtaining IF's, i.e., by low and high-pressure injection molding, extrusion, and rotamolding [1–5]. Compositions are foamed using PBA's (nitrogen) or CBA's, the latter frequently being employed as concentrates (see Chapter 2). In the '*Phillips Petroleum*' *process*, integral foam items are manufactured on standard injection molding machines by feeding the concentrates and other additives, mixed with pelletized PO, into the plasticator. High counterpressure in the plasticator reduces the plasticization time. Injection is carried out at a low pressure and without additional feeding. The use of the concentrated BA differs mainly in that the composition is injected into a cold mold, both the female and male dies being subjected to deep cooling. The distribution channels and inlets have to be 2 or 3 times larger than those used when molding unfoamed polymers. The thickness of the mold walls should not be less than 2.5 to 3.8 mm. The most uniform core structure is achieved by using linear PE with a narrow molecular-mass distribution (MMD) [2,3].

Low-pressure injection molding can be implemented using both special and standard molding machines. In any case, however, the concentration of the BA has to be fairly high (Fig. 16.1), some 50 to 100% higher than for integral PS. It is good practice to use a high injection rate. Melt temperatures for polyolefin foams vary between 200 and 300 °C; the cooling time is 30 to 40% longer than for integral PS foams; the

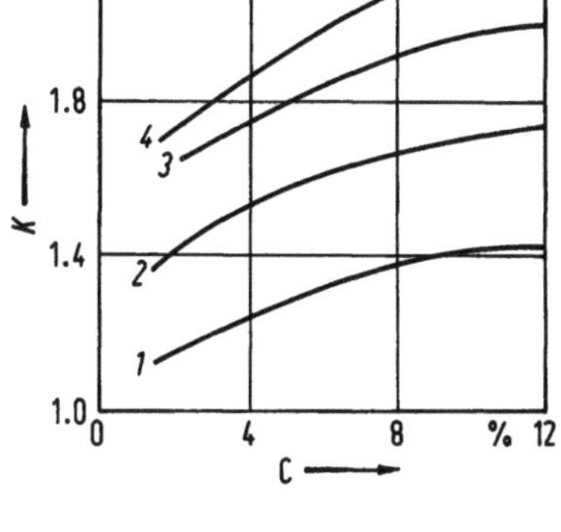

Fig. 16.1. Relationship between the foaming coefficient K for integral PP foam and the concentration of the blowing agent (nitrogen) C, for IF articles of varying thickness; 1, 2, 3, 4: respectively 3, 5, 7.5, and 10 mm [4]

minimum thickness of fabricated items is 6 or 7 mm; and the shrinkage of the items in the mold (after cooling) is very high, amounting to 2.6% for PE foam at $\varrho_a = 800$ kg per m^3, compared to 0.7 and 2.0% for integral PS and PP foams, respectively, with the same density [6].

As reported by Tamura [7] the larger densities of the polyolefins (PO) call for low injection rates. The temperature decreases and, therefore, the viscosity of the melt rises considerably because of the long flow time. Increasing the injection rate from 200 to 500 g/s leads to a rise of the foaming coefficient of PE at low pressure from 1.8 to 2.7 ($\delta_a = 5$ mm). At high injection rates (over 500 g/s), the density of the foam is practically unaffected by either the thickness or the mold temperature (between 40 and 100 °C). The high injection rate leads to a laminar flow of the melt and eliminates the "multi-layer" character of the items due to turbulence in the front of the melt flow. It should be borne in mind that the density of IF's based on PE can only be changed within narrow limits, since an increase in the degree of foaming to 35–40% rises the number of voids in the core and sharply lowers the flexural modulus of elasticity. The drop in flexural modulus E_f is severe when the length of travel of the melt in the mold exceeds 70 times the distance between mold walls; an increase in the amount of BA is then no longer an effective means of lowering the density.

The *multi-station injection molding machines* made by Desma-Werke (FRG). are an example of specialized equipment for producing integral PO foams. They combine several molds and presses with a single worm plasticator [8]. The mobile injector part moves from one mold to another. By accurately controlling the volume of the melt and the injection rate, up to 15% of the raw material may be saved.

To obtain *two-component IF's based on HDPE*, a technique developed by Bayer AG (see Chapter 7) uses a composition with the following proportions (mass parts): pelletized PE, 100; freon-113, 5; magnesium stearate, 0.25; and nucleating agents, 0.5 to 1%. Items in the form of disks (diameter 120 mm and thickness 15 mm) have $\varrho_a = 450$ kg/m^3, $\varrho_c = 250$ kg/m^3, and $\delta_s = 2$ mm [9].

16.1.2 Extrusion

Extrusion methods can be used for obtaining integral PO foams (see above, Chapter 9). Profiled items from 5 to 30 mm thick, as well as pipes can be produced using methods similar to those for processing integral PS foams. Techniques for producing extrusion materials based on prefoamed PO pellets and glass-filled HDPE have been disclosed in a number of patents [1,2].

IF items based on *cross-linked PO* can be manufactured according to the same principles as those underlying the production of non-integral PO foam [10]. Japanese scientists have developed methods for producing chemically and physically cross-linked integral foams, based on PE, PP, and their mixtures [11].

The extrusion method for manufacturing chemically cross-linked PE consists of the following steps: A double-worm extruder, having three heating zones, receives a mixture of PE, CBA, 4,4'-oxybis-(benzosulfonylhydrazide), dicumyl peroxide and a filler. The temperature in the first heating zone equals the melting point T_m of the polymer, but remains below the decomposition temperature T_{dec} of the peroxide; freon is injected into the melt as it exits from this zone. In the second zone, the melt is heated to T_{dec} of the CBA and the peroxide, the residence time of the melt in this zone corresponds to 1 or 2 half-lives of the peroxide. On exit from the third zone, the temperature of which is somewhat lower than that of the second zone (to avoid decomposition of the polymer), the material is extruded with a cross-section measuring 1.3 × 1.3 cm, foamed ($\varrho_c = 80$ kg/m^3) and then cooled.

The manufacture of a physically cross-linked integral PE foam is based on cross-linking the unfoamed matrix (Furukawa process) [10]. The polymer sheet containing a CBA (ACA) is irradiated with an electron beam, the angle of incidence of the beam being varied to adjust the degree of cross-linking, and therefore the properties, of different zones of the IF items. For example, varying the angle of incidence from 60 to 90° changes the content of the gel fraction in the finished material from 67 to 43%, ϱ_s from 95 to 35 kg/m^3, and the diameter of the core cells from 0.105 to 0.085 mm. This method makes it possible to manufacture IF items with a "cross linked" core and an "non-cross linked" skin (ϱ_a = 200 kg/m^3; δ_a = 5 to 20 mm) [14].

16.2 Properties

The readily adjustable thickness and density of items made from integral PE foams mean that the properties of the final items can be varied over a wide range (Fig. 16.2).

The data in Table 16.1 and Table 15.1 (Chapter 15) are evidence that the *absolute strengths* of PE integral foams are lower than those of the unfoamed polymers. However, these results concern IF items, the thickness of which was approximately half that of the monolithic items. It is known that the modulus of elasticity of an IF increases with the cube of its thickness. As a result, the stiffness of an IF item is 3.4 times higher than that of a monolithic plastic item of the same mass.

The strength of commercial integral PP foams is very high. For example, at densities of ϱ_a = 600–650 kg/m^3, the specific strength of these materials is three times higher than that of isotropically foamed PP, 12 times higher than that of aluminum and 14 times higher than that of steel [19]. IF items based on PP have a high resistance to thermal ageing; they can be used at 80 to 100 °C for several years[6]. Their resistance to UV is enhanced by introducing lamp black into the starting composition.

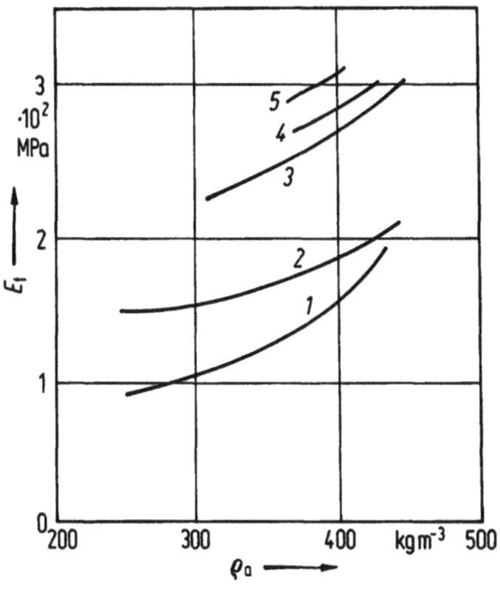

Fig. 16.2. Relationship between the flexural modulus and the density ϱ_a of integral PE foam, for varying skin thickness; 1, 2, 3, 4, 5: respectively 0.5, 1.0, 1.5, 2.0, and 2.5 mm [12]

Table 16.1. Comparison of the physical properties of integral foam and unfoamed HDPE materials [13)]

Parameter	HDPE-based IF	Unfoamed HDPE
Apparent density, kg/m³	550	960
Tensile strength (along the skin), MPa	7.0	21–35
Compressive strength (across the skin), MPa	4.9	17
Shear strength, MPa	8.3	21
Flexural strength, MPa	12.0	36
Flexural modulus, MPa	670	1,050
Linear expansion, °C⁻¹		
along the skin	4×10^{-5}	4×10^{-5}
across the skin	$3 \cdot 10^{-5}$	—
Process shrinkage, %	1.8–2.0	2.2

Table 16.2. Mechanical properties and costs for injection molded HDPE parts with various contents of asbestos filler[a] [16)]

Material	Specific density, kg/m³	Asbestos content, wt.%	Flexural modulus, MPa	Flexural strength, MPa	Coeff. expansion, 1/ °C·10⁻⁴	Mass cost, cts/kg	Volume cost, $/m³
Unfoamed part[b]	962	0	1,120	27.3	1.20	63.9	618
	1,027	10	2,100	42.7	—	67.8	701
	1,101	20	3,500	55.3	—	71.8	795
	1,186	30	5,180	70.0	0.23	75.8	904
Integral foam[b]	700	0	770	23.8	1.24	63.9	450
	730	10	1,960	33.6	0.68	67.8	498
	760	20	2,590	40.6	0.53	71.8	549
	780	30	3,360	49.0	0.40	75.8	594
Integral foam[c]	650	0	560	12.3	1.30	63.9	424
	660	10	1,295	18.6	1.05	67.8	443
	670	20	2,170	34.3	0.92	71.8	481
	680	30	2,940	38.5	0.85	75.8	518

[a] high purity, high aspect ratio colloidal chrysotile asbestos, grade RG-600;
[b] impact grade homopolymer with melt index of 0.72;
[c] copolymer with a density of 0.953 g/cm³ and a melt index of 2.0

The coefficient of *thermal conductivity* of integral PO is higher (in particular, for heavier items) than that of integral PS foam (see Fig. 3.7).

The *fire resistance properties* of PP parts will be presented in Table 22.1.

The *impact strength and resistance* to variable bending loads of integral PO foams are higher, and their stiffness lower, than those of integral PS foams. To minimize creep and raise stiffness, integral PE foams (Table 16.2) and integral PP foams (Table 16.3) are filled with glass fiber, talc, mica, or asbestos (up from 10 to 40%) [16a)]. It has been shown, in particular, that filling with 20% (mass) of glass fiber, 3 to 6 cm long, doubles the modulus of elasticity; it also raises σ_t by a factor of 1.8, and σ_f by a factor of 3; the thermal resistance increases from 40 to 70 °C, and the fire resistance

Table 16.3. Properties of talc filled polypropylene integral foam [17]

Content of talc, %	Density, kg/m³	Flexural Modulus, MPa
0	730	910
10	770	1,505
20	880	1,890
40	980	2,870

by a factor of 3. However, the impact strength drops by a factor of almost 2 (see Figure 14.7).

A process developed by Solvay (Belgium) produces integral PE foam, *filled with glass fiber* (20%) and having a density of $\varrho_a = 800$ kg/m³. The properties of this material are much better than those of the unfilled IF: the creep drops by 30%, whereas the stiffness and E_f increase 2 to 4 times [18,19].

Harris [15] reports the development of new types of high-strength integral PP foams *filled with silicone microspheres* (20%). These materials may be called integral-syntactic gas-filled plastics. Another achievement in PO technology is the creation of one-sided IF's, with a compact skin on one side only [2].

16.3 Applications

The main areas of application of PO integral foams are the same as those for any other IF's: substitution for wood, metal, and solid plastics, especially for complex-shape and many-sub-unit parts.

PO integral foams are well established and successful materials for trays, handling containers, and crates. These parts are light, durable and hygienic, they won't rot, rust, or splinter.

An all-plastic, in-ground swimming pool has been developed for leisure use. The swimming pool uses a 0.75 mm PVC liner supported by a modular PE extruded wall system which is assembled with interlocking dovetail joints [18].

As far as new uses of extruded IF articles are concerned, the trend is toward the design and production of large complex parts for exterior and interior applications, such as panels, doors, and cabinets.

Recently, a rotamolded IF camper top based on a cross-linked PE outer skin and a foamed HDPE core has been developed using the "two-drop" process (see Chapter 10) [20]. Part weight was cut by a total of 4.5 kg (originally it weighed 61.3 kg) and the switch in CBA agents to Celogen AZ translated into annual savings of more than $ 6,300 in production costs.

An integral foam based on PO is also being used for battery compartment lids for commercial vehicles and mobile construction plants, particularly when the batteries are enclosed in housings mounted under the floors of the bodies or cabs. For this application HDPE integral foam is used as it is a good electric insulator, resistant to acid attack, and its impact resistance remains high at very low temperatures (Fig. 16.3).

A good example of an under-the-hood application for an IF based on HDPE is a battery housing for protecting the battery case from damage by vibration; a complementary lid of the same material adds to the appearance of the application. A 12-volt, 135 ampere-h, battery, weighing 7.2 kg in a hard-rubber case, weighs only 2.5 kg clad in HDPE integral foam, and at the same time the cost is lower and the specific power characteristics are improved [1].

An outstanding IF product is a bedstead produced entirely as an IF polypropylene molding. The structure is rigid and provides a firm base to support the mattress. Optional headboard and sidetables, also molded from PP integral foam, can be clipped into position, not requiring fastenings nor tools for the assembly.

Fig. 16.3. Waste water collector made from Vestolen A integral PE foam (Courtesy of Chemische Werke Hüls AG, Marl, FRG)

Another outstanding product is a washing machine outer tank, which replaces 16 metal parts, and is the largest component so far made in glass-reinforced PP integral foam [21].

A unique and highly styled instrument panel based on filled PP (40% talc) IF has been developed by the Ford Motor Company for a heavy truck. It helped reduce panel complexity, and improved routing of electrical wiring and air pipes. The panel, size 2,180 × 790 × 400 mm, weighs 16 kg. Talc filled PP integral parts, weighing up to 10 kg, have been molded for seat frames, door panels, load floors and roof-tops for the automobile industry [17].

An IF based on flame-retardant PP is in service in form of airflow-guiding cowls for radiators and fan installations, on road vehicles and in stationary construction equipment.

Four-wheel "wheelbarrows" produced in England have a base molded from PP integral foam [22] that can carry loads up to 50 kg, and comes in sizes (areas), 1,000 by 600 mm, and 1,000 by 700 mm. The new bases replace conventional steel or timber platforms. The barrows are 8–12% more expensive than those of conventional design. But they have passed many types of impact tests, including drop tests at subzero temperatures. The surface hardness and good frictional properties of the integral PP foam prevent excessive scratching and scuffing in service, while keeping loads from sliding.

Another recent development is a line of glass reinforcement (10%) integral PE foam transom platforms; they are more economical to produce and easier to maintain than the teak platforms traditionally used in luxury yachts [3]. A new mechanized process reportedly [23] produces automobile seats that are lighter and less expensive than conventional types; it uses vinyl/fabric laminates, foam backfill, and PP integral foam for the seat shell.

16.4 Technical and Economic Analysis

Integral PE foams have been used extensively in the USA and Europe, particularly for material handling applications, because until recently PE raw material was cheaper than PP. But over the last few years the cost of PE has increased[24-27] (see Chapter 23).

Integral PP foams are comparatively recent entrants in the IF market, but *their use is increasing* rapidly because of the extensive range of grades and properties available, as well as the favorable price trend compared with other thermoplastics and products (Table 16.4). From these data it is clear that PP should feature strongly in the future development of IF's. Many exciting developments have already led to higher output growth rates compared to other plastics such as high density PE, PS, modified PPO, etc. [25].

Integral PP foam with a 20% *mica loading* has the same weight as unfilled solid PP, but is 15% lower in material costs, has three times the stiffness, 30% greater flexural strength, and 15% higher heat deflection. When a 40% talc/solid PP is compared with a 40% mica/PP integral foam, the advantages of the filled IF include a 33% material cost saving and a 50% greater stiffness [25].

Table 16.4. Relative price increase of various materials referred to polypropylene, from 1963 to 1980

Material	Relative price increase[a]
Wood	3.8
Glass	3.2
Paper and board	2.7
Steel	2.6
Aluminum	2.6
Polystyrene, high impact	2.4
ABS-plastic	1.9
Poly(vinyl chloride)	1.8
Polyethylene, low density	1.7
Polyethylene, high density	1.4
Polypropylene	1.0

[a] (Price of material 1980/price of PP 1980)/(price of material 1963/price of PP 1963)

16.5 References

1. Semerdjiev, S.: Introduction to Structural Foam. Brookfield Center: Society of Plastics Engineers, 1982
2. Berlin, A. A., Shutov, F. A.: Strengthened Gas-Filled Plastics. Moscow: Khimia, 1980 (in Russian)

3. N.N.: Mod. Plast. Int. *11*, No 3, 24 (1981)
4. Shibata, O. et al.: Japan Plast. Age *11*, No 2, 15 (1973)
5. Caritat de Peruzzis, C., Lempereur, C.: Rev. Gen. Plast. Caoutch. *52*, 601 (1975)
6. Schleith, O.: Plastverarbeiter *26*, 703 (1975)
7. Tamura, S.: Sange Kikay No 238, 4 (1970)
8. N.N.: Kunststoffe *66*, 81 (1976)
9. Menges, G., Schramm, K.: Kunststoff-Rundschau *20*, No 4, 165 (1973)
10. Berlin, A. A., Shutov, F. A.: Chemistry and Technology of Gas-Filled High Polymers. Moscow: Naika, 1980 (in Russian)
11. N.N.: Polym. Age *6*, No 3, 92 (1973)
12. Simmonds, R. C.: Mod. Plast. Encycl. *47*, No 10-A, 258 (1971)
13. Gonzalez, H.: J. Cell. Plast. *12*, 49 (1976)
14. Pramuk, P. F.: Polym. Eng. Sci. *16*, 559 (1976)
15. Harris, C. A.: Plast. Mod. Elast. *29*, No 7, 60 (1977)
16. Michno, M. J.: 32nd Ann. Techn. Conf. of the SPI, Washington USA, 1977
16a. Burton, R. H., Hornsby, P. R.: J. Mater. Sci. Letters *2*, 195 (1983)
17. Dominick, G.: J. Elastom. Plast. *11*, No 2, 133 (1979)
18. N.N.: Mod. Plast. Int. *9*, No 11, 46 (1979); *10*, No 8, 22 (1980)
19. N.N.: Plast. Design. Process. No 4, 9 (1981)
20. Kravitz, H. et al.: Plast. Technol. *25*, No 11, 63 (1979)
21. N.N.: Europ. Plast. News *8*, No 2, 32 (1981)
22. N.N.: Europ. Plast. News *5*, No 5, 58 (1978)
23. N.N.: Kunststoffe *72*, No 2, 101 (1982)
24. N.N.: Plast. World *38*, No 5, 11 (1980)
25. Norgan, M. R.: Cell. Polym. *1*, No 2, 161 (1982)
26. Semerdjiev, S., Popov, N., Tuleshkov, N., Sgureva, I.: Injection Molding of Structural Foams. Sofia: Technika. 1983 (in Bulgarian)
27. Ahmadi, A. A., Hornsby, P. R.: Plast. Rub. Proc. Appl. *5*, No 1, 35—49, 51—59 (1985)

17 Integral Foam Based on ABS-Copolymers

ABS-plastics, i.e. plastics based on copolymers of styrene and acrylonitrile (15–30%) and blended with butadiene or butadiene-styrene rubber (5–25%), differ from high-impact PS by their higher tensile strength and stiffness, and from PE and PVC by their greater hardness and flexural strength. These properties, along with good water and chemical resistance, are retained by foamed materials, including integral foams. This explains why the proportion of ABS in total world IF output amounted to 30% until recently. Today, however, this figure is somewhat lower because of the increasing manufacture of IF's based on polyurethanes and PO's. The industrial production of ABS began in 1967, and at present several dozens of commercial grades are available in industrially developed countries[1–3].

17.1 Technology Specifications

Integral ABS foams are manufactured with a wide range of ϱ_a, δ_a and δ_s, using practically all methods of fabricating integrally foamed structures, i.e. injection molding, extrusion, rotation molding, and casting.

We shall not deal in detail with the operating parameters of the equipment involved, as they are similar to those for obtaining PS and PO-based IF's (see Table 9.2). We shall only consider those parameters specific to manufacturing ABS foams.

The starting formulation contains *a pelletized copolymer ABS* (20:15:65) *and low-temperature pelletized CBA* (0.1 to 15%) the CBA may also be in concentrate form (matrix mixtures)[4]. In some instances, *cut glass fiber* (up to 15%) is added as filler (see Table 17.1 below). The composition is held in the plasticator at 200 to 230 °C, achieving complete CBA decomposition, to lower the density of the final item. A holding time of 4.5 min reduces ϱ_a from 1,010 to 700 kg/m^3. The cooling time of an item is determined by its density and size, and by the mold temperature; it may range from 30 to 300 s.

To obtain large items (measuring 2.2 × 1.1 m) with complex configurations, Milner Corporation (USA) has developed a *semi-automatic molding process* (the Marban process). It requires no external pressure, but a specially designed molding machine[5]. This machine consists of two cylindrical chambers connected by a closed conveyor system. A mobile mold containing the polymer mixture passes through the first chamber (the heating chamber) where the polymer is foamed, then enters the second chamber (the cooling chamber). The throughput is very high, up to 725 kg/h, with the unit mass of items of up to 70 kg, with a density of 400 to 800 kg/m^3.

An interesting extrusion process manufactures pencils by extruding an integral ABS foam directly onto a graphite rod[6]. As a rule, the molding time of extruded integral ABS items is 10 to 25% longer than that for the production of solid ABS plastics; the density is not below 500 kg/m³.

A detailed analysis of the rheological and process parameters of the composition and their effect upon the structure and properties of integral ABS foams was made by Throne [7].

17.2 Properties

17.2.1 Molding Parts

The strength properties of ABS integral foams depend on the method of manufacture as well as on density and size of the items (Fig. 17.1). Strength and stiffness of integral ABS foams are greater than the respective values for IF's based on PE and high-impact PS [1,6,7] (see Fig. 14.7).

A comparison of the physical properties of an unfoamed ABS part versus a foamed part with a 10% density reduction (Table 17.1) shows a deterioration of these properties as a result of foaming. However, improved tensile, flexural and heat distortion

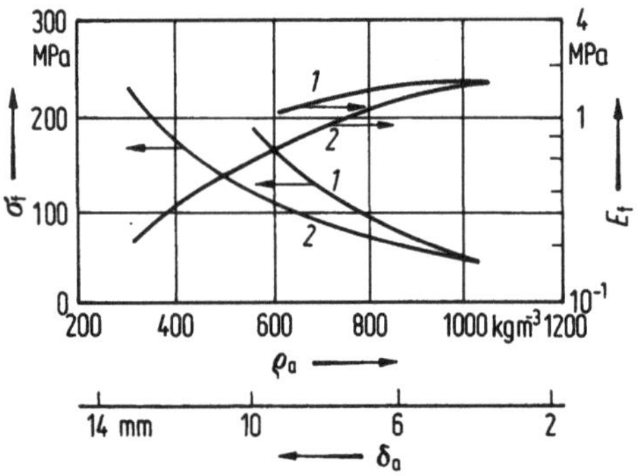

Fig. 17.1. Variations of the flexural strength and flexural modulus as functions of the density ϱ_a and thickness δ_a of an integral ABS component produced by (1) injection molding casting and (2) the casting molding process [13]

Table 17.1. Properties of unfoamed material, conventional non-integral foam, and glass fiber reinforced integral foam, based on the same ABS-copolymer (10% density reduction in the foams)

Property	Unfoamed	Foamed	Integral foams			
			Un – reinforced	5% GF	10% GF	15% GF
Tensile strength, MPa	44.1	32.2	28.7	35.0	37.8	46.2
Tensile modulus, MPa	2,240	1,820	2,030	2,380	3,010	3,570
Elongation, %	25	10	3.1	2.9	1.6	1.2
Flexural strength, MPa	77.0	61.6	63.0	64.4	67.9	90.3
Flexural modulus, MPa	2,310	2,100	2,240	2,310	3,080	3,290
Heat distortion at 1.85 MPa, °C	94.4	85.0	80.5	86.1	87.8	92.2

Table 17.2. Physical and mechanical properties of an ABS integral foam, Cycolac® FBK, Borg-Warner Corp. [9]

Property	Value	ASTM Test
Specific density, kg/m^3	750–850	D-792
Tensile strength, MPa	17.5–21.0	D-638
Flexural strength, MPa	31.5–45.5	D-790
Flexural modulus, MPa	1,575–1,890	D-790
Falling-dart impact, J	13.5	D-3029
Shore hardness, Scale D	75–85	D-785
Mold shrinkage, cm/cm	0.005–0.008	D-955
Deflection temperature at 455 kPa load, °C	77–83	D-648
Flammability	V-0	UL Subject 94

properties can be obtained by filling with short glass fibers; as the content of glass fiber increases, the properties improve [8].

The most popular ABS integral foams belong to the '*Cycolac*' *family*, which is produced by Borg-Warner Corporation, by a LP injection molding process. These materials have very good mechanical properties, carry a flame rating of 94 V-O (at 6.4 mm thickness and a density of 750 kg/m^3), and have good performance characteristics (Table 17.2) [9].

A drop of the temperature from 23 to 0 °C is accompanied by an appreciable increase (by 25 to 30%) in the strength properties of integral ABS foams; a rise of the temperature to 74 °C sharply deteriorates the properties:

Temperature, °C	74	23	0
Tensile strength, MPa	5.6	10.5	14.0
Flexural strength, MPa	8.4	17.5	24.5
Modulus of elasticity in flexure, MPa	420	630	770

The coefficient of thermal conductivity of these materials is quite low ($\lambda = 0.07$ to 0.09 W/(m · K)) and does not depend much on the ambient moisture content, because of a small water absorptivity (lower than that of wood) [6]. The high dielectric values of IF's based on ABS are the reason why they are widely used as electric insulating elements in electric appliances and radios. Some comparative data on the electric insulating properties of solid ABS and integral foam ABS (both dry and moist) are given below [6,9]:

Property	Solid ABS	IF ABS, dry	IF ABS[a], moist
E_{ds}, kV/mm	—	90	70
$\varkappa_s$, Ohm	—	1×10^3	4×10^{10}
$\varkappa_v$, Ohm · cm	1×10^{13}	4×10^{14}	4×10^{14}
ε at $f = 10^3$ Hz	3.5	2.2	2.2
tg δ at $f = 10^3$ Hz	0.020	0.014	0.014

[a] after exposure for 4 days at 23 °C and 80% relative humidity.

17.2.2 Coextruded Sheets

Coextruded ABS foam-core sheet has engineering properties similar to those of solid ABS, although somewhat reduced according to the composite density. Table 17.3 compares rigidity, flexural properties, dart impact, and heat distortion under load, of three ABS sheets having different composite thickness, but the same skin thickness. The major advantage of increasing sheet thickness (by increasing the foam core) is that it greatly improves the flexural rigidity. Doubling the sheet thickness and reducing the specific density to 700 kg/m^3, increases the stiffness 6- to 7-fold.

Table 17.4 compares the properties of three ABS coextruded foam sheets having a constant overall thickness with skin thickness varying from 0.5 to 1.5 mm. As the specific density of the sheet increases, the flexural strength increases proportionally. The flexural modulus and sheet stiffness also increase, but the change in rigidity is small in comparison with that obtained by increasing the overall sheet thickness.

Table 17.3. Effect of thickness on properties of a foam-core ABS sheet, with equal 1 mm skin [11]

Properties	Sheet thickness, mm		
	6.35	9.5	12.7
Specific density, kg/m^3	800	730	700
Flexural rigidity, kg/cm	13	39	87
Flexural strength, MPa	47.1	39.1	34.4
Flexural modulus, GPa	1.8	1.5	1.45
Falling-dart impact, J			
at 23 °C	22–25	27–30	38–41
at −40 °C	11–14	14–16	19–22
Deflection temperature, °C under load			
1820 kPa	79	82	92
455 kPa	89	89	92

Table 17.4. Effect of skin thickness on the properties of a 9.5 mm ABS coextruded foam sheet [11]

Properties	Skin thickness, mm		
	0.5	1.0	1.5
Specific density, kg/m^3	700	730	770
Flexural rigidity, kg/cm	35	39	44
Flexural strength, MPa	35.1	39.1	44.6
Flexural modulus, GPa	1.4	1.5	1.7
Falling-dart impact, J			
at 23 °C	11–16	27–30	52–55
at −40 °C	8–11	14–16	25–27
Deflection temperature, °C			
1820 kPa	82	82	84
455 kPa	89	89	90

The important result of increasing the skin thickness is the substantial improvement of the toughness with no increase in overall thickness, and only a small increase in density [11].

The *fire resistance properties* of ABS parts are presented in Table 22.1.

17.2.3 Coextruded Pipes

Table 17.5 shows the falling-dart impact resistance and the stiffness of ABS foam-core pipes with different wall thickness. By increasing the skin thickness, at 3.17 mm wall thickness, the impact and stiffness criteria (ASTM 2751 sewer ABS specification) can be approached with the overall specific gravity remaining at less than 850 kg/m^3. The versatility of the coextrusion feedblock (see Chapter 9) is such that changes can be readily made to either the inner or outer skins, to achieve the skin thickness necessary to give the required properties [12].

Studies have been done to identify the design required for the best overall performance concerning burst-pressure loading, stiffness, and impact resistance in foam-core ABS DWV pipe [11]. Based on this analysis, the inner skin and the foam core should each represent 40% of the total pipe wall thickness, and the outer skin should make up the remaining 20%. Such ABS foam-core pipe has overcome a significant practical hurdle towards its acceptance for DWV's.

Table 17.5. Stiffness and impact test results on 100 mm diameter foam core ABS sewer and drain pipe at 23 °C [11]

Specific gravity, kg/m^3	Skin thickness		Impact[a], J	Stiffness at 5% deflection, kPa	Min. wall thickness, mm	Wt/meter, kg
	inner, mm	outer, mm	mm			
770	0.51–0.76	0.02–0.51	71	3,519	3.9	0.89
740	0.38–0.64	0.02–0.51	71	6,693	4.7	1.00
790	0.76–1.00	0.76–1.00	57	3,036	3.4	0.86
770	1.00–1.27	0.89–1.14	89	9,315	4.8	1.15
790	1.00–1.27	0.89–1.14	97	9,729	5.0	1.18
660	0.51–0.76	0.38–0.64	30	8,142	5.3	1.04
1,040	Solid		62	3,105	3.2	1.26

[a] Falling-dart impact (ASTM D2444), 52 mm (2 in) rad. tip

17.3 Applications

Integral ABS foams are ideal in applications requiring lightweight, rigid qualities, such as domestic appliance bases, housings, and internal component, business machine housings, computer bases, covers and other components, television cabinetry and internal framework [1,6,9].

In the recreational market, applications such as camper trailer tops, boats, and golf carts are likely targets for ABS foam-core sheet, probably laminated on the exposed side with an acrylic layer for improved weatherability.

As the proper resonance frequency of integral ABS foam is substantially higher than that of wood (140 as against 100–110 Hz), these materials can be used for manufacturing casings for TV and radio receivers, tape recorders and loudspeakers.

The agricultural, transportation and recreational markets offer many opportunities for thermoformed coextruded ABS foam-core sheets as they compete favorably on a cost-performance basis with FRP, aluminum, steel, and solid plastics. Applications such as exterior body parts on tractors and seeders are being pursued in the USA as a possible replacement for sheet steel [9,10].

Flooring, roofing, international transmission covers, tool boxes for cars and trucks, and motorcycle luggage covers are potential applications where low weight, rigidity, load-bearing capability, corrosion resistance and low cost make ABS foam-core sheet an attractive option.

Coextruded ABS foam-core structures should play their part in helping automobile manufacturers to make their vehicles lighter in order to meet, for example, the USA federal energy standard of an average 18 miles per US gallon (28.8 km per 3.79 l) for 1978 models rising to 20 miles (32 km) per gallon by 1980 and to 27.5 miles (44 km) per gallon by 1985 [11,12].

ABS foam-core pipes have been applied as sewer and drain pipes, utility ducting, carpet winding cores, and for tubular furniture [13–15].

An interesting application of extruded ABS integral foam is for pencils, with an overall density 400–800 kg/m^3; these pencils can be easily cut and sharpened if the solid skin is thin and the cellular core uniform [6].

17.4 Technical and Economic Analysis

Table 17.6 compares the cost of various materials, for the same stiffness. It can be seen that coextruded ABS foam-core sheet is the most effective utilization of material in achieving stiffness. Sheet steel and sheet or die-cast aluminum offer the best stiffness per unit raw material cost, but the steel product would be more than twice as heavy as the coextruded ABS foam-core product, and the processing, fabricating and high finishing costs of aluminum or steel can add appreciably to the raw material depending on the size, complexity, and finish required for the product. Coextruded sheet may be preferred for thicker articles where the advantage of greater rigidity

Table 17.6. Comparison of material costs for the same stiffness at 23 °C

Material and cost/kg	Modulus, GPa	Specific density, kg/m^3	Thickness for same stiffness	Cost, $	Relative stiffness to weight ratio	Relative stiffness to cost ratio
Coextruded ABS, 1 mm skin, $ 1.32	1.5	730	9.53	0.85	0.70	1.18
Solid ABS, $ 1.20	2.3	1,040	8.25	1.38	0.40	0.72
Sprayed FRP, $ 1.32	5.5	1,700	6.58	1.38	0.43	0.72
Match-mold FRP, $ 1.32	8.3	1,800	8.56	1.28	0.47	0.78
Die-cast/sheet Al $ 1.1	69	2,600	2.82	0.76	0.66	1.32
Steel sheet, $ 0.31	207	7,800	1.96	0.44	0.32	2.27
Die-cast Zn, $ 0.9	16	6,600	4.55	2.54	0.16	0.39

Table 17.7. Economics of extruded plastic pipes (76.2 mm diameter) for drain, waste and vent piping

	Solid ABS	Coextruded expanded ABS		Solid PVC
Specific density, kg/m^3	1,400	830	730	1,400
Resin cost, ¢/kg	88.1	88.1	88.1	52.9
Resin cost, ¢/unit[a]	88.1	70.5	61.7	74.0
CBA cost, ¢/unit[b]	–	0.50[c]	0.66[d]	–
Unit cost, ¢	88.1	71.0	62.36	74.0[e]

[a] This particular type of pipe requires 0.454 kg (1 lb) of solid ABS to make 30.48 cm (1 ft). Other types of pipe will have proportionately higher or lower costs/unit length
[b] the chemical blowing agent is azodicarbonamide, $ 3.3/kg
[c] 0.15% CBA
[d] 0.20% CBA
[e] conservative estimate, does not include 2.2 ¢/kg that industry sources say should be added for in-plant labor/mixing cost

is important. ABS foam-core sheet will require less trimming; it also offers integrated color and smooth or textured surfaces. Moreover, die-cast aluminum or zinc, as well as aluminum or steel sheet require buffing, degreasing and priming before the finish coat can be applied, and this adds appreciably to the cost of the finished article [11].

The major advantage of *foam-core pipe* is the polymer usage, which provides weight and cost benefits over both solid ABS and PVC pipe. Because of the density reduction of the cellular core, foam-core pipe is 20% lighter than solid ABS, and about 40% lighter than solid PVC. Translated into polymer savings, the density reduction gives the ABS foam-core construction about a 9% cost edge over solid PVC, and a 20% saving over solid ABS (Table 17.7) [11].

ABS integral foam is an economic engineering material. Molded products can be finished immediately after molding with no surface preparation such as filling or sanding. For example, compared to PC parts, ABS reduces unit costs by 7 to 15%, due to reduced material cost and a smaller number of finishing operations [1].

ABS IF's have the following advantages over PC and modified PPO integral foams: lower cost, less finishing, reduced degassing cure time. The limitations of ABS IF's are: low heat deflection temperature, poor physical and mechanical properties, and a single source of manufacture[1,6,17].

17.5 References

1. LaPlaca, J. P.: J. Cell. Plast. *16*, 36 (1980)
2. N.N.: Plastverarbeiter *31*, No 12, 731 (1980)
3. Wehrenberg, R. H.: Mater. Eng. *89*, No 4, 50 (1979)
4. Kramer, A.: Kunststoffe *64*, 350 (1974)
5. Freund, R. W., Ludwig, C. T.: Plast. Technol. *18*, No 7, 35 (1972)
6. Berlin, A. A., Shutov, F. A.: Strengthened Gas-Filled Polymers. Moscow: Khimia, 1980 (in Russian)
7. Throne, J. L.: J. Cell. Plast. *12*, 264 (1976)
8. Dominick, G.: J. Elast. Plast. *11*, No 2, 133 (1979)
9. Cycolac® FBK, ABS Flame Retardant Structural Foam: Technical Bulletin. Borg-Warner Chemical Europe, Brussels, Belgium

10. Young, R. H.: SPE Techn. Papers *19*, 631 (1973)
11. Malpass, V. E.: Plast. Rubber Mater. Appl. *3*, No 4, 149 (1978)
12. Ahnemiller, J.: Plast. World *38*, No 6, 82 (1980)
13. Krumm, S.: 31th Ann. Techn. Conf. SPE, Montreal Canada, 1973
14. N.N.: Mod. Plast. Int. *10*, No 1, 58; No 10, 38 (1980)
15. Structural Foam: Brochure Union Carbide Corporation, New York USA, F-44669B, 11/74-10M (1981)
16. Wehrenberg, R. H.: Mater. Eng. *96*, No 6, 34 (1982); *97*, No 2, 26 (1983)
17. Shutov, F. A.: 4th Intern. Conf. on Composites, Varna Bulgaria, 1985

18 Integral Foams Based on Poly(phenylene oxide)

Plastic materials, as well as the IF's, based on poly(phenylene oxide) (PPO) are characterized by good impact strength, high thermal resistance and incombustibility. All commercial grades of PPO are modified with PS, enhancing appreciably the flexural modulus of the material. The most popular commercial grade of IF based on PPO is the "Noryl" family, which was developed and produced by General Electric, both in USA and in West Europe, at the beginning of 1970's.

Commercial PPO integral foams are produced by injection molding, usually at low pressure.

18.1 Raw Material Specifications

18.1.1 Blowing Agents

PPO sheets foamed with different CBA's show very small differences in density, but this difference is substantial in sheets produced using nitrogen (Table 18.1). The differences between the mechanical properties correspond approximately to the differences in density, which means that with respect to mechanical properties none of the blowing agents is better than the other [1].

According to data obtained by General Electric, USA, substituting Freon-22 for N_2 when making integral PPO foam by a LP injection molding process can reduce swirl, upgrade physical properties, and save cycle time [2,3]. As a result, the cost of a finished IF part may be cut by 40%. These conclusions are based on trials with flat

Table 18.1. Properties of PPO IF sheets (dimension 520 × 520 mm) foamed by a chemical or a physical blowing agent

Property	CBA	PBA (N_2)
Specific density, kg/m^3	840	940
Tensile strength, MPa	25	32
Elongation, %	12	17
Flexural strength, MPa	53	62
Flexural modulus, MPa	1,950	2,300
Impact strength, N/m^2	20,000	32,000
Heat deflection temperature[a], °C	85	90

[a] ISO R 75, method A

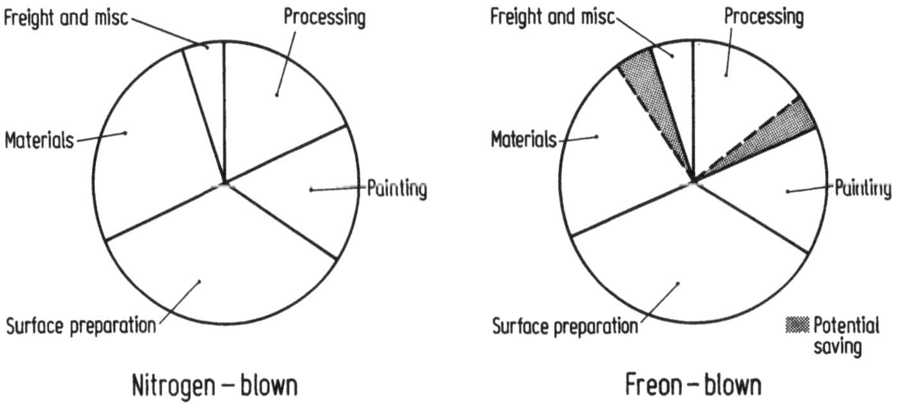

Fig. 18.1. PPO integral part costs. Comparison between foaming with N_2 and with Freon-22 [2)]

molded panels, 610 × 915 × 6.4 mm, and with an average density reduction of 15%. Freon-foamed panels have finer cells, fewer voids and more uniform skin thickness than N_2-foamed panels. Moreover, less outgassing occurs after demolding. The flexural and tensile strengths are improved by about 20%, while the flexural modulus and heat deflection temperature are improved by about 10%. The use of Freon reduces the mold-fill time by 30% and decreases the total time by up to 25%, compared to a N_2-PBA. Despite Freon's higher cost (57.1 ¢ per liter versus 35.7 ¢ per liter for N_2), the results suggest that the extra material cost will be largely offset by savings in the manufacturing process (Fig. 18.1).

18.1.2 Freon Ecology Problem

We would like to take this opportunity to discuss briefly the "Freon question". An inevitable protest arises whenever Freon is used, that fluorocarbons would degrade the Earth's UV-filtering ozone layer. The answer is that not all Freon compounds are alike. For example, Freon-22, which is the particular compound that is proposed for foaming, has the lowest "ozone-depletion index" of all known Freon compounds, being only about 20–25% that of the most objectionable Freons. Because of its relatively low potential effect on the ozone layer, Freon-22 is not likely to encounter any restriction [2)].

18.2 Properties

The main physical and mechanical properties of two PPO integral foams, together with those of a polycarbonate integral foam, are presented in Table 18.2 [4)]. The high specific strength properties of PPO IF are very clearly illustrated by the following data [5)]:

Material	PPO	Aluminum	Steel	Zinc
Specific stiffness	110	90	45	30
Specific flexural modulus	133	45	20	5

Table 18.2. Properties of PPO (Noryl[a]) and PC (Lexan[a]) integral foams

Property	Noryl FN 215	Noryl FN 5110	Lexan FL 900
Specific density, kg/m^3	800–850	850–900	850–950
Tensile strength, MPa	23	32	37
Tensile modulus, MPa	1,600	1,700	1,900
Elongation, %	5–10	2–4	3–5
Flexural strength, MPa	45	58	70
Flexural modulus, MPa	1,800	2,100	2,300
Compressive strength, MPa	35	36	36
Charpy impact, N/m^2	20,000	20,000	–
Creep 300 hr., 23 °C at 7 MPa, %	0.6	0.5	0.4
Heat deflection temperature at 1.85 MPa, °C	83	110	126
Mold shrinkage, %	0.6–0.9	0.5–0.8	0.5–0.8
Thermal conductivity, W/Km	0.123	0.128	0.151
Inflammability, 6.4 mm thick.	UL 94 Vo	–	UL 94 Vo
Oxygen index	28.5	–	>35
Dielectric strength, v/mm	216	220	220
Dielectric constant			
100 Hz	2.27	–	2.22
10^6 Hz	2.18	–	2.18
Dissipation factor			
100 Hz	0.0047	–	0.0012
10^6 Hz	0.0039	–	0.0061

[a] Commercial grades of General Electric Company USA and General Electric Plastics B.V. Holland

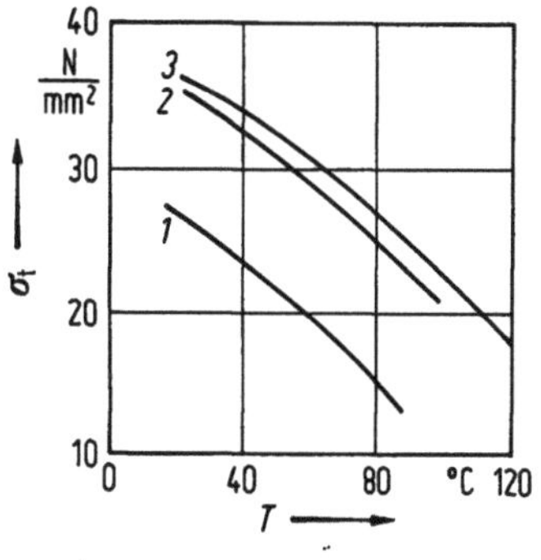

Fig. 18.2. Tensile strength versus temperature for IF's with a density of 800 kg/m^3; 1: PC (Lexan FL 900), 2: PPO (Noryl FN 215), 3: PPO (Noryl FN 5110) (Courtesy of General Electric Plastics B.V., Bergen op Zoom, Holland) [4)]

Because of its cellular core, an IF part is not generally recommended for pure tensile loading. However, an outstanding feature of IF's is their *retention of tensile strength over a wide temperature range.* For example, a PPO integral foam retains 70% of its room-temperature tensile strength at 66 °C, while a PC foam retains 65% of its tensile strength at 95 °C (Fig. 18.2).

Excellent *flexural properties* of IF's result from the natural sandwich-like distribution of polymer throughout the molded foam product. Unfoamed polymer is situated at the edges of the cross-section where the flexural stress is at its maximum. This material distribution guarantees that a high flexural modulus is maintained even at elevated temperatures. For example, a PPO integral foam retains 80% of its room-

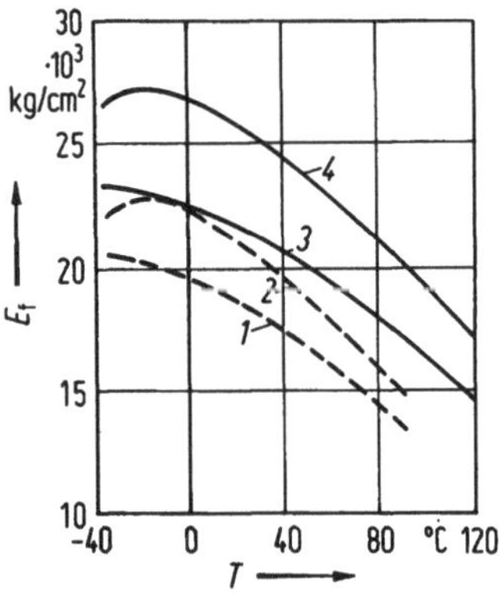

Fig. 18.3. Flexural moduli versus temperature for an IF based on PPO (Noryl FN 215) with density of (1) 800 kg/m³ and (2) 900 kg/m³, and for an IF based on PC (Lexan FL 900) with density of (3) 720 kg/m³ and (4) 950 kg/m³ (Courtesy of General Electric Plastics B.V., Bergen op Zoom, Holland)[4)]

temperature flexural modulus at 66 °C, while a PC foam retains 70% of its room-temperature modulus at 95 °C (Figure 18.3)[4–6)].

All IF's have a definite advantage over metals in the *reduction of noise levels,* thanks to the ability of foam to damp large vibrations much faster than metal (Table 18.3). When the actual source of noise is housed in an IF casing, the noise is dampened up to 16 times more efficiently than it would be by a metal[4)].

Integral PPO foams are characterized by their excellent *hydrolytic stability.* They are therefore unaffected by hot and cold water, detergents, weak and strong acids and alkali; however, PPO will soften or dissolve in many halogenated and aromatic hydrocarbons (see Chapter 19). The fire resistance properties of PPO foams are presented in Tables 13.10 and 22.1.

Table 18.3. Sound damping characteristics of IF parts and metals[4)]

Material	Thickness, mm	Decay rate, dB/sec
Integral PPO foam (Noryl FN 215)	6	70
Integral PPO foam (Noryl FN 5110)	6	85
Integral PC foam (Lexan FL 900)	6	118
Aluminum	3	12.8
Steel	3	7.4

18.3 Applications

The main applications are based on the following features of PPO integral foams: hot water and detergent resistance, heat and sound insulation, low flammability, electric safety, versatility in application, and unique property balance. Hence integral PPO foams are applied in tumble-driers, dishwashers, washing machines, housings for photocopiers, cash registers, minicomputers, and TV's, medical instruments, radar and navigational equipment, electrical control gear, recreational and playground equipment, film spools, etc.[7)].

Metal is still the preferred material for housings of electronic systems, whenever manufacturers have to retain a flexibility in producing a large number of products with relatively low production runs for each individual item. This need for flexibility, combined with the relatively high tooling costs for molded plastics, has been considered a barrier to the use of plastics [8,9].

The recent trends towards the adoption of standard printed circuit-card sizes (19 inches or 482.6 mm), and the related adoption of national and international standards for the height divisions within internal rack space, have created greater opportunities for plastics. This has promoted the use of PPO IF's for housings, especially where modular design and part integration are necessary [10-13].

As the miniaturization of electronic circuits advances, it becomes easier to provide complete systems, such as a Private Branch Exchange (PBE) system for a small-to medium-size office, in a small desk-top or under-desk cabinet. IF's have an important role to play in this area, as electronic equipment itself becomes integrated with furniture as "Electronic Furniture'. The key factor in favor of IF molding is then its ability to provide aesthetic styling more cost-effectively than metal fabrication [11,12].

IF parts based on modified PPO are today's leader in business machine industry. Almost three-quarts of all material usage is PPO; PC holds a strong second psotion, followed by ABS and PS [14].

But IF foams are not only used for housings. For example, integral PPO foam was designed into a one-piece, unitized instrument base for a gas chromatograph: the main frame was the main support onto which other components, such as printed-circuit-board or "card" guides, were assembled [14].

A critical aspect in molding such large integral PPO parts (weight up to 10 kg) is holding dimensional tolerances. Electric contacts (prongs) must match with mating connectors. Deviations of a few thousandths of an inch could result in a malfunction, tolerances for the card guides were hold to within 0.010 in. (0.254 mm). According to Wehrenberg [14], the cost of the IF base was $ 39 ($ 70 with inserts), versus about $ 200 for its aluminum counterpart.

A cost analysis of integral PPO parts was presented in Chapter 14, where it was compared with integral PS parts.

18.4 References

1. Eckardt, H.: Plast. World *38*, No 4, 66 (1980)
2. N.N.: Plast. World *39*, No 8, 64 (1981); *40*, No 8, 10 (1982)
3. N.N.: Mod. Plast. Int. *10*, No 11, 20 (1980); *11*, No 6, 39 (1981)
4. Engineering Structural Foam Resins: Brochuer of General Electric Plastics B.V., Bergen op Zoop, Holland, 1976
5. Meyer, W.: J. Cell. Plast. *14*, 50 (1978)
6. Stencel, J.: J. Cell. Plast. *17*, 152 (1981)
7. N.N.: Plast. World *38*, No 4, 70 (1980); *39*, No 9, 27 (1981)
8. N.N.: Des. Eng. (Gr. Brit.) No 9, 71 (1979)
9. N.N.: Mod. Plast. Int. *10*, No 6, 17 (1980); *11*, No 5, 14 (1981)
10. Swinstead, N.: Engineerg. No 9, 711 (1981)
11. Vink, D. A.: Internat. Conf. Plastics in Telecommunication, London England, 1982
12. N.N.: Mod. Plast. Int. *12*, No 6, 64 (1982); Plast. World *41*, No 6, 68 (1983)
13. N.N.: Mod. Plast. Int. *12*, No 9, 28 (1982); *13*, No 6, 82 (1983)
14. Wehrenberg, R. H.: Mater. Eng. *96*, No 6, 34 (1982)

19 Integral Foam Based on Polycarbonates

The good physical and mechanical properties of the polycarbonates (PC) — maintenance of strength under the action of impact loads, shape stability, low creep, and resistance to UV irradiation — have been the major factors causing the rapid development of manufacturing methods for obtaining IF items based on polycarbonates. Today, quality integral foams based on PC are being manufactured in many countries, for applications in various industries and construction.

The most popular commercial PC integral foams are the "Lexan" family (General Electric Plastics B.V.) and the "Merlon" family (Mobay Chemical Corporation) [1,2].

19.1 Technology Specifications

Integral PC foams are manufactured on standard machines by high and low-pressure injection molding techniques, using high-temperature CBA-s, for example, 5-phenyltetrazole. The use of a complex BA in the form of two types of CBA (one low-temperature, and the other high-temperature) lowers ϱ_a, the molding time and decreases the dimensions of the cells, while improving the "smoothness" of the skin.

The *high viscosity of the melt* makes it impossible to manufacture lightweight IF's as the degree of foaming cannot be higher than 50 or 60% (0.15 to 0.25% CBA). Hence ϱ_a of the final items usually ranges between 600 and 1,100 kg/m^3 [3].

The IF's are made from pelletized polymers which are heated in a plasticator to 270 to 300 °C and quickly injected into a mold (mold injection time is not longer than 2–3 s). The mold temperature governs the thickness of the skin and must range between 82 and 93 °C for δ_s = 6.2 mm and between 65 and 77 °C for δ_s = 9.2 mm [3]. Despite the high melt temperature, molding times are short because of the high deformation stability of the starting polymer during heating. For the same reason, the finished items posses a negligible shrinkage.

The effect of the process parameters on the properties of an integral PC foam (high-pressure injection molding and 5-phenyltetrazole as BA) was investigated in detail by Matonic and Kaminski [3]. They showed that the strength values of these materials increase as the injection rate, evaluated in terms of the mold filling time, fell (Fig. 19.1 a). This effect is due to two causes, viz. the decrease of the thickness of the skin, and the deterioration of its quality. However, the relationship is reversed for thicker items, where an increase in mold filling time improves the strength. The strengths vs. mold-temperature relationship (Fig. 19.1 b) is fairly complicated; it is not only different for each parameter, but also for items of different thicknesses. The molding time (τ_d) and the temperature of the composition (T_c) affect σ_t more

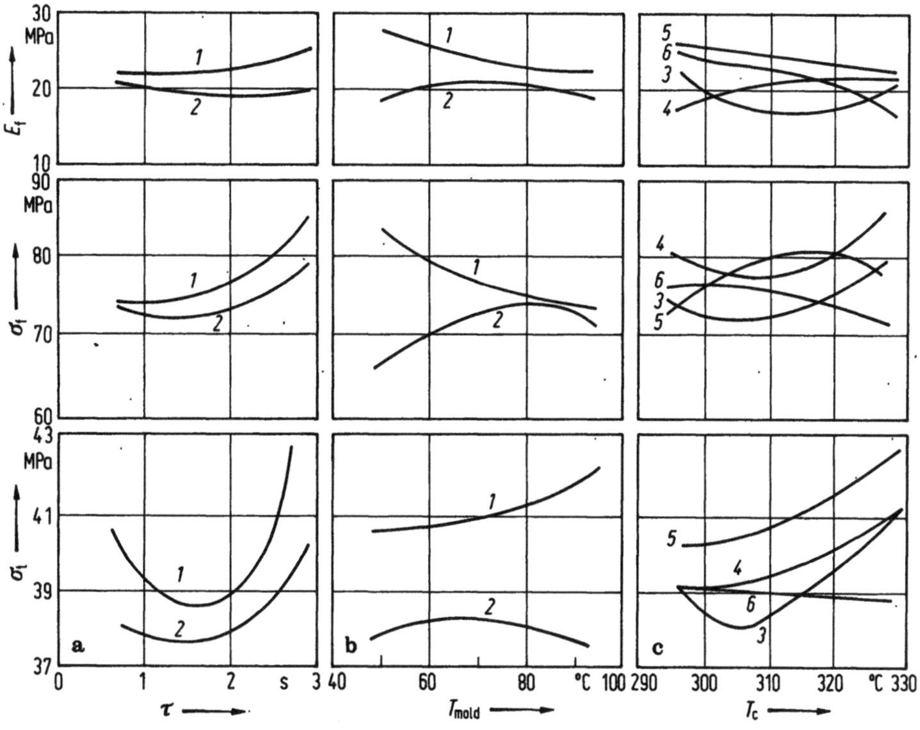

Fig. 19.1 a–c. Dependence of the flexural modulus and strength, and tensile strength of an integral PC upon (a) the mold filling time τ, (b) the mold temperature T_{mold}, and (c) the temperature of the composition T_c. 1 and 2: IF parts respectively 6.4 and 9.2 mm thick; 3, 4, 5, and 6: molding time τ_d respectively 10, 15, 20, and 40 min [3)]

than they affect σ_f and E_f (Fig. 23.1c). However, τ_d must never be shorter than 15 min., and T_c not less than 310–320 °C for molded items 6.35 mm thick, and 293–305 °C for molded items 9.25 mm thick [3)]. These data demonstrate conclusively that the process parameters of molding an IF item must be determined in advance, according to the concrete application (service conditions) of the item.

19.2 Properties

The physical and mechanical properties of integral PC foams, as well as their temperature resistance, have already been discussed (Chapter 18, Tables 18.2 and 18.3, and Fig. 18.2 and 18.3) It should be mentioned once more that a PC foam is characterized by its excellent behavior at sustained *high temperatures*, and its *ageing properties* allow it to be used under load at temperatures closer to its heat deformation than almost any other thermoplastic.

Industrially produced integral PC foams have higher impact strengths than any other IF and this property is practically independent of the temperature between −40 and +20 °C. The clear-cut macrostructure anisotropy of these materials deter-

Table 19.1. Properties of glass fiber reinforced Merlon SF-810 (10% GF) and Merlon SF-820 (20% GF) polycarbonate IFs, produced by Mobay Chemical Corp., USA [10]

Properties[a]	ASTM Test Method	Merlon SF-810	Merlon SF-820
Density, kg/m^3 ($Ib./in.^3$)	D792	950 (0.037)	1,000 (0.039)
Mold shrinkage, mm/mm (in./in.)	D955	0.004–0.006	0.003–0.005
Water absorption (immersion),			
24 hours (%)	D570	0.10	0.10
lond term (%)	D570	0.30	0.30
Tensile stress at break, MPa (psi)	D638	41 (6,000)	45 (6,500)
Tensile strain at break, %	D638	4.0	3.5
Tensile modulus, MPa (psi)	D638	2,170 (310,000)	3,150 (450,000)
Flexural stress at	D790	76 (11,000)	83 (12,000)
Flexural modulus, MPa (psi)	D790	2,840 (406,000)	3.850 (550,000)
Shear strength, MPa (psi)	D732	33.1 (4,800)	34.5 (5,000)
Deflection under load,			
1.82 MPa, °C (264 psi, °F)	D648	134 (273)	136 (277)
0.46 MPa, °C (66 psi, °F)	D648	144 (292)	146 (295)
Flammability	UL 94	V-0/5V	V-0/5V
Dielectric strength, kV/mm (V/mil)	D149	16 (400)	16 (400)

[a] All properties were measured at 6.4 mm thickness, with the exception of electric properties measured at 3.2 mm thickness

Table 19.2. Chemical resistance of PPO (Noryl) and PC (Lexan) integral foams

Substance	Lexan	Noryl	
Mineral acids (dil.)	good	good	excellent
Mineral acids (cons.)	fair-good	fair	
Alkali	fair	excellent	
Solvents:			
alcohols	good[a]	excellent	
ketones	poor	poor	
Chlorinated hydrocarbons	poor[b]	poor[b]	
Aromatic hydrocarbons	poor	poor	
Detergents sol.	good	excellent	
Greases and oils	excellent-good[c]	fair-good[c]	

[a] Except mathly alcohol; [b] Methylene chloride, ethylene dichloride and trichloroethylene are good solvents for these polymers; [c] Highly dependent on additives used.

mines the substantial differences in the mechanical properties throughout the thickness of the items (see Fig. 1.1). The highest values of σ_f amount to 70.3 and 84.0 MPa for densities of 900 and 1,100 kg/m^3, respectively [1,3].

A widely used means for improving strength properties is the use of *glass fiber fillers* [4,5]. Two glass-fiber reinforced, "second-generation" foamable grades of Merlon PC, with the exceptional stiffness and dimensional stability required for the most demanding business machine and automotive applications, have been introduced by Mobay Chemical Corp. (USA) [10]. Coupled with their high glass content (up to

10–20%), Merlon SF-810 and Merlon SF-820 integral foams offer excellent impact strength, flame retardance and heat resistance (Table 19.1). In addition to business machine and automotive uses, the new grades are also suitable for injection molding of large IF parts for the fields of materials handling and recreational products [4,5]. The mechanism of failure of integral PC foams has been analyzed in detail by Hobbs [6].

It is interesting to compare the stiffness-to-weight ratio of PC and PPO integral foams, for equivalent material weights, with other materials, i.e. Lexan FL 900 (PC) — 124, Noryl FN 5110 (PPO) — 112, aluminum — 105, Noryl FN 215 (PPO) — 100, steel — 13, zinc — 4.5 [1].

The fire resistance properties of PC integral foams are presented in Tables 13.10 and 22.1.

The chemical resistance of an IF is similar to that of the polymer of which it is composed. On the other hand, IF parts are relatively stress-free in comparison with compact injection molding parts. This aspect is important with respect to resistance to stress-corrosive environments. Environments which may be completely harmless to unstressed parts may cause stress-corrosion problems with highly stressed parts [4].

As to chemical resistance (Table 19.2), a PC IF is resistant to dilute mineral and organic acids, insoluble in aliphatic hydrocarbons, ether and alcohol, partially soluble in aromatic hydrocarbons, and soluble in many halogenated hydrocarbons [1].

19.3 Applications

PC integral foams are mainly applied where their rigidity, impact performance, heat resistance, and low flammability are needed. Accordingly, integral PC foams are applied in cargo pallets, cable spools, tractor cabs, train and aircraft components, truck and bus seat shells, luggage racks, institutional furniture, office machines, and telecommunications equipment [7,8].

In most European countries a cross-connection cabinet will be found at roadside locations to aid local telephone companies to control traffic lights, and in certain countries to distribute cable television. Traditionally the upper cabinet has been constructed from either metal or SMC, and combined with a concrete base passing into the ground, through which the underground cables are fed into the upper cabinet to the distribution connection blocks. Recently the traditional concrete base has been replaced by PC (Lexan) parts (FRG). A typical base consists of 6 parts totalling 28–33 kg, compared with the 331 kg of an equivalent concrete base. Although the IF base costs 3.7 times more than the concrete base, it cuts installation costs by 80% [9].

The widespread use of IF parts in *automobile design* started as a means of reducing costs. However, as the performance of engineering thermoplastics, and specifically IF thermoplastics, has improved, new designs not possible with any other material are now being made. In addition, the IF components often have better properties than those made by the traditional methods. For example, a Jeep roof top originally conceived as a multi-piece metal assembly was found to be impracticable. A polyester resin design was considered instead, but the number of special tools needed to form the complex shape would have been too expensive. The solution was PC (Lexan FL 900) integral foam, which allowed the roof to be produced as an integrated one-piece molding weighing 36 kg; it had relatively low tooling costs, yielded a profile exactly meeting the requirements, and ensured excellent reproducibility from part to part. In addition, the material has high mechanical and dimensional stability between −40° to 120 °C, and easy removability of the roof is ensured by the light weight. Further cost savings over conventional production methods result from the integration of bolting points and door hinge fixing points.

A tooling comparison for producing the roof from PC (Lexan) foam or by sheet molding a polyester resin showed the costs to be, respectively, $ 165,000 and $ 380,000 for the CJ7 Jeep top, $ 384,000 and 1,318,000 for the GN Blazer top [1].

19.4 References

1. Engineering Structural Foam Resins: Brochure of General Electric Plastics B. V., Bergen op Zoom, Holland, 1976
2. N.N.: Mod. Plast. Int. *11*, No 2, 14 (1981); Plast. Eng. *37*, No 12, 7 (1981)
3. Matonic, D. E., Kaminski, A.: Plast. Technol. *22*, No 10, 57 (1976)
4. Jones, H. L., Yu, R.: J. Cell. Plast. *17*, 328 (1981)
5. McQuinton, H.: Plast. Eng. *36*, No 6, 18 (1980)
6. Hobbs, S. Y.: J. Cell. Plast. *12*, 258 (1976); J. Appl. Phys. *48*, 4052 (1977)
7. N.N.: Mod. Plast. Int. *12*, No 5, 58; No 7, 64 (1982)
8. Berlin, A. A., Shutov, F. A.: Strengthened Gas-Filled Polymers. Moscow: Khimia, 1980 (in Russian)
9. Vink, D. A.: Internat. Conf. Plastics in Telecommunication, London England, 1982
10. N.N.: J. Cell. Plast. *19*, 138 (1983)

20 Other Types of Integral Foams

Over the past few years the output of new types of thermoplastic and thermosetting IF's has enlarged rapidly. We shall discuss briefly the recent laboratory and industrial developments and the achievements of these new IF materials.

20.1 Polyamide

Integral polyamides based on nylon-6 and nylon-6,6 are manufactured industrially in the USA (Zytel and DuPont), in France (Technyl and Rhône-Poulenc) and in FRG (see Fig. 7.7) [1].

These materials are manufactured by the *high-pressure injection molding process* (the USM-process) using CBA's, for example, 5-phenyltetrazole. The introduction of glass fiber considerably enhances the stiffness and heat resistance. Carbon fiber reinforcement also has a positive effect (Table 20.1, see also Table 20.3) [2].

The fire resistance of nylon IF's will be presented in Chap. 22 (Table 22.1). Glass fiber reinforced nylon, containing an internal foaming agent, has a specific density of 1,380 kg/m^3 and costs 3,975 \$/m^3 in its unfoamed form. Foaming to a specific density of 1,100 kg/m^3 (80% solids content), reduces the actual end product material cost to 3,178 \$/m^3, foaming to 60% solids content to 2,386 \$/m^3 [3].

Table 20.1. Properties of fiber reinforced Nylon-6 integral foam

Property	Un-reinforced	Reinforced	
		5% glass	10% carbon
Specific density, kg/m^3	800	880	860
Tensile strength, MPa	49	70	91
Elongation, %	10	2–3	2–3
Flexural strength, MPa	85	116	159
Flexural modulus, MPa	2,100	4,550	7,000
Heat distortion at 1.85 MPa, C	73.4	199	249

20.2 Thermoplastic Polyester

Rapid advances are being made in the industrial production of IF's based on a crystalline polymer, thermoplastic polyester [4]. In the USA, General Electric is producing the Valox family of IF's by injection molding and extrusion using CBA's [5]. Several commercially available IF's are filled with up to 30% glass fiber. These materials probably possess *the highest flexural strength of all industrial thermoplastic IF's* produced thus far (see Table 20.3) [6]:

density, kg/m³	800	900	1,000	1,100
flexural strength, MPa	108	120	139	143
flexural modulus, MPa	510	560	660	700
tensile strength, MPa	49	56	67	77

According to Breitenfeller and Leidig [7], the relative stiffness of an IF based on reinforced thermoplastic poly(butylene terephtalate), or PBT, is higher than that of any other IF material or metal (Table 20.2).

A glass fiber reinforced polyester IF is being used to mold a lean burn modular housing, and General Motors is currently molding the roof frame of the Corvette Moon from the same material [3].

The fire resistance properties of these materials are shown in Table 22.1.

Table 20.2. Comparison of the relative stiffness of some metals and thermoplastic IF materials [7]

Material	Specific density, kg/m³	Relative stiffness, A[a]	Relative weight, B[b]	A:B
Metals:				
Steel	7,800	1.0	1.0	1.0
Aluminum	2,700	8.1	0.49	2.04
Zinc	2,100	0.6	1.17	0.85
Plastics:				
PBT, unfoamed	1,310	2.5	0.74	1.35
PBT, unfoamed + 30% glass fiber	1,530	5.4	0.57	1.75
PBT, integral foam + 30% glass fiber	1,050	11.3	0.47	2.12
PC, integral foam + 30% glass fiber	850	4.8	0.59	1.69
PPO mod., integral foam	750	8.0	0.50	2.00

[a] stiffness at the same thickness;
[b] weights relative to steel

20.3 Polyacetal

IF's based on polyacetals and polyketals are being manufactured industrially, e.g. Celcon (Celanese Plastics, USA), and Minton and Delrin (DuPont, USA) [8]. These materials are produced by *low-pressure injection molding* (the UCC process), with azodicarbonamide as the CBA. Items containing from 13 to 33% of glass fiber filler (Minton and Delrin of DuPont) combine elasticity and high stiffness (see Table 20.3). In particular, the σ_t and E_f values of this material filled with 25% glass fiber (30% less dense than the solid plastic) are, respectively, 46 and 28% lower than the σ_t and E_f of the solid polymer.

Table 20.3. Properties of various thermoplastic IF materials [11]

Property	Copolymer of polyacetals	Copolymer of polyacetals + 30% glass fibre	Thermoplastic polyester + 30% glass fiber	Nylon-6 + 33% glass fiber
Density, kg/m³	990	1,130	1,170	970
Tensile strength, MPa	38	74	71	94
Flexural strength, MPa	64	133	120	143
Flexural modulus, MPa	185	530	675	450
Thermoresistance, °C	154	164	220	255

20.4 Polyimide

A new material developed by Ciba-Geigy AG (Switzerland) is an integral polyimide foam (IMB 1907) [9,10]. Its special features are good mechanical and electrical properties at service temperatures up to 240 °C. The material is supplied as a single-component powder, pellets, or tablets containing a CBA. It can be foamed to densities up to 700 kg/m³, using a closed heated mold (direct molding process), with a wall thickness of 2 cm. The main application areas of this IF are in the electrical, machinery, and aerospace industries. The price is expected to be about SFr 40/kg.

20.5 Epoxy Resins

Recently, integral epoxy resin foams were developed by Ciba-Geigy AG, with densities between 300 and 800 kg/m³ [10]. They are supplied as a two component system (resin and hardener), with a PBA to be premixed with the resin. Foaming and cross-linking proceed simultaneously, and the skin thickness is determined by the mold temperature, the normal range of which is 40 to 55 °C. The material's special feature is its very high modulus of elasticity; moreover it is not hydrolized even at high temperatures. It is aimed at applications where a combination of good mechanical and electrical properties is needed. In addition, it can be processed by RIM [11].

In the USSR integral epoxy foam is used in radio/TV applications and, in the construction industry, for window frames [12].

20.6 Phenolic Resins

In recent years, the production of IF's and multi-layer materials based on phenol formaldehyde resins (PF) has been developed [13,14]. Manufacturing these materials is not different, in principle, from obtaining IF's based on other polymerizable reactive oligomers, for example the urethanes. However, the large-scale production of IF's based on PF's is limited by equipment corrosion and the high adhesion of the starting compositions and finished items to the metallic surfaces of the equipment. The latter problem can be overcome, although at the cost of a substantial increase in molding time, by coating the mold walls with wax or silicone lubricants.

When manufacturing an IF based on a resol type of PF resin, the blowing agents are *carbonates of alkali metals*, and freons. Foaming is carried out in the presence of an acid catalyst, such as benzenesulfonic, phenolsulfonic or toluenesulfonic acid. The subsequent neutralization is carried out in special chambers by *gaseous ammonia* at 60–70 °C. The physical and the mechanical properties of a resol IF ($\varrho_a = 100$ kg per m^3) are: $\sigma_f = 0.65$ MPa, $\sigma_c = 0.78$ MPa, and a modulus of elasticity of 20 MPa. To reduce the brittleness of an item based on novolak PF, 20 to 30 mass % of a synthetic rubber is added. The stiffness of the foam can be enhanced by the introduction of up to 40 mass % of glass fiber [15, 16].

A considerable achievement in the technology of integral foams was the development, by Xentex Corporation (USA), of a material based on a hybrid (urethane and phenol) matrix and a filler (40 to 90% glass fiber) [13]. This material, manufactured by low-pressure injection molding (the UCC process), has a ϱ_s of 1,040 and a ϱ_c of 48 kg/m^3, whereby its stiffness and impact strength are more than 4 times those of the non-reinforced IF's.

20.7 Other

Another recent achievement in IF technology is the manufacture of items based on highly thermally resistant *cross-linked polymers*. For example, Phillips Petroleum (USA) manufactures industrial IF's based on polysulfones (Udel) and poly(phenylene sulfide) [9]. An addition of 20 to 40% of glass fiber to the latter makes it possible to retain 91% to 83.5% of the σ_t value, and 53 to 50% of the E_f value, of the unfoamed plastic, while decreasing the density as much as 34% [9].

Molding processes have been used to produce IF's based on poly(methyl methacrylate) [9], on polyacrylates [17], on copolymers ethylene-vinyl acetate [18] and its mixtures with PE, as well as on ionomer resins[19], and many others[20–24].

20.8 References

1. N.N.: Europ. Plast. News *6*, No 12, 27 (1979)
2. Dominick, G.: J. Elast. Plast. *11*, No 2, 133 (1979)
3. Vaccari, J. A.: Prod. Eng. (USA) *50*, No 3, 29 (1979)
4. N.N.: Plastic World *38*, No 7, 38 (1980)
5. N.N.: Mod. Plast. Int. *9*, No 1, 53 (1979); *10*, No 5, 74 (1980)
6. N.N.: Plast. Technol. *21*, No 3, 11; No 4, 7; No 8, 61 (1975)
7. Breitenfellner, F., Leidig, K.: Kunststoffe *67*, 487 (1977)
8. N. N.: Mod. Plast. Int. *10*, No 2, 61; No 3, 20 (1980)
9. Litman, A., et al.: AIChE Symp. Ser. *73*, No 170, 163 (1977)
10. N.N.: Mod. Plast. Int. *10*, No 3, 18 (1980)
11. N.N.: J. Cell. Plast. *11*, 171 (1975)
12. All-Union Res. Inst. Synth. Resins: Technical Bull., Vladimir: VNIISS, 1982
13. Bonfiglio, G.: Poliplast. Plast. Reinforz. *24*, No 218, 21 (1976)
14. N.N.: Plast. Ind. News. No 7, 99 (1981)
15. N.N.: Mod. Plast. Int. *6*, No 6, 8 (1976)
16. Shutov, F. A., Ivanov, V. V.: Building Matter. Constr. No 1, 21 (1982) (in Russian)
17. N.N.: Mod. Plast. Int. *3*, No 1, 16 (1973)
18. N.N.: Mod. Plast. Int. *5*, No 3, 32 (1975)
19. N.N.: Mod. Plast. Int. *12*, No 4, 41 (1982)

20. Berlin, A. A., Shutov, F. A.: Strengthened Gas-Filled Polymers. Moscow: Khimia, 1980
21. Shutov, F. A.: USSR Conf. Gas-Filled Plastics. Suzdal USSR, 1982
22. Wang, C. S.: J. Appl. Polym. Sci. 27, 1205 (1982)
23. Shutov, F. A.: Intern. Sympos. Characterization and Analysis of Polymers, Melbourn Australia, 1985
24. Shutov, F. A.: 2nd Intern. Sympos. Roofing Technology, Chicago USA, 1985

Section D: Calculation, Design and Marketing Fundamentals

21 Calculation of Strength and Other Properties

21.1 Basic Morphological Parameters of Integral Foams

21.1.1 Apparent Density

A necessary condition for quality control, as well as for the correct calculation of the specifications of gas-filled materials (including polymeric foams), is a rigorous quantitative evaluation of the apparent density distribution over the bulk. In the case of classical foams, a consideration of the density distribution in a plane, i.e., a two-dimensional distribution of the ϱ-values, would be quite sufficient. Evidently, with the integral foam structures (see Fig. 1.1–1.3), one has to consider a volume density distribution, i.e., a three-dimensional variation of the ϱ-values. Since a graphic representation of the behavior of any IF property in three-dimensional space is not always clearly comprehended, a two-dimensional "property *vs.* density" plot for a particular cross section, most often one normal to the surface skin, is used in practice.

On the other hand, the term "apparent density" employed for integral structures needs clarification, because it may have different meaning and numerical values depending upon which portion of an article is considered, and which evaluation method is used.

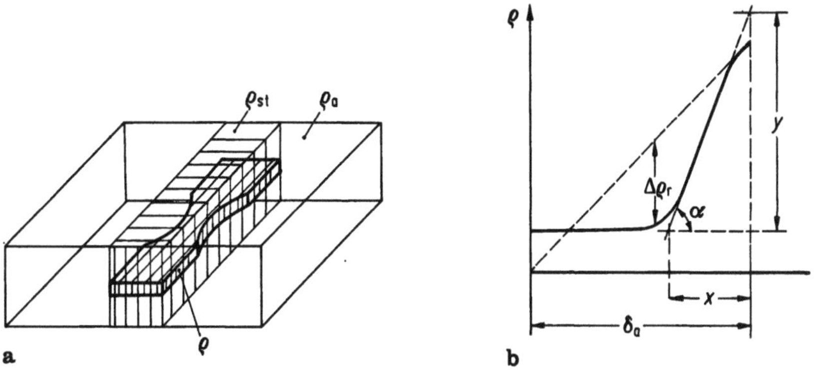

Fig. 21.1a and b. Morphological parameters of integral polymer foam according to Hubeny [1]; **a)** apparent density of an article, ϱ_a; apparent density of the structure, ϱ_{st}; apparent density of a specimen for testing, ϱ; **b)** relative apparent density $\Delta\varrho_r$; apparent density gradient $D_\varrho = \text{tg}\,\alpha = y/x = \Delta\varrho/\Delta\delta$

According to the classification suggested by Hubeny [1], a detailed analysis of integral foam structures necessitates, in addition to the apparent density of the core (ϱ_c) and that of the surface skin (ϱ_s), the definition of some more morphological parameters (Fig. 21.1):

1) the apparent, or average, density of an article (ϱ_a), which is expressed as the ratio of the whole mass of an article to its volume;

2) the apparent density of a "structure" (ϱ_{st}) which is expressed as the ratio of the mass of a particular portion of an article to its volume, including the surface skin; an IF article of a certain ϱ_a value may contain a number of structures with different ϱ_{st} values; the ϱ_{st} value is sensitive to the part of the article from which a given structure is cut out;

3) the apparent density of a "specimen" ($\varrho_{sp} = \varrho$) which is the mass-to-volume ratio for that portion of a structure which has been cut out for testing purposes (for example, to determine the ultimate tensile strength, as shown in Fig. 21.1a). There are cases where $\varrho_a = \varrho_{st} = \varrho$.

21.1.2 Density Distribution

These three parameters are necessary, but not sufficient characteristics of the IF morphological structure. It is essential to introduce additionally a quantitative estimate characterizing the behavior of the density on a transition from the surface skin, across the intermediate zone, to the core of an article. The need for such an estimate follows from the frequently observed fact, that a sudden change of density between the skin and the core is just as detrimental for the strength of an article as a completely gradual transition[1-3,29,30].

A simple, though rather approximate quantitative estimate of the transition concerned, is the ϱ_s/ϱ_a or the ϱ_c/ϱ_a ratio, i.e. the "current" density of a specimen [6].

A more accurate characteristic is the φ_ϱ value, suggested by Essipow [4] for an estimate of IF density distributions:

$$\varphi_\varrho = \frac{\varrho_s + \varrho_c}{\varrho_a} \tag{21.1}$$

To characterize the "curvature" of the density distribution in the intermediate zone, Throne [5,6] has introduced the "shape factor" (*C*) in terms of which the apparent density of a specimen, ϱ, is expressed as follows:

$$\frac{\varrho}{\varrho_p} = \frac{\varrho_c}{\varrho_s} + \left(1 - \frac{\varrho_c}{\varrho_s}\right)[(C+1)\,Z^C - CZ^{C+1}] \tag{21.2}$$

where ϱ_p is the density of the unfoamed polymer ($\varrho_p \geqq \varrho_s$);
$Z = x/L$, x = the distance between the central axis and the point being determined;
L = the specimen height.

For all known IF patterns, $C = 3$–9; for an unfoamed polymer $C = 0$ ($\varrho/\varrho_p = 1$); for a uniformly foamed polymer (conventional foam) $C = \infty$. The difficulty and inaccuracy with regard to a graphical determination of C, on the one hand, and the vague physical meaning of this factor, on the other hand, impose certain limitations, however, on its practical application.

In estimating the heterogeneity of the apparent density distribution in the IF bulk, the "density gradient" (D_ϱ) is extensively used [5-8] According to Hubeny [1] it is determined as the tangent of the angle α between the slope of the steepest portion of the ϱ versus specimen thickness curve, and the abscissa (Figure 21.1b):

$$D_\varrho = \mathrm{tg}\,\alpha = y/x = \Delta\varrho/\Delta\delta \tag{21.3}$$

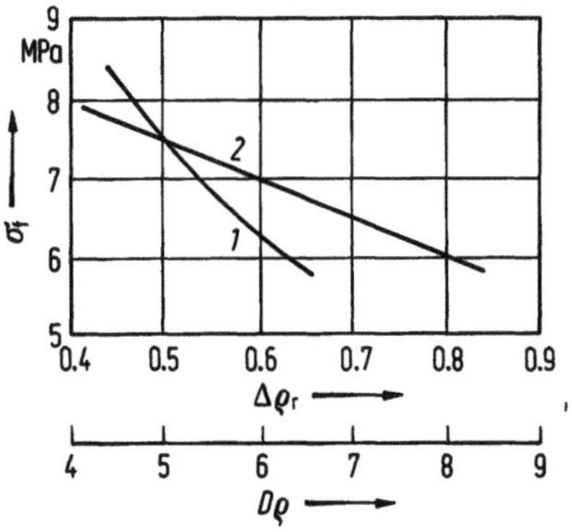

Fig. 21.2. Flexural strength of integral PUR foam versus density gradient D_ϱ (1) and relative apparent density $\Delta\varrho_r$ (2) [1)]

Along with the paramter D_ϱ, use is also made of such indicators as the "differential" apparent density ($\Delta\varrho$) found as the difference between the densities of two adjacent layers (cross-section) $\Delta\varrho = \varrho_1 - \varrho_2$, and the relative apparent density $\Delta\varrho_r$ defined as the maximum distance between the curve $\varrho = f(\delta)$ and the straight line corresponding to $D_\varrho = 1$ (Fig. 21.1 b).

The significance of introducing these parameters is clearly illustrated by the following example. Integral PUR specimens exhibiting a similar overall density of $\varrho_a \simeq 220$ kg/m³, but different density distributions in the intermediate zone, were tested for bending [1)]. It turned out that the higher the values of $\Delta\varrho_r$ and D_ϱ, i.e., the sharper the change of the density from skin to core, the lower the value of σ_f (Fig. 21.2). Note that the dependence $\sigma_f = f(D_\varrho)$ is more distinct than $\sigma_f = f(\Delta\varrho_r)$.

The nature of the density distribution in the intermediate zone has also a considerable bearing on IF strength. Indeed, if the transition is sudden (high values of D_ϱ, $\Delta\varrho$, and $\Delta\varrho_r$) skin and core may separate when exposed to thermal or mechanical stresses because of greatly differing strengths, coefficients of thermal conductivity, and moduli of elasticity. On the other hand, a continuous transition (low values of D_ϱ, $\Delta\varrho$, and $\Delta\varrho_r$) would increase the density of the IF [6)].

The question of the effect of a density gradient within the skin upon the strength indicators of an IF has not been thoroughly studied as yet.

Thus, even with equal average apparent density and the same values of ϱ_c and ϱ_s, the strength properties of an IF material may be quite different, if the density behavior in the intermediate zone differs [3)].

The parameters D_ϱ, $\Delta\varrho$, and $\Delta\varrho_r$, however, are not the only characteristics of the IF macrostructure. Of great importance for the whole set of physico-mechanical properties are also the absolute values of the core (δ_c) and skin (δ_s) thickness, with respect to the thickness of the article as a whole (δ_a), as well as the ratios δ_c/δ_s, δ_c/δ_a, and δ_s/δ_a. The effect of these morphological parameters on IF properties will be treated in a later discussion.

21.2 Calculation of Strength Properties

21.2.1 Problems

The development of a reliable calculation scheme, relating the specific features of the IF morphological structure with the IF strength properties and the boundary conditions of service, requires a rigorous quantitative evaluation of the morphological parameters, as well as unified methods for the physicomechanical tests of these materials[2,3,31,32)].

In determining the physico-mechanical properties of integral foams, one is faced with considerable methodical difficulties. None of the known methods for measuring the technical characteristics of conventional, i.e. isotropic, foams can take into account the specific effect produced by the nonuniform density of the integral foam, specially the density and thickness of the skin and core of the material. In order to compare the strength properties of various nonintegral foams, it is quite sufficient to screen the specimens with respect to a single morphological parameter, i.e., the apparent density. On the contrary, in the case of integral foams having the same overall density, their strength properties may greatly vary because of the difference in their other morphological parameters, i.e. ϱ_s, ϱ_c, δ_s, δ_c, δ_{iz}, D_ϱ, $\Delta\varrho$, etc. (see Sect. 21.1)[29,30].

The most important strength characteristics of IF's are stiffness (or rigidity) and flexural strength, since their specific values (calculated per unit weight) exceed those of the non-integral foamed and unfoamed polymers, as well as those of many woods and metals. Another important strength indicator is the compression strength. The value of this indicator is mainly determined by the value of the apparent density of the core of the foamed material. Certain difficulties do also arise while considering the effect of macrostructure nonuniformities on IF thermophysical and electric properties.

Currently, in many countries, there are established government standards for the determination of physical and mechanical properties of the IF materials, while in some countries only provisional recommendations have been made for the specimen preparation and for carrying out the corresponding tests. Currently, the Society of Plastics Industry's Structural Foam Standardization Subcommittee has recommended certain necessary standard tests [9].

After these general remarks we proceed now to the consideration of the main theoretical concepts interrelating the morphology and the properties of integral foam structures. To begin with, note that an analysis of the existing correlations between structure and properties of conventional (non-integral) foams has revealed that they are not useful for correctly calculating the strength properties of the integral structures [3].

21.2.2 Semerdjiev's Approach

The main use of IF parts being in engineering and load-bearing applications, it is very important to find correct methods for the computation and prediction of the mechanical strength properties of these materials. It is known that the classical elastic theory and the theory of strength of materials can be applied only if the change in density across the IF parts is continuous. On the other hand, IF's might be regarded as three-layer structures. For sandwich structures, the shear stresses at the interface

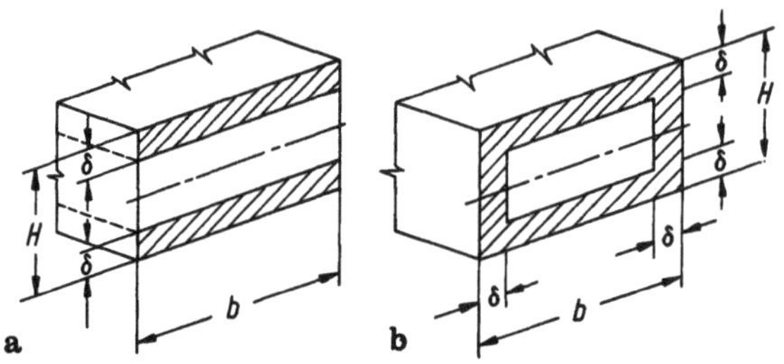

Fig. 21.3. Cross section of beams loaded in bending (see Table 21.1 for explanation of symbols) [7]

Table 21.1. Formulas for determining the deflection of IF beams under flexural load, according to Semerdjiev [7)]

Parameter	Section of IF beam according to Fig. 21.3	
	A	B
Reduced thickness, δ	$\frac{H}{2}\cdot\frac{\varrho_m}{\varrho_p}$;	$\frac{H+b}{4}-\frac{1}{2}\left(\frac{H+b}{2}\right)^2-Hb\cdot\frac{\varrho_m}{\varrho_p}$
Reduced moment of inertia, T	$\frac{b\delta}{6}\left[1+3\left(\frac{H}{\delta}-1\right)^2\right]$;	$\frac{bH^3}{12}-\frac{(b-2\delta)(H-2\delta)^3}{6H}$
Moment of resistance, W	$\frac{b\delta^3}{3H}\left[1+3\left(\frac{H}{\delta}-1\right)^2\right]$;	$\frac{bH^2}{6}-\frac{(b-2\delta)(H-2\delta)^3}{6H}$
Static moment, s_y	$\frac{b\delta}{2}(H-\delta)$;	$\frac{1}{8}[bH^2-(b-2\delta)(H-2\delta)^2]$
Shear stress, τ_{max} (Q = shear force)	$6Q\frac{\delta(H-\delta)}{b[H^3-(H-2\delta)^3]}$;	$\frac{3Q}{4}\cdot\frac{bH^2-(b-2\delta)(H-2\delta)^3}{bH^3-(b-2\delta)(H-2\delta)^3}$.

of the skin and the core are determining factors, but yet such a boundary is considered physically impossible to define for IF's. Therefore, the classical laminate-beam theory, if applied to calculate the deformation of a given IF body, results in incorrect data.

Semerdjiev [7)] suggested a simplified method to calculate the strength of IF parts. It assumes that the entire mass of the polymer material of the IF body is concentrated in the skin layers, and that the core strength can be completely ignored. The bending strength can than be calculated, using classical formulas of the theory of strength of materials. These formulas, applied to the cross sections of beams shown in Fig. 21.3, are summarized in Table 21.1. However, since this method is not based on the real structure of the IF body, the calculated moments of resistance are greater than the experimental ones. Nevertheless the method is accepted for a simplified analysis of any IF material having a simple configuration.

21.2.3 Gonzalez's Approach

The stiffness of an article with uniform density is known to be the product of the modulus of elasticity (E) and the moment of inertia (I). In the particular case of a material with a square cross-section, the stiffness is given by

$$EI = \frac{EP^3}{12B^2\varrho_a^3} \tag{21.4}$$

where B is the linear size (width), P is the weight, and ϱ_a is the density of the article.

A corresponding IF body can be treated, for theoretical purposes, as a two-component beam having a skin with a modulus E_s and a core with a modulus E_c. According to Gonzalez [10)], the stiffness of such a beam $(EI)_a$ is the sum of the stiffnesses of the core and the skin

$$(EI)_a = (EI)_c + (EI)_s \tag{21.5}$$

This model assumes that $\delta_{iz} \ll \delta_c$, i.e., it represents the IF structure as a simple three-layer panel. The terms $(EI)_c$ and $(EI)_s$ are defined as

$$(EI)_c = \frac{E_c \delta_c^3 B}{12}; \qquad (EI)_s = \frac{E_s(\delta_a^3 - \delta_c^3)\, B}{12} \tag{21.6}$$

Assuming $\delta_c/\delta_a = r$, and transforming equations 21.5 and 21.6, we obtain

$$E_a = E_s - r^3(E_s - E_c) \tag{21.7}$$

The weight of the IF part (per unit length) is

$$P = \varrho_a \delta_a B \tag{21.8}$$

where $\delta_a = 2\delta_s + \delta_c$.

On the other hand two equations are known relating the morphological parameters of the core, skin and IF article proper[10]:

$$\delta_s = \frac{\delta_a(\varrho_a - \varrho_c)}{2(\varrho_s - \varrho_c)} \tag{21.8a}$$

and

$$\varrho_a = \varrho_s - \frac{\delta_c}{\delta_s}(\varrho_s - \varrho_c) \tag{21.8b}$$

These equations 21.8a and 21.8b are valid for "thick" IF articles, i.e. for the case where the thickness of the intermediate zone δ_{iz} is many times smaller than $\delta_c(\delta_{iz} < \delta_c < \delta_a)$.

Then, substituting equations 21.6 into 21.5, and using equations 21.8, 21.8a and 21.8b we obtain

$$(EI)_a = \frac{P}{12 B^3}\left[\frac{E_s}{\varrho_s^3} \cdot \frac{1 - r^3(1 - E_c/E_s)}{1 - r(1 - \varrho_c/\varrho_s)^3}\right] \tag{21.9}$$

By differentiating equation 21.9, and using the condition $d(EI)/dr = 0$, we can find the value of r_0 corresponding to maximum rigidy of the IF part:

$$r_0 = \left(\frac{\delta_c}{\delta_a}\right)_0 = \sqrt{\frac{1 - \varrho_c/\varrho_s}{1 - E_c/E_s}} \tag{21.10}$$

at given values of B, δ, and ϱ.

Table 21.2 lists the values of stiffness of the model IF structure, based on equations 21.9 and 21.10 for various values of δ_c/δ_a and E_c/E_s, at $\varrho_c/\varrho_s = 0.6$ [10].

Table 21.2. Optimum stiffness $(EI)_{max}$ of model integral structures for various E_c/E_s and δ_c/δ_a ratios, at $\varrho_c/\varrho_s = 0.6$, calculated according to Equation 21.9[10]

E_c/E_s	δ_c/δ_a				
	0.20	0.40	0.60	0.80	1.00
0	1.27	1.58	1.79	1.55	0
0.20	1.28	1.60	1.88	1.88	0.93
0.40	1.28	1.62	1.98	2.20	1.85
0.60	1.28	1.64	2.08	2.53	2.78
0.80	1.28	1.67	2.18	2.85	3.70
1.00	1.28	1.69	2.28	3.18	4.63

21.2.4 Hartsock's Approach

Hartsock[11] gives another variant that can be used for calculating the extent of deflection, y, of a model IF structure having the form of a three-layer panel, with a foam filler:

$$y = \frac{PL^3}{48E_sI} + \frac{PL}{4GB\delta_a} \tag{21.11}$$

In this equation, G is the shear modulus of the foam filler and

$$I = \frac{B\delta_a^2 \delta_{s_1} \delta_{s_2}}{\delta_{s_1} + \delta_{s_2}} \tag{21.12}$$

where δ_{s_1} and δ_{s_2} are the thickness of the upper and lower layers.

When $\delta_{s_1} = \delta_{s_2} = \delta_s$ we have

$$I = \frac{B\delta_a^2 \delta_s}{2} \tag{21.13}$$

On the other hand, an approximate empirical dependence relating the shear modulus of a foam with its apparent density has been given by Benning[12]:

$$G = G_0(\varrho/\varrho_0)^n \tag{21.14}$$

where G_0 is the shear modulus for a foam with density ϱ_0; G is the shear modulus for a foam with density ϱ; n is an empirical constant equal to $n = I \div 2$.

In the case of the three-layer model of density ϱ_a, the density of the foam filler ϱ_c is

$$\varrho_c = \frac{\delta_a\varrho_a - 2\delta_s\varrho_s}{\delta_a - 2\delta_s} \tag{21.15}$$

Then, the shear modulus of the panel is

$$G = G_0\left[\frac{\delta_a\varrho_a - 2\delta_s\varrho_s}{10(\delta_a - 2\delta_s)}\right]^n \tag{21.16}$$

Thus, the expression 21.11 can be transformed as follows

$$y = \frac{PL^3}{24E_sB\delta_a^2 + \delta_s} + \frac{PL}{4G_0B\delta_a}\left[\frac{10(\delta_a - 2\delta_s)}{\delta_a\varrho_a - 2\delta_s\varrho_s}\right]^n \tag{21.17}$$

21.2.5 Hobbs's Approach

A two-component beam problem can be analyzed in terms of equivalent beams of a material whose cross-sections are altered perpendicularly to the axis of symmetry, for accounting local variations in stiffness. Thus, the modulus differences in the real beam are accounted for by constricting or broadening of the cross-sections of the model beams. Such an approach was developed by Hobbs[13]; the cross-sections considered in his study are shown in Fig. 21.4.

Cross-section *b* represents the well-known I-beam approximation, commonly

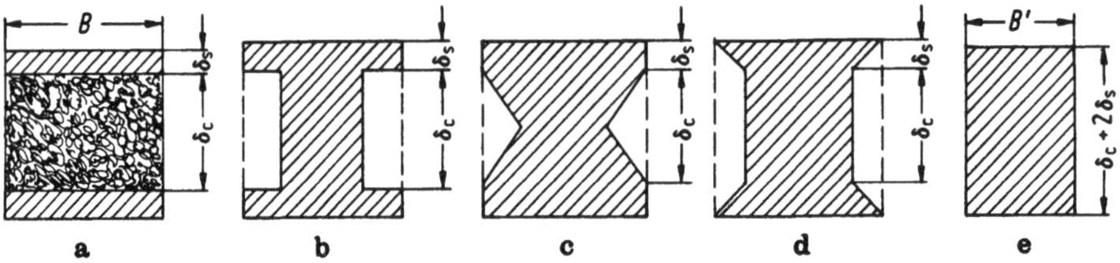

Fig. 21.4a–e. Calculation of the strength properties of an IF part by means of model structures: three-layered two-component beam (a) and one-component beams equivalent to the latter (b–e) [13]

used for modelling sandwich moldings, and is recommended for use with the IF parts. It assumes a skin of thickness δ_s, modulus E_s, and a core thickness reduced by a factor E_c/E_s. Cross-section *c* differs from *b* assuming that the foam modulus reaches a minimum near the center of the sample (a linear gradient was chosen for the sake of convenience). The minimum sample width is such that the total area in cross-section *c* equals that of *b*. Cross-section *d* is similar to *b* except that the modulus of the skin is assumed to decrease linearly over its thickness δ_s. Here, note that the width of the core is somewhat greater than that in *b* to account for the material displaced from the skin. The last cross-section *e* represents an oversimplified model in which the entire cross-section is reduced by a factor E_{avg}/E_s (where E_{avg} is the "average" modulus of the foamed bar) and the skin-core effects are not considered at all.

The central deflection of a beam in a three-point bending is given by:

$$y = \frac{PL^3}{48EI} \tag{21.18}$$

where I is the moment of inertia of the beam and $E = E_s$.

To calculate the properties of three-layer (sandwich) panels with foamed fillers (Figure 21.4a), Ogorkiewicz and Sayigh[14] have suggested a linear relationship between the elasticity moduli of skin and core:

$$E_c = (1 - n)\,E_s \tag{21.19}$$

and a — more accurate — nonlinear relationship:

$$E_c = \frac{1 - \sqrt{n}}{1 - \sqrt{n} + n}\,E_s \tag{21.20}$$

where $1 - n = \varrho_c/\varrho_s$.

The moments of inertia of the equivalent beams (Figure 21.4b–e) are, respectively,

$$I_2 = \frac{B_s\delta_c^3}{12} + \frac{B\delta_s^3}{6} + \frac{B\delta_s(\delta_c + \delta_s)^2}{2} \tag{21.21}$$

$$I_3 = \frac{B_s\delta_c^3}{12} + \frac{B\delta_s^3}{6} + \frac{B\delta_s(\delta_c + \delta_s)^2}{2} + \frac{(B - B_s)\,\delta_c^3}{18} \tag{21.22}$$

$$I_4 = \frac{B_s\delta_c^3}{12} + \frac{B_s\delta_s^3}{6} + \frac{B_s\delta_s(\delta_c+\delta_s)^2}{2} + \frac{(B-B_s)\,\delta_s^3}{18} + (B-B_s)\,\delta_s\left(\frac{\delta_c}{2}+\frac{2\delta_s}{3}\right)^2 \quad (21.23)$$

$$I_5 = \frac{B_s(\delta_c+2\delta_s)^3}{12} \quad (21.24)$$

where B is the width of a specimen; $B_s = B(E_c/E_s)$.

The extent of deflection of an isotropic three-layered beam is determined [13,14] by the following expression

$$y = \frac{PL^3}{48\delta_c}\left(1+\eta\frac{\delta_a^2}{L^2}\right) \quad (21.25)$$

Here,

$$\delta_c = \frac{B}{6\alpha_s}\left(3\delta_s\delta_c^3 + 6\delta_s^2\delta_c + 4\delta_s^3 + \frac{\alpha_s}{\alpha_c}\cdot\frac{\delta_c^3}{2}\right);$$

$$\alpha_s = \frac{1}{E_s(1-\nu_s^2)}; \quad \alpha_c = \frac{1}{E_c(1-\nu_c^2)};$$

$$\eta = \frac{3\delta_c}{6\alpha_s\delta_a(1-\delta_c^2/\delta_a^2)\,G}; \quad \delta_a = \delta_c + 2\delta_s$$

where ν_s and ν_c are the Poisson numbers for the skin and the core, respectively; the shear modulus is $G = E_c/3$.

The validity of the different models and of the corresponding equations have been studied by Hobbs [13], for integral foams based on PPO and PC. A comparison of the calculated with the measured deflection data has revealed unexpected results: it is model *d* (Fig. 21.4) that has proved to be the most accurate, rather than model *b* which is most often used for calculations. The discrepancy between the calculated (using the non-linear equation 21.20) and the measured results was below 1% for PPO, and below 5% for PC. Moreover, model *e* was found to yield more accurate results than the models *b* and *c* (with use of equation 21.19).

The approach of Hobbs is an excellent first choice for quickly assessing the stiffness of many IF parts, over a broad density range.

21.2.6 Throne's Approach

The studies of Throne [5,15] represent a more accurate quantitative approach to the calculation of the mechanical properties of IF parts, taking explicitely into account the density gradient.

Let us briefly outline the main ideas of this approach. Assuming that the foam in a model three-layer panel is not exposed to shear stresses, the values of deflection and moment of inertia are:

$$y = \frac{PL^3}{48\,E'I}; \quad I = \frac{B\delta_a^3}{4} \quad (21.26)$$

where E′ is the average modulus of elasticity of the panel, which is equal to

$$E' = \int_0^Y E(y)\,dy \Big/ \int_0^Y dy \tag{21.27}$$

where $Y = 0.5\delta_a$; the modulus of elasticity of the foam according to Ham [16], is given by:

$$E = E_0(\varrho/\varrho_0)^m \tag{21.28}$$

where m is similar to n in Eq. 21.14.

In an actual integral foam structure, the density changes rather smoothly from the skin to the core, and not in steps (as in a three-layer model). This change of ϱ over the height (y) of an IF body can be expressed mathematically in the form of a polynomial:

$$\varrho = a + by + cy^2 + dy^3 + \ldots \tag{21.29}$$

with the following boundary conditions for the density

$$\varrho = \varrho_c \quad \text{at} \quad y = 0\,; \qquad \varrho = \varrho_s \quad \text{at} \quad y = Y \tag{21.30}$$

and the density gradient:

$$d\varrho/dy = d^2\varrho/dy^2 = \ldots = 0 \qquad \text{at} \qquad y = 0, Y \tag{21.31}$$

Assuming that the distribution curve of ϱ is symmetrical with respect to the central axis of the IF specimen, we can write

$$\varrho/\varrho_s = R + (1 - R)\,[(C + 1)\,Z^C - CZ^{C+1}] \tag{21.32}$$

where $R = \varrho_c/\varrho_s$; $Z = y/Y$; C is the shape factor of the ϱ distribution curve (see eq. 21.2). For the investigated specimes of integral PUR foam, $R = 0.2$ and $C = 0.7$ [5].

From Eq. 21.32 follows an important relation:

$$\varrho_a = \left(\frac{2 + CR}{2 + C}\right) \cdot \varrho_0 \tag{21.33}$$

By substituting equation 21.32 into equation 21.28 and by integrating equation 21.33 one obtains two analytical expressions for two limiting cases, m = 1 and m = 2:

$$E_1' = E_s \frac{2 + CR}{2 + C} \tag{21.34}$$

$$E_2' = E_s\left[R^2 + \frac{4R(1 - R)}{C + 2} + (1 - R^2) + \frac{(C + 1)^2}{2C + 1} + \frac{C^2}{2C + 3} - C\right] \tag{21.35}$$

Thus, the explicit dependence of the extent of deflection upon the magnitude of load, density distribution, and IF dimensions is, according to Throne[5,15], as follows (for m = 1):

$$y = \frac{PL}{48E_1'I} = \frac{PL^3(2 + C)}{12B\delta_a^3E_s(2 + CR)} \tag{21.36}$$

One can easily see that equation 21.36 does not include the thickness of the surface skin, nor the shear modulus of a foam. This equation provides, nevertheless, a more accurate calculation of the IF strength properties than equation 21.17, which was based on a three-layer panel. For integral PUR, the experimentally measured values of E′ and ϱ_a are 539 MPa and 394 kg/m^3, respectively, whereas the calculated values are 926 MPa and 422 kg/m^3, respectively. This means, however, that the extent of deflection calculated by equation 21.36 differs from the experimental values by a factor of about 10.

21.2.7 Other Calculation Formulas

The rigidity enhancement (ΔEI, %) on substituting an integral foam article for one made of an unfoamed plastic of the same mass, is expressed, according to Fugate [17], by the following equation:

$$\Delta(\mathrm{EI}) = \left[\frac{E_c}{E_s} \cdot \frac{\varrho_s^3}{\varrho_c^3} - 1\right] \cdot 100 \tag{21.37}$$

Vice versa, the reduction of mass (ΔP, %) on substiting an integral foam article for an unfoamed one of equal rigidity is:

$$\Delta P = \left[1 - \frac{\varrho_c}{\varrho_s} \cdot \left(\frac{E_s}{E_c}\right)^{1/3}\right] \cdot 100 \tag{21.38}$$

For the sake of qualitative estimates, the rigidity of integral foam bodies having a rectangular cross-section can be assumed to be proportional to ϱ_a and δ_s^3 [7-10]. An unfoamed plastic is certainly stronger than an IF of indentical chemical nature and thickness. But, if comparing bodies of identical mass, the specific strength of the integral foam is 30–60% higher. Hence, for an identical specific strength, the polymer consumption in the case of an IF article is 30–60% less, as compared with an unfoamed article. Bright prospects for the application of IF's as construction materials are evident from the following estimates. For a constant skin thickness, a 25% reduction in the core apparent density results in almost a two-fold increase of the article's strength. Similarly, a 25% increase of δ_s leads to a two-fold enhancement of the IF's rigidity (with the same mass) [17].

In the case of cylindrical IF articles, the stiffness is proportional to ϱ_a^2; if ϱ_a is reduced by 20%, the value of δ_a should be increased by 16% in order to maintain the same rigidity; this will save 7% of feedstock, but, at the same time, increase the cooling time by 10% (due to the increased δ_a) [5].

For determining the stiffness requirements for an IF part, which is to replace a part of another material, the flexural modulus of the original material is frequently taken as the standard. This can, however prove to be very misleading. For example, if a plastic with a modulus of 1725 MPa is compared to aluminum with a modulus of 68950 MPa, it would appear that the plastic part should be forty times thicker. A more realistic value is obtained using an equation suggested by McBrayer [18] for equivalent rigidity in bending of simple beams:

$$\delta_1 = \sqrt[3]{E_2(\delta_2)^3/E_1}\,, \tag{21.39}$$

where δ and E are the thickness and the flexural modulus. According to this equation, the required thickness of the plastic part is only 3.4 times that of the aluminum part.

To have an approximate qualitative estimate of the strength of an IF article, it is convenient to apply an Eq. suggested by Throne [5]

$$\sigma = \frac{\delta_a}{\delta_s}\sigma_p + \left(1 - \frac{\delta_a}{\delta_s}\right)\left(\frac{\varrho_a}{\varrho_p}\right)^2 \sigma_p \tag{21.40}$$

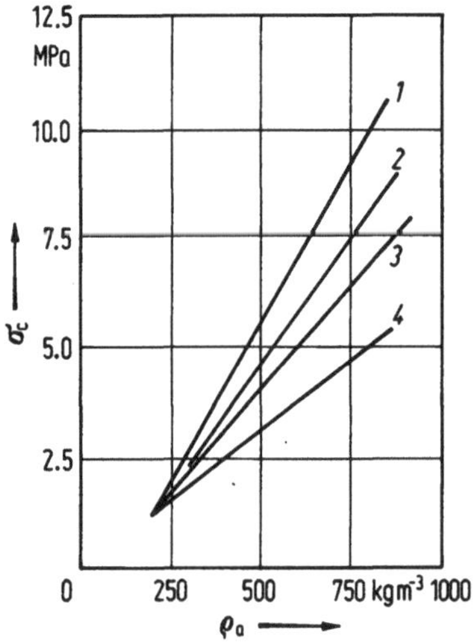

Fig. 21.5. Relationship between the compression strength and the density of integral PUR foam parts, for various shapes (area-to-volume ratio) of the parts; 1, 2, 3, 4: respectively 1.5, 1.8, 2.5, and 4.5 cm^2/cm^3 [19]

where σ and σ_p are any strength characteristics of the IF and of the corresponding unfoamed plastic.

The strength properties of IF parts are influenced not only by density and thickness of the skin, but also by the shape of the item (Fig. 21.5). The stiffness of an IF item is governed, as shown previously, by the moment of inertia of the item, which can be calculated according to known Eqs., provided the transversal distribution of the density is known. To a first approximation, an IF item may be divided into two zones — skin and core — and it may be assumed that the density of the material is the same throughout each zone. The relationship between the stiffness of an integral PUR foam part and the thickness of its skin, presented in Fig. 21.6, was calculated with this assumption.

To increase the stiffness of an IF part, use is made not only of mineral fillers, but also of surface reinforcement with glass fiber and glass fabric, with polymer films and sheets, and with metallic foils [19-22].

Other important equations for calculating various mechanical properties of IF parts were developed by Semerdjiev and his school [7].

For design engineers some other relevant equations are presented in Chap. 22 (Table 22.2).

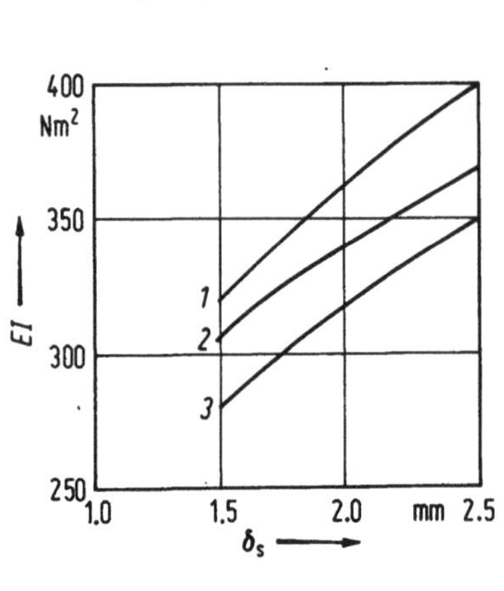

Fig. 21.6. Stiffness EI versus skin thickness δ_s, for integral PUR foam parts; thickness of parts 21 mm (1) and 20 mm (2, 3); elasticity modulus of core 5,500 MPa (1, 3) and 7,000 MPa (2); elasticity modulus of skin is the same for all parts, 2,500 MPa [20]

21.3 Calculation of Thermal Conductivity

Not only the strength, but also the thermal-insulation properties of integral foam structures are directly related to skin thickness. In particular, the coefficient of thermal conductivity of an integral foam is always higher than that of the corresponding isotropically foamed material (for similar apparent densities). This is accounted for by the fact that the thermal conductivity of the surface skin (not containing a gas-phase) is much higher than that of the core. On the other hand, under actual service conditions, say, at a high moisture content in the environment, the thermal conductivity of an IF increases less than that of a conventional foam, which is attributed to the protecting (screening) effect of the surface skin [3].

According to Skinner [23], the coefficient of thermal conductivity of a three-layer structure λ' is given by:

$$\frac{1}{\lambda'} = \frac{\delta_s + 2\delta_c}{(\delta + 2\delta_s)\,\lambda_s} + \frac{\delta_c}{(\delta + 2\delta_s)\,\lambda_c} \tag{21.41}$$

For an isotropic cellular structure, the coefficient λ is determined, according to Kerner [24], as follows:

$$\lambda = \frac{\lambda_0(1 - \vartheta_p)\dfrac{3\lambda_p}{\lambda_0 + 2\lambda_p} + \lambda_p\vartheta_p}{(1 - \vartheta_p)\dfrac{3\lambda_p}{\lambda_0 + \lambda_p} + \vartheta_p} \tag{21.42}$$

where λ_0, λ_p are the coefficient of thermal conductivity for air and unfoamed polymer, respectively; ϑ_p is the volume fraction of the polymer.

A comparison of the measured values of λ with those calculated by Eqs. 21.41 and 21.42, for integral PS and ABS foams, indicates fairly good agreement (Table 21.3).

At this point, it should be noted that the significance of the introduction by Throne [5,15] of the ratio $\varrho_a/\varrho_p(\varrho_p/\varrho_s)$ into the calculation of the stiffness (Eq. 21.40), is supported by the validity of another Eq., suggested by Progelhof [25] for calculating coefficients of thermal conductivity for integral foam structures:

$$\lambda_a = \lambda_g\left(1 + A\frac{\varrho_a}{\varrho_p}\right)^B \tag{21.43}$$

Table 21.3. Calculated and experimental values of thermal conductivity, for various integral polymer foams [23]

IF type	ϱ_a, kg/m³	δ_c/δ_a	λ, W/m · °K experimental values	calculated, Eq. 21.42	calculated, Eq. 21.41
PS	210	0.93	0.044	0.038	0.036
	318	0.87	0.047	0.045	0.041
	506	0.84	0.049	0.059	0.055
ABS	430	0.87	0.056	0.055	0.053
	500	0.84	0.062	0.062	0.059
	640	0.72	0.068	0.075	0.069
	660	0.72	0.076	0.076	0.071

where λ_a and λ_g are the coefficients of thermal conductivity for the IF sample and the foaming gas respectively; A and B are empirical coefficients, numerical values of which are given below for some integral foams:

Starting polymer	Foaming gas	A	B
Polyurethane	CO_2	5.37	0.6
Polyurethane	freon	9.44	0.7
Polystyrene	N_2	4.00	0.8

21.4 Further Problems

Even while testing unfoamed plastics, many variables are encountered: processing and atmospheric conditions, molding temperature and pressure, and part configuration. As to IF articles, additional to all these variables, there is the influence of a cellular core structure with varying density, surrounded by a skin which can vary in thickness from 10 to 50 mm. With this in mind, one can easily understand the problem of obtaining accurate and reproducible test results from a group of IF test samples. Nevertheless, the picture is not so pessimistic, because for constant process conditions, the physical properties are in most cases also constant.

Certain trends in the engineering properties of IF's may be mentioned [26]. For example, the reduction of the tensile strength is usually about 10% less than the density reduction; e.g. a 30% density reduction yields a tensile strength about 40% less than that of the solid (unfoamed) injection molded plastic. The flexural modulus usually decreases parallel to the density reduction, i.e. a 30% density reduction yields a flexural modulus about 30% less than that of the unfoamed injection molded plastic. The HDT (heat deflection temperature) is affected far less than the mechanical properties; the HDT values of IF's are close to injection molded values.

For a development engineer, it is sometimes difficult to conduct a reliable stress analysis on IF parts, because of the lack of physical property data on these materials. There is considerable work being undertaken to develop this type of data and to establish suitable test methods to accurately determine the properties of an IF part. It should be noted, however, that the properties of IF's involve additionally those parameters normally used to describe the properties of unfoamed molding, as well as of non-integral foams.

During the past few years, by the efforts of many scientists, and particularly of Bürger [27], Progelhof [25], Semerdjiev [7] and Throne [5,6,15] there has been considerable progress in the calculation methods for IF parts. For an advanced, accurate method of calculation of the mechanical properties of these materials it is necessary to determine quantitatively the parameters characterizing the main morphological sub-units of the integral polymer foams: the skin, the core, and the intermediate zone.

Any further development in the field of "calculation" trends should, in our opinion, involve the quantitative consideration of additional fine details of the IF morphology (such as anisotropy of the structure, dispersity of cells in size and shape, fraction of open cells, quality of skin surface) as well as the introduction of parameters characterizing the chemical nature of the polymeric matrix [28-32].

21.5 References

1. Hubeny, H.: J. Cell. Plast. *11*, 256 (1975); Kunststoffe *66*, 746 (1976); *68*, 479 (1978)
2. Shutov, F. A.: Adv. Polym. Sci. *51*, 155 (1983)
3. Shutov, F. A.: USSR Conf. Chem. and Technol. of Polyurethanes, Vladimir USSR, 1979
4. Jessipow, J. L.: IFL-Mitt. *14*, 335 (1975)
5. Throne, J. L.: Mechanical Properties of Thermoplastics Structural Foam. In: Engineering Guide to Structural Foam. Wendle, B. C. (ed.). Westport: Technomic, 1976, pp. 90–114
6. Throne, J. L.: J. Cell. Plast. *8*, 208 (1972); *12*, 264 (1976)
7. Semerdjiev, S.: Introduction to Structural Foam. Brookfield Center: Soc. Plastics Engineers, 1982
8. Berlin, A. A., Shutov, F. A.: Strengthened Gas-Filled Polymers. Moscow: Khimia, 1980 (in Russian)
9. Nutter, D. A.: 6th Structural Foam Conf., Bal Harbour USA, 1978
10. Gonzalez, H.: J. Cell. Plast. *12*, 49 (1976)
11. Hartsock, J. A.: Design of Foam-Filled Structure. Westport: Technomic, 1969
12. Benning, C. J.: J. Cell. Plast. *3*, 62 (1967)
13. Hobbs, S. Y.: J. Cell. Plast. *2*, 258 (1976)
14. Ogorkiewicz, R. M., Sayigh, A. A. M.: Plast. Polym. *40*, 64 (April 1972)
15. Throne, J. L.: Structural Foams. In: Mechanics of Cellular Plastics. Hilyard, N. C. (ed.). London: Applied Science, 1982, pp. 263–322
16. Ham, S. J.: J. Cell. Plast. *14*, 42 (1978)
17. Fugate, D. A.: Plast. Eng. *31*, No 4, 213 (1975)
18. McBrayer, R. L.: J. Cell. Plast. *17*, 332 (1980)
19. Thompson, E. J., et al.: J. Cell. Plast. *9*, 35 (1973)
20. Dunkley, C. D.: Ingeniersblad *42*, No 7, 173 (1973)
21. Vaccari, J. A.: Prod. Eng. *50*, No 3, 29 (1979)
22. Siggelkow, H.-J.: Ifl-Mitt. *19*, No 1, 9 (1980)
23. Skinner, S. J.: VDI-Berichte No 182, 155 (1972)
24. Kerner, E.: Proc. Phys. Soc. *69*, 802 (1956)
25. Progelhof, R. C.: SPE Techn. Papers *24*, 678 (1978); *27*, 863 (1981)
26. Wendle, B. C.: Materials. In: Engineering Guide to Structural Foam. Wendle, B. C. (ed.). Westport: Technomic, 1976, pp. 76–89
27. Börger, H.: Kunststoffe *69*, 865 (1979); *72*, 359 (1982)
28. Shutov, F. A.: USSR Conf. Gas-Filled Plastics, Suzdal USSR, 1982
29. Shutov, F. A.: 4th Conf. Mechanics and Technology of Composites, Varna Bulgaria, 1985
30. Shutov, F. A.: 28th Microsymp. Macromolecules, Prague Czechoslovakia, 1985
31. Shutov, F. A.: 4th Intern. Conf. Polymer Science and Technology, Mar-tel-Plata Argentina, 1985
32. Shutov, F. A.: Intern. Symp. Characterization and Analysis of Polymers, Melburn Australia, 1985

22 Design Concepts

22.1 Design Possibilities

The virtually unlimited design freedom IF's offer has given the industrial design engineer many new and exciting possibilities. Engineering IF's means that creativity in design and technical ingenuity are no longer restricted by the properties of metals and the limitations of the metal fabrication processes (see Fig. 1.8) [1-4].

The larger an IF component is, and the greater its structural complexity, the greater are the savings to be made by using IF instead of other materials.

Light weight and manufacturing economy open up many new applications for IF technology. In vehicles and transportation, the combination of rigidity and light weight with good appearance allows IF parts to meet the requirements for both visual and structural applications [5-14a].

It should be mentioned here that part complexity and part size are limited for IF's only by the process and equipment used. There are no principle obstacles to the production of complete housings for large business machines and appliances, and of major vehicle assemblies and structural groups like truck cabs, roofs, doors, etc.

22.2 Thin-Wall Integral Foams

In earlier chapters it was shown, by many examples, that any change in wall thickness will affect the physical properties of the final IF part. As the molding pressures increase, the trend to thinner wall sections also increase. The term "thin wall integral foam" has become common, and techniques have been developed by material suppliers to fit their materials into this mode[14,15].

Thin-wall IF parts produce tensile strengths, flexural strengths, and flexural moduli that are higher than those of their thick-wall counterparts — primarily because the density is higher in a thin wall. Density reduction in a 4 mm section, for instance, is less than that of a 9 mm section.

During the last two years several commercial grades of thin-wall IF parts were developed: Noryl-150 modified PPO and Lexan FL-410 polycarbonate (General Electric Co., USA), Styron 6087 SF polystyrene (Dow Chemical, USA), Baydur 726 (Mobay Chemical, USA), and RIM 160 polyurethane (Union Carbide, USA), phenylene oxide copolymer (Borg Warner, USA) [10-13]. The properties of the thin-wall IF parts were presented above in Table 13.10.

At the new thickness limits — 3.2 mm as opposed to the former 6.4 mm standard — material savings of 50% are being widely realized, along with significant cycle-time reductions. A very important factor for market growth has been that the size of thin-wall IF parts is limited only by the pressure equipment [12].

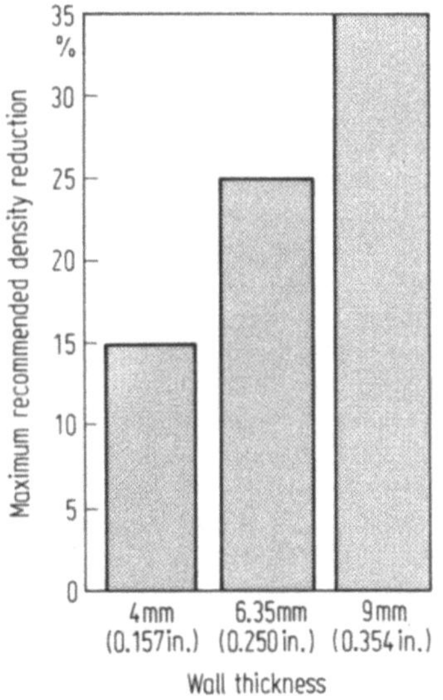

Fig. 22.1. Maximum recommended density reduction for varying wall-thickness of IF parts, according to General Electric Co.[10)]

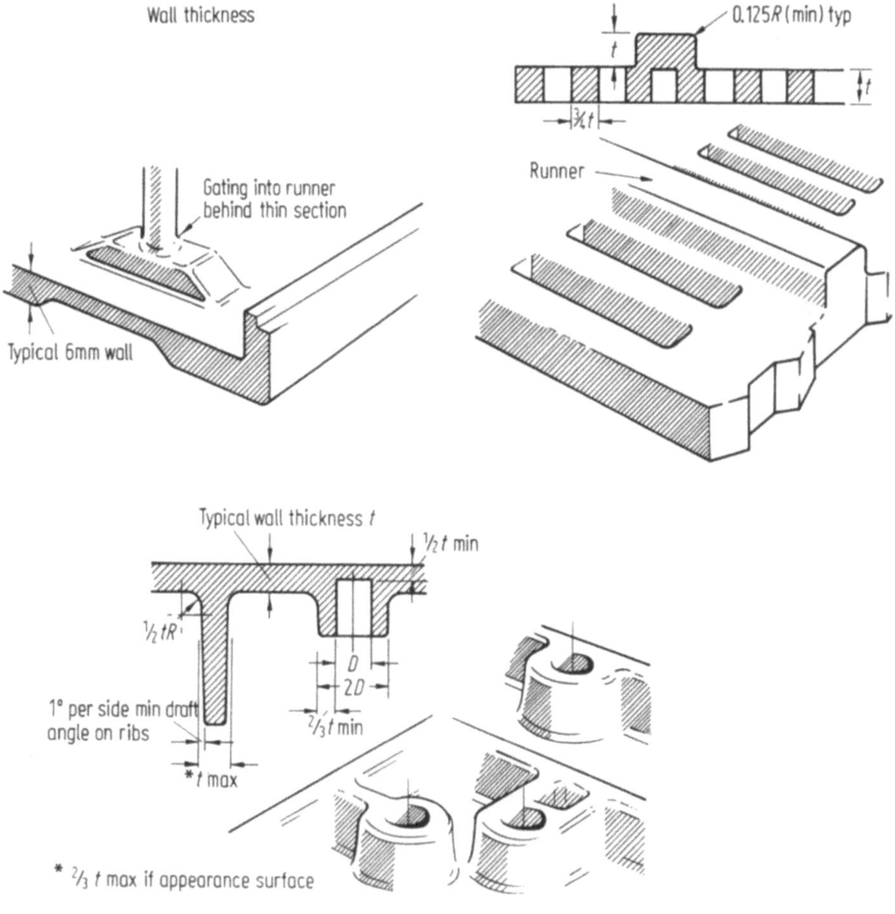

Fig. 22.2. Gating, runner, rib and boss recommended for IF parts design (Courtesy of General Electric Plastics B.V., Bergen op Zoom, Holland)

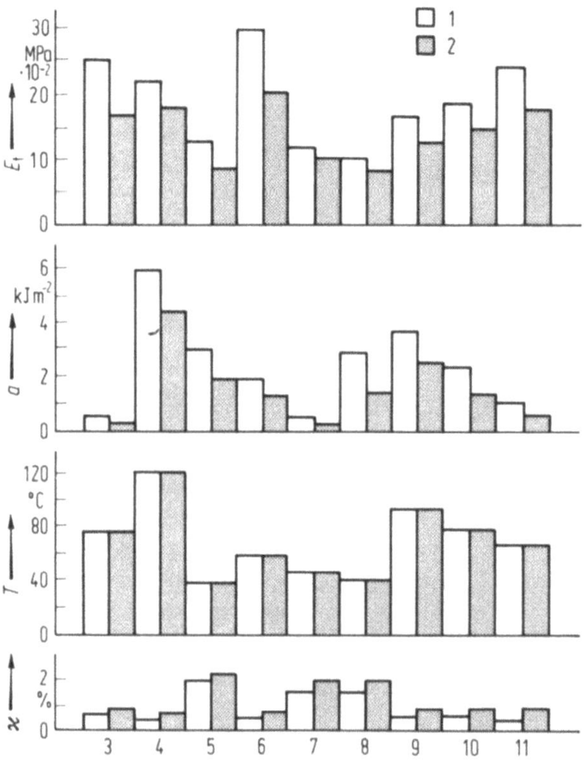

Fig. 22.3. Comparison of the physical and mechanical properties of IF's based on various polymers with foaming coefficient of (1) 20% and (2) 40%; the polymers are: (3) ABS, (4) PC, (5) PE, (6) PE + 20% glass fiber, (7) PP, (8) PP copolymer, (9) PPO, (10) high-impact PS, and (11) general-purpose PS; E_f = flexural modulus, a = impact resistance, T = thermostability, and ϰ = shrinkage [6]

22.3 Wall-Thickness Selection

The following recommendations, developed by General Electric Co, USA [10], indicate which material – modified PPO, 5%, 10%, or 30% glass fiber reinforced PC – and which wall thickness yield the best material property:

In applications where *impact strength* is a prime concern, 5% glass fiber reinforced PC and a thick wall provide the best performance; with a 9 mm wall, for example, the falling-ball impact strength for an IF based on PC is as high as 67.8 J (50 ft-lb).

Rigidity is proportional to wall thickness to the third power (see Chapter 21). Therefore, thick walls yield the most rigid IF parts. Where space is not limited, ribbing can be used to improve the rigidity of thin-wall designs (Fig. 22.1). Sink marks are usually the limiting factor on the thickness of a rib at its base; sink marks appear more readily in thinner sections.

The *flexural modulus* is directly proportional to the glass fiber content. Since the density is reduced less in thinner walls, moduli are correspondingly higher than in thick walls.

Load-bearing capabilities increase with higher wall thickness. Because deflection and stress for a simply supported beam are proportional to thickness, greater wall thickness increase stiffness and load-bearing ability.

Flexural strength, too, is greater as glass fiber loading goes up. Higher glass content and thinner walls yield the highest strength.

Tensile strength is affected by wall thickness and type of material. Higher strength results from thinner wall sections because, similar to flexural strength, lower density reductions are attained in thin parts. Also, increased glass loading boosts strength.

Table 22.1. Fire resistance of various types of integral foam and wood, according to US Standards (thickness of all test parts is 6.4 mm)

Type of IF	Flammability rating, UL-94	Oxygen index, ASTM-D-2863	Radiant panel test, ASTM-E-162
ABS	HB	18.8	—
FR-ABS[a]	V-0, 5V	27.0	150
Polyacetal	HB	—	—
Poly(phenylene oxide)	V-0, 5V	28.5	110
Nylon	HB	—	—
PBT	HB	—	—
FR-PBT	V-0	—	23
Polycarbonate	V-0, 5V	35.0	79
FR-Polycarbonate	V-0, 5V	42.0	18
Polypropylene	HB	17.5	—
FR-Polypropylene	V-0	—	—
Polystyrene	HB	17.5	—
FR-Polystyrene	V-2 to V-0	25.0	240
FR-Polyurethane	V-0	—	—
PVC	—	40.0	—
Red Oak	—	24.6	100

[a] FR indicates that flame retardant materials have beed added to the polymer matrix

Table 22.2. Design criteria for a low pressure IF part [7)]

Mechanical property	Equation[a]	Comments
Tensile strength	$\sigma_a^t = \sigma_0^t(\varrho_a/\varrho_0)^2$	Coefficient can increase with temperature
Compressive modulus	$E_a^c = E_0^c(\varrho_a/\varrho_0)^2$	Most data on short columns
Shear modulus	$E_a^s = E_0^s(\varrho_a/\varrho_0)^2$	Limited data
Flexural modulus	$E_a^f = E_0^f(\varrho_a/\varrho_0)^{3/2}$	Empirical — uncontrolled skin thickness, low foam density
	$E_a^f = E_0^f(\varrho_a/\varrho_0)$	Law of mixtures — thick skins, high foam density
Impact strength	$\sigma_a = \sigma_0^i(\varrho_a/\varrho_0)^4\ (\delta_s/\delta_0)^2$	Very few data from many test procedures (tentative)
Fatigue strength	$\sigma_a^{ft} = \sigma_0^{ft}(\varrho_a/\varrho_0)^2$	Very tentative, rule-of-thumb
Creep strength	$\sigma_a^{cr} = \sigma_0^{cr}(\varrho_a/\varrho_0)^2$	Very tentative, rule-of-thumb

[a] The "a" and "0" subscripts indicate the IF and unfoamed polymer matrix respectively; δ_s and δ_0 are thickness of the skin of the IF and the thickness of the unfoamed polymer matrix, respectively

Necessary features like mounting points, card entry slots, paper trays, holes, gating, ribs and bosses can be designed into the main structure, saving time and reducing the cost that would otherwise incur in secondary manufacturing operations and assembly (Fig. 22.2).

22.4 Polymer Type Selection

Correctly designing an IF part depends on the correct selection of the polymer type. Fig. 22.3 presents a comparison of the main technical properties of widely available commercial IF's based on different polymers.

In addition to taking into account the mechanical properties, the best choice of the polymer matrix cannot be made without knowing the fire resistance properties (Table 22.1) [8,9]; see also Tables 13.10, 15.2, 18.2, and 19.1.

22.5 Design Criteria

A summary of approximate design criteria for IF parts produced by low pressure injection has been given by Throne (Table 22.2) [7]. These criteria should be used with care, because only few of them are based on sound data. A design engineer should remember that the data are extracted from stress-strain curves, and because an IF part may be anisotropic the data should be considered to be apparent values. It should also be assumed that these design criteria are provisional, and that extensive structure-property analyses will lead to more sophisticated and accurate formulations.[16,17].

22.6 References

1. Belmont, C., Hester, F.: 7th Structural Foam Conf., Norfold USA, 1979
2. Wendle, B. C.: 7th Structural Foam Conf., Norfolk USA, 1979
3. Turner, N. B.: 7th Structural Foam Conf., Norfolk USA, 1979
4. Stencel, J., et al.: J. Cell. Plast. *17*, 152 (1981)
5. Engineering Structural Foam Resins: General Electric Plastics B.V., Bergen op Zoom, Holland, 1976
6. Caritat de Peruzzis, C., Lempereur, C.: Rev. Sen. Caoutch. Plast. *52*, 601 (1975)
7. Throne, J. L.: Mechanical properties of thermoplastic structural foams. In: Engineering Guide to Structural Foams, Wendle, B. C. (ed.). Westport: Technomic, 1976, pp. 91–114
8. Nelson, G. L.: 6th Structural Foam Conf., Bal Harbour USA, 1978; J. Cellular Plast. *18*, 36 (1982)
9. Fire Experiments on Structural Foam Plastics Equipment Enclosures: Rep. Structural Foam Division, Soc. Plastics Ind., New York USA, 1981
10. Wehrenberg, R. H.: Mater. Eng. *97*, No 6, 34 (1982)
11. Wigotsky, V.: Design Eng. Febr., 15 (1982)
12. N.N.: Plast. World *41*, No 6, 68 (1983)
13. N.N.: Modern Plast. Intern. *13*, No 6, 82 (1983); *14*, No 7, 56 (1984)
14. Wendle, B. C.: Intern. Conf. on Foamed Plastics, Düsseldorf FRG, 1983
14a. Wendle, B. C.: Structural Foam, a Purchasing and Design Guide. New York: M. Dekker, 1985
15. Loar, D.: Plast. World, *42*, No 2, 40 (1984)
16. Shutov, F. A.: 28th Microsymp. Macromolecules, Prague Czechoslovakia, 1985
17. Shutov, F. A.: 30th IUPAC Intern. Symp. Macromolecules, Hague Holland, 1985

23 Current and Future Trends

23.1 Industry Position of Integral Polymer Foam

The production of IF's is growing extremely fast. Thus, according to some estimates[1,2], the annual growth rate of IF production in USA is about 50%, whereas in Western Europe, which started producing IF's on a commercial scale somewhat later, it is 75–100%. The absolute figures are: in 1971 the United States produced 10,000 tons of IF, in 1974 40,000 tons, in 1976 110,000 tons, while for 1984 a production of 900,000 tons was predicted [3]. In 1975 West European countries produced 10,000 tons, and in 1984 they were predicted to produce 230,000 tons [4]. Integral foams comprised about 10% of the total plastic production in USA in 1975, whereas for 1984 25% were predicted [5,6].

According to Business Communication Co. (USA) [7], between 1976 and 1980 the annual growth in USA for thermoplastic IF's was 33%. This makes IF production the fastest growing area in the whole plastics industry.

For IF resin consumption in the 1982–1987 period, a 6.7% annual growth rate is forecast – from 82 to 114 million kg [8].

On the other hand, it is known [7] that 56% of the IF consumed in USA in 1975 was PS-based, but currently, the production of PS foams is decreasing, while the output of the PUR foams is growing very quickly, both for rigid and flexible foams. This is in spite of the higher raw material cost for PUR foams, which is almost twice as high (per unit mass) as that for an IF based on PS or PE (see also Chapter 13).

Table 23.1. Market projection for engineering thermoplastic resins

Type of resin	1981		1986	
	Quantity, 10^6 kg	Value, 10^6 \$	Quantity, 10^6 kg	Value, 10^6 \$
ABS	680	887	1,130	1,400
Acetale	72.5	145	118	234
Fluoroplastics	9.5	72.5	15.4	117
Nylon	158	335.4	281	589
Poly(phenylene oxide)	138	290	278	581
Polycarbonate	142	342	283	691
Poly(phenylene sulfide)	13.1	58	32.7	144
Polysulfone	10.9	53	27.2	132
SAN	27.2	25	38.6	36

In Western Europe the situation is appreciably different. In fact, a very rapid growth of the output of IF's based on ABS, modified PPO and PS was confidently predicted for 1985, with transportation being singled out as the fastest growth sector. By 1985, the usage of thermoplastic foam in Europe was expected to be 700,000 t [2].

According to Wendle [7], the forecast for the IF industry in USA until 1985 ca be summarized as follows:

Year	All Resins, metric tons	Commodity resins, %	Engineering resins, %
1976	45×10^6	75	25
1978	220×10^6	76	24
1980	350×10^6	78	22
1985	450×10^6	80	20

This includes all types of IF materials based on thermoplastic and thermosetting plastics; the "commodity resins" considered are ABS, PE, PP, PS, PUR and PVC.

The changing distribution between the commodity and engineering resins reflects the expected sharp increase of the cost of engineering plastics over the next years (Table 23.1) [9].

The development of high-performance plastic materials, ranging from reinforced commodity thermoplastics to very high-performance polymers, will remain a focus of interest in the 1980's. The state-of-the-art in plastic processing includes off-the-shelf machinery and equipment for high-volume processes. As shown in Table 23.2, the material cost of a thermoplastic IF for a business-machine housing is only 40% of the total manufacturing cost, making obvious the thrust toward process improvement, as the business-machine market becomes price competitive. This is one reason for the importance accorded to coinjection processes, where the improvement in surface quality holds the potential of reduced finishing costs [10].

Manufacturers need to consider which percentage of their total plastic requirement they plan to process in-house compared to what they will buy from custom processors. Table 23.3 indicates which respondents anticipate becoming involved in which processes, within the next few years. It is interesting to note that the order of importance for the six processes is virtually identical for both custom and

Table 23.2. Estimated conversion costs for different plastic processes

Process/product	Manufacturing cost, %	
	Materials	Conversion
Polyethylene film extrusion — bag making	60	40
Integral foam — computer housing	40	60
RIM[a] — part (no finishing)	50	50
SMC[b] — part (such as housing)	45	55
Pultrusion[c] — compression molding	35	65

[a] Reaction Injection Molding; [b] Sheet Molding Compound; [c] Process by which thermosetting liquid resins and fiber reinforcements are combined and pulled through a die rather than being pushed or extruded

Table 23.3. Anticipated involvement by 1985 in newer plastics technologies [11]

Technology	Captive plants[a], %				Custom plants, %
	Anticipated involvement	100% in-house	Partially in-house	Will buy 100% from custom	
Integral foam	44	30	27	36	45
Reaction injection molding	41	42	34	20	24
Scrapless forming	25	60	24	12	17
Stretch-blow molding	18	50	17	6	10
Cold forming	12	55	36	9	11
Foam reservoir molding	7	43	43	14	6

[a] Results indicate only a pattern of thought, and the percentages do not total 100% because often a respondent indicated more than one approach he might take

"captive" plants. With the exception of integral foam, the percentages of plants anticipating involvement in captive plants are slightly higher, but not by a large margin. As to IF technology, the anticipated involvement — both for captive and custom plants — is the highest of all the new plastic technologies [11].

The future of IF processing lies with machinery manufacturers who are able to produce "fool proof" equipment, i.e. equipment that can be handled in a three-shift production shop without the need of many high technology personnel [12]. The fact that this type of molding requires substantially less clamp pressure and hence is less expensive (per unit of mass of processing capability), makes it a natural replacement candidate for existing foam machines.

Therefore, the prospects for the development of IF have considerable promise [2,6,12,14]. Further increases in the output of these materials will be stimulated by cost reduction due to a shortening of the molding time achieved by using high-speed mixing heads, nested molds and compositions of greater reactivity.

23.2 Marketing Position of Integral Polymer Foam

The growth rate of IF output is presently estimated at better than 50% annually, with predictions indicating this will continue. No particular application, or special feature, can explain this growth; it is rather the wide flexibility of IF technology that has brought about this success. The trend will continue, because new materials with better properties — especially heat and flame resistance — are being developed, and larger capacity machines are being designed. Even though present application possibilities are far from being saturated, these developments will provide further impetus. Additionally, several industries are looking at IF more closely because of the increasing appearance of weight as an important design parameter [2,15,16].

A breakdown of the marketed applications of IF materials is given in Table 23.4 [17]. The IF industry was and is a very fragmented industry covering and utilizing many different resins in different types of machinery. Moreover, the IF problem has been complicated by a marketplace that is changing daily.

Table 23.4. Market potentials for engineering integral foams [17)]

Industry	Applications	Estimated annual volume, 10^{12} kg	
		1976	1982
Aerospace	seating, containers	–	0.45
Major appliances	central air conditioners, vending machines, video equipment, consoles	0.55	7.26
Portable appliances	floor care, room air conditioners, humidity control, power tools	0.02	0.23
Building and construction	underground junction boxes, underground pipe connectors	0.02	0.18
Business machines	mini-computers, terminals, copiers, mailing equipment, editing typewriters	6.81	24.9
Electrical/communications	modems, signal repeaters, connectors, junction boxes	0.18	2.27
Furniture	desks, hospital equipment, mail room furniture, storage equipment	1.22	3.20
Instrumentation and control	spectrophotometers, chromatographs, traffic signal housings	0.07	0.45
Laboratory/medical	centrifuges, angiographs, cardiographs, x-ray developers	0.07	0.45
Material handling	trays, totes, specialty pallets, conveyors	0.68	4.54
Military	ammunition containers, temporary shelters	0.005	0.98
Photographic	film processors, projectors, micrographic equipment	0.09	0.98
Recreation	snowmobiles, campers, trailers, private aircraft interiors	0.02	1.54
Transportation	automobile interiors, off-road vehicles, agricultural equipment, doors, seating, panels, roof tops	0.68	13.60
Miscellaneous	textile bobbins, compressor covers, ladders, etc.	1.59	7.26
	Total	12.0	68.3

Today in USA the IF market, while still very fragmented, may be grouped into two major types of industries [12)]. Those molders who do custom work, whereby the equipment manufacturer owns the tooling, tend to supply the business machine enclosure market. The "captive" molder (an integral part of a larger manufacture) tends to supply the material handling market (including trays, waste recepticals, mop buckets, large bins, etc.).

The first business machine housing to use IF was a calculator base molded from foamed PS in 1969; the first all-structural foam housing was a mini-computer unit produced in the early 1970's. Today over 50% of all table top or larger electronic housings are produced from IF [12, 18)]. Current and future positions for the IF process in the business machine housing industry are excellent (Table 23.5) [19)]. Business machine applications represented 30% of the activity in 1982 and are expected to rise to 35% by 1987. The growth rate of that sector will be more than twice that of the IF as a whole [18)].

In the material handling area, IF was an immediate success for specialty type large applications: pallets, a banana tray, and a line of trays and boxes were the first applications in what was to become a long line of heavy duty handling units. Strong merchandising has also helped to open up this market; the products include such items as mop buckets and a trash receptical with a dome top.

Table 23.5. Processes for 1982's and for 1987's business machines[a]

Process	1982 %	Next five years % More	Less	Same
Sheet metal	47	5	21	16
Die casting	19	5	10	5
Hand lay up	4	0	2	0
Sheet molding compound	14	9	5	7
Thermoforming	14	5	4	5
Injection molding	47	32	2	7
Polyurethane RIM	4	4	0	4
Structural integral foam	56	32	12	12

[a] Data shown are percentages of total OEM respondents in a Mobay Chemical and Business Communications Corporations study

Other existing markets include the home construction industry where window frames, bifold doors and shutters have been produced from integral foam; complete cabinets of IF are also in use in the hospital industry [12]. Over the last few years new applications of IF's have been developed such as beehives, civil band radios, earth-moving equipment, and leisure products.

Extruded IF's are a new family of plastic materials, and can be substituted for wood and metal, as well as for unfoamed and non-integral foamed plastic. For example, during the last 10 years, IF profiles for window frames have accounted for an ever-increasing proportion of the market, although the market share of plastic window frames varies considerably between different countries. For example, in West Germany the market share has increased from less than 5% in 1970 to 35% in 1979. Recently rigid extruded PUR integral foams have been applied to window frames, too [20].

A challenge to the IF industry will be to *develop new polymers and material compositions* incorporating more sophisticated fillers and reinforcement systems to provide a better balance of mechanical properties. Another promising area of study would be to marry aluminum with an IF, or a non-foamed RIM with an IF [21,22]. Another challenge in the 1980's is to discover other non-traditional fabrication areas (like printed circuit card guides) where IF's may out-perform metal in both cost and quality [23]. Design and processing innovations along with technological advances will keep the IF industry expanding beyond the 1980's.

Finally, as our experience with the different processes grows, the many economic factors — material and energy savings, tooling and equipment costs — are necessitating investigations of integral foam opportunities all throughout the manufacturers' product-lines (see Chapters 8, 9, 12, 15) [24,25].

As we conclude this book we want to recall that only a few years ago IF technology was considered so "troublesome" that it could never become a universal method for processing polymers. The data we have presented in our work are conclusive evidence of the incorrectness of this view. In fact, integral foams can be obtained today from any kind of synthetic polymer. Future development in IF technology will involve not so much the expansion of the range of polymers, but mainly the refinement of the processing methods for manufacturing integral foams in large tonnage and from cheap polymers[26-29]. The development of efficient manufacturing methods will open up new fields of application of plastics, and also substantially change our ideas on the potentialities of synthetic materials and, in particular, of polymers.

23.3 References

1. N.N.: Polym. News *2*, No 11–12, 36 (1976)
2. N.N.: Europ. Plast. Neus *7*, No 2, 8 (1980); *11*, No 10, 12 (1984)
3. Ham, S. J.: J. Cell. Plast. *14*, 42 (1978)
4. Litman, A., et al.: AIChE Ser. *73*, No 170, 163 (1977)
5. Meyer, W.: J. Cell. Plast. *14*, 50 (1978)
6. N.N.: Mater. Eng. *97*, No 5, 39 (1983)
7. Wendle, B. C.: Future of Structural Foam. In: Engineering Guide to Structural Foam. Wendle, B. C. (ed.). Westport: Technomic, 1976, pp. 225–228
8. Colangelo, M.: Plast. Technology *29*, No 6, 41 (1983)
9. Mock, J. A.: Plast. Eng. *38*, No 1, 17 (1982)
10. Wark, D. T., Eller, R.: Plast. Eng. *38*, No 1, 31 (1982)
11. Riley, M. W.: Plast. Technology *25*, No 1, 62 (1979)
12. Wendle, B. C.: Intern. Conf. Foamed Plastics, Düsseldorf FRG, 1983
13. Berlin, A. A., Shutov, F. A.: Strengthened Gas-Filled Polymers. Moscow: Khimia, 1980 (in Russian)
14. Wilder, R.: Plast. Technology *28*, No 5, 61 (1982)
15. N.N.: Plast. World *38*, No 1, 11; No 9, 29 (1980)
16. N.N.: Mod. Plast. Int. *11*, No 5, 17 (1981); 12, No 9, 28 (1982)
17. Rager, R.: Engineering Resins. In: Engineering Guide to Structural Foams. Wendle, B. C. (ed.). Westport: Technomic, 1976, pp. 15–20
18. N.N.: Plast. World *41*, No 7, 5 (1983)
19. Wehrenberg, R. H.: Mater. Eng. *96*, No 6, 34 (1982)
20. Kleiman, H.: Cell. Polym. *1*, No 2, 105 (1982)
21. Domenick, G.: J. Elast. *11*, No 2, 133 (1979)
22. Hendry, J.: 6th Structural Foam Conf., Bal Harbour USA, 1978
23. Brooks, R. L., Colbert, S. J.: J. Cell. Plast. *17*, 94 (1981)
24. N.N.: The Use of Polymer Materials in the Construction Industry. United Nations, Economic and Social Council, ECE/Chem/26, 1980
25. N.N.: Japan Plast. Age *20*, No 7, 32 (1982)
26. Shutov, F. A.: USSR Conf. Gas-Filled Plastics, Suzdal USSR, 1982
27. Shutov, F. A.: Intern. Symp. Polyurethane-83, London England, 1983
28. Shutov, F. A.: Intern. Symp. Plastics in Building, Liége Belgium, 1984
29. Shutov, F. A.: 30th IUPAC Intern. Symp. Macromolecules, Haque Holland, 1985

Firms, Firm-Processes and Grade Index

T-2.1 = Table 2.1; F-17.3 = Figur 17.3

Subject Index

T-1.2 = Table 1.2

Bei Fragen zur Produktsicherheit wenden Sie sich bitte an:
If you have any questions regarding product safety,
please contact:

Walter de Gruyter GmbH
Genthiner Straße 13
10785 Berlin
productsafety@degruyterbrill.com